THE APPLICATION OF FRACTURE MECHANICS TO POLYMERS, ADHESIVES AND COMPOSITES

Other titles in the ESIS Series

EGF 1	*The Behaviour of Short Fatigue Cracks* Edited by K.J. Miller and E.R. de los Rios
EGF 2	*The Fracture Mechanics of Welds* Edited by J.G. Blauel and K.-H. Schwalbe
EGF 3	*Biaxial and Multiaxial Fatigue* Edited by M.W. Brown and K.J. Miller
EGF 4	*The Assessment of Cracked Components by Fracture Mechanics* Edited by L.H. Larsson
EGF 5	*Yielding, Damage, and Failure of Anisotropic Solids* Edited by J.P. Boehler
EGF 6	*High Temperature Fracture Mechanisms and Mechanics* Edited by P. Bensussan and J.P. Mascarell
EGF 7	*Environment Assisted Fatigue* Edited by P. Scott and R.A. Cottis
EGF/ESIS 8	*Fracture Mechanics Verification by Large Scale Testing* Edited by K. Kussmaul
ESIS/EGF 9	*Defect Assessment in Components - Fundamentals and Applications* Edited by J.G. Blauel and K.-H. Schwalbe
ESIS 10	*Fatigue under Biaxial and Multiaxial Loading* Edited by K. Kussmaul, D.L. McDiarmid and D.F. Socie
ESIS 11	*Mechanics and Mechanisms of Damage in Composites and Multi-Materials* Edited by D. Baptiste
ESIS 12	*High Temperature Structural Design* Edited by L.H. Larsson
ESIS 13	*Short Fatigue Cracks* Edited by K.J. Miller and E.R. de los Rios
ESIS 14	*Mixed-Mode Fatigue and Fracture* Edited by H.P. Rossmanith and K.J. Miller
ESIS 15	*Behaviour of Defects at High Temperatures* Edited by R.A. Ainsworth and R.P. Skelton
ESIS 16	*Fatigue Design* Edited by J. Solin, G. Marquis, A. Siljander and S. Sipilä
ESIS 17	*Mis-Matching of Welds* Edited by K.-H. Schwalbe and M. Koçak
ESIS 18	*Fretting Fatigue* Edited by R.B. Waterhouse and T.C. Lindley
ESIS 19	*Impact of Dynamic Fracture of Polymers and Composites* Edited by J.G. Williams and A. Pavan
ESIS 20	*Evaluating Material Properties by Dynamic Testing* Edited by E. van Walle
ESIS 21	*Multiaxial Fatigue & Design* Edited by A. Pineau, G. Gailletaud and T.C. Lindley
ESIS 22	*Fatigue Design of Components.* ISBN 008-043318-9 Edited by G. Marquis and J. Solin
ESIS 23	*Fatigue Design and Reliability.* ISBN 008-043329-4 Edited by G. Marquis and J. Solin
ESIS 24	*Minimum Reinforcement in Concrete Members.* ISBN 008-043022-8 Edited by Alberto Carpinteri
ESIS 25	*Multiaxial Fatigue and Fracture.* ISBN 008-043336-7 Edited by E. Macha, W. B,edkowski and T. Lagoda
ESIS 26	*Fracture Mechanics: Applications and Challenges.* ISBN 008-043699-4 Edited by M. Fuentes, M. Elices, A. Martin-Meizoso and J.M. Martinez-Esnaola
ESIS 27	*Fracture of Polymers, Composites and Adhesives.* ISBN 008-043710-9 Edited by J.G. Williams and A. Pavan
ESIS 28	*Fracture Mechanics Testing Methods for Polymers Adhesives and Composites.* ISBN 008-043689-7 Edited by D.R. Moore, A. Pavan and J.G. Williams
ESIS 29	*Temperature-Fatigue Interaction.* ISBN 008-043982-9 Edited by L. Rémy and J. Petit
ESIS 30	*From Charpy to Present Impact Testing.* ISBN 008-043970-5 Edited by D. François and A. Pineau
ESIS 31	*Biaxal/Multiaxial Fatique and Fracture* ISBN 008-044129-7 Edited by A. Carpinteri, M. de Freitas and A. Spagnoli
ESIS 32	*Fracture of Polymers, Composites and Adhesives II* ISBN: 008-044195-5 Edited by B.R.K. Blackman, A. Pavan and J.G. Williams

For information on how to order titles 1-21, please contact MEP Ltd, Northgate Avenue, Bury St Edmonds, Suffolk, IP32 6BW, UK. Titles 22-32 can be ordered from Elsevier (http://www.elsevier.com).

THE APPLICATION OF FRACTURE MECHANICS TO POLYMERS, ADHESIVES AND COMPOSITES

Co-ordinating Editor:
D.R. Moore

ESIS Publication 33
The book is the work of the European Structural Integrity Society (ESIS)
through its Technical Commitee 4 (TC4)

2004

Elsevier

Amsterdam – Boston – Heidelberg – London – New York – Oxford
Paris – San Diego – San Francisco – Singapore – Sydney – Tokyo

ELSEVIER Ltd
The Boulevard, Langford Lane
Kidlington, Oxford OX5 1GB, UK

First edition 2004

Library of Congress Cataloging-in-Publication Data
A catalog record from the Library of Congress has been applied for.

British Library Cataloguing in Publication Data
A catalogue record from the British Library has been applied for.

ISBN : 0-08-044205-6
ISSN : 1566-1369

♾ The paper used in this publication meets the requirements of ANSI/NISO Z39.48-1992 (Permanence of Paper).
Printed in The Netherlands.

The papers presented in these proceedings have been reproduced directly from the authors' 'camera ready' manuscripts. As such, the presentation and reproduction quality may vary from paper to paper.

EDITORIAL TEAM

CONTENTS

Editorial Team Affiliations v

INTRODUCTION
P.E. Reed 1

APPLICATION OF FRACTURE MECHANICS TO POLYMERS

Polymer Performance

Ductile-Brittle Fracture Transitions in Polyethersulphone 7
D.R. Moore

The Application of Fracture Mechanics to the Fatigue Crack Propagation of Toughened Thermosets 17
H. Sautereau

Fracture Mechanical Behavior of Thermoplastic Polymers as a Function of Molecular and Supermolecular Variables 25
J. Karger-Kocsis

Failure of Polymers by Rupture versus Disentanglement of Molecules 31
F. Ramsteiner

Rate Dependent Fracture of Polymers – Applicability and Limitations of Force Based Fracture Mechanics Approaches 39
Z. Major and R.W. Lang

Creep Crack Growth in High Density Polyethylene 47
G. Pinter and R.W. Lang

Plastic Zone Corrected LEFM: Benefits and Limits 55
C. Grein, H.-H. Kausch and Ph. Béguelin

Why Ductile Thermoplastics Sometimes Fracture Under Impact 63
P.S. Leevers

Rate and Temperature Effects on Fracture Toughness of Polystyrene at Varying Molecular Mass 71
L. Castellani, S. Colombarini and R. Frassine

Applicability of Linear Elastic Fracture Mechanics to Fatigue in Amorphous and Semi-Crystalline Polymers 83
R.W. Lang, W. Balika and G. Pinter

Polymer Design

Comparison between Material Toughness and Component Toughness 93
D.R. Moore

The Engineering Design of Water Storage Tanks 97
D.R. Moore

Critical Pressure for Rapid Crack Propagation in Thermoplastic Water Pipe 103
P.S. Leevers and C.J. Greenshields

The Use of Fracture Mechanics to Assess Plastics Pressure Pipe Materials 109
G.P. Marshall and J.G. Williams

Materials Selection Using Fracture Mechanics 121
D.D. Huang, S. Li and J.G. Williams

APPLICATION OF FRACTURE MECHANICS TO ADHESIVES

Adhesives Performance

Peel Strength and Adhesive Fracture Toughness 131
D.R. Moore

Prediction of Ductile-Brittle Transitions in Coating Materials 137
D.R. Moore

The Durability of Adhesive Joints in Hostile Environments 143
B.R.K. Blackman and A.J. Kinloch

Adhesives Design

The Influence of Adhesive Bond Line Thickness on the Toughness of Adhesive Joints 149
A.J. Kinloch and D.R. Moore

Adhesive Joint Strength Prediction for Composite Design 157
J.P. Sargent and P. Davies

The Role of the Interphase and its Residual Stresses in the Determination of Fracture Toughness 163
J. Bouchet and A.A. Roche

Scratch Resistance and Embrittlement of Coated Polymeric Systems 171
F. Ramsteiner, T. Jaworek, M. Weber and St. Forster

Evaluation of the Interfacial Strength in Polymeric Systems 179
F. Ramsteiner

The Use of Fracture Mechanics Techniques to Predict the Service Life of Adhesive Joints 187
A.J. Kinloch and A.C. Taylor

Residual Stress Effects in Fracture of Composites and Adhesives 193
J.A. Nairn

APPLICATION OF FRACTURE MECHANICS TO COMPOSITES

Composites Performance

Toughness Comparisons With and Without Linear Elastic Fracture Mechanics 201
D.R. Moore

Finite Fracture Mechanics of Matrix Microcracking in Composites 207
J.A. Nairn

Fracture Mechanics of the Microbond and Pull-Out Tests 213
J.A. Nairn

Fracture Anisotropy 219
D.R. Moore

Determination of Interface Adhesion in Single-Fibre Pull-Out and Microbond Experiments 227
B. Lauke, T. Schüller and W. Beckert

Microstructure Dependent Fracture and Fatigue Crack Propagation Behavior in Injection-Molded Short Fiber-Reinforced Thermoplastics 233
J. Karger-Kocsis

Composites Design

A Stiffener Debonding Study Using Fracture Mechanics Data 241
P. Davies

The Influence of Epoxy-Metal Interphase Properties on Fracture Toughness 249
A.A. Roche and J. Bouchet

Fracture Tests to Optimise Marine Composite Manufacturing 257
P. Davies and H. Loaec

Suppression of Initiation of Delamination Cracking in Unidirectional Composites by Self-Same Resin Interleaving 265
I.K. Partridge and D.D.R. Cartié

Suppression of Propagation of Delamination Cracking in Unidirectional Composites by Z-Fiber® Pinning 273
D.D.R. Cartié and I.K. Partridge

Author Index 279

Subject Index 281

Author Affiliations 285

The Application of Fracture Mechanics to Polymers, Adhesives and Composites
D.R. Moore (Ed)

INTRODUCTION

P. E. REED

WORK of ESIS TC4

Technical Committee 4 of the European Structural Integrity Society (ESIS) was formed in 1985. The objective of TC4 is to further research and understanding of fracture mechanics for use in the development of a broad range of polymer materials and products manufactured from polymers. The first urgent need identified was that of developing standard, fracture mechanics based test methods to measure the fundamental fracture toughness parameters required for material development or use in design. A good deal of academic work had already been done when TC4 started, but the usefulness to industry was limited at that time by the lack of agreed standards. Over the past 17 years, TC4 has devoted much of its effort to the development of fracture mechanics based test methods for determining fracture toughness parameters.

An initial draft protocol is produced for each test method, which is then subjected to examination by means of round-robin laboratory testing by members of TC4. TC4 meets twice a year in Les Diablerets, Switzerland, to review progress of all aspects of its work, including the results of the round robin testing. The adequacy of the test method set down in the protocol to produce accurate, reproducible fracture toughness parameters within narrow limits, consistent from all sources, is considered and the protocol then revised accordingly. Development continues through further rounds of testing and rewriting of the protocol until a satisfactory test method is established. When considered appropriate, the developed protocol is submitted for consideration as an ISO standard.

Establishment of the international standards has naturally involved increasing collaboration between TC4 and other organisations similarly engaged around the world. A list of the TC4 activities leading towards ISO test standards is given in Table 1, together with information on the current state of development of ongoing programmes of work. An overview of the development of the testing methods activities by TC4 is published in '*Fracture Mechanics Testing Methods for Polymers, Adhesives and Composites*' [1].

APPLICATION of FRACTURE MECHANICS

A second need identified at the outset of TC4 was that of exploring the use of data obtained from fracture mechanics test methods in the design of plastic components. The need for such assessment in the design of polymer materials and of items fabricated from plastic and polymer composite materials arises from the increased use of these materials in critical load bearing situations.

Any component under load contains stored (potential) energy. The fundamental concept of fracture mechanics is that fracture will occur when the release of the stored energy is sufficient to propagate an existing crack or flaw, i.e. sufficient to overcome the fracture resistance of the material. As loading on a component is increased, so the stored energy in that system is increased, until a level is reached which causes crack growth. In the case of elastic systems, the crack growth can be unstable, leading to catastrophic failure. The energy released at this point to propagate the crack is the *critical strain energy release rate,* G_c, or, alternatively, the *J-fracture toughness*, J_c, for non-elastic dissipative materials. The parameter G_c, or J_c in the case of ductile polymers, is considered as a material property: each different polymer having its own unique value. Consequently knowledge of the fracture toughness parameter, G_c or J_c, enables the prediction of the loading conditions that will lead to crack growth in any situation i.e. it lays the foundation for the design of polymer materials or polymer components where cracks or flaws exist. Such design depends on having reliable values for the fracture toughness parameters. It is for this reason that TC4 has expended so much energy on developing reliable test methods for the measurement of G_c and J_c. A detailed introduction to the theory of fracture mechanics for polymers is given by Williams [2].

Flaws and defects at which failure might initiate and propagate exist at all levels of structure, from the smallest continuum structure of the material itself, through the compounding or construction of a material from several components and through the processing of material to form a product. Flaws can exist after manufacture or develop in service, especially where cyclic loading occurs. Cracks can also be introduced in the construction of assembled products where two or more parts are joined together. Fracture mechanics can therefore be applied both in the design and assessment of the polymer material itself, be it homopolymer or composite, as well as in the design of large products or assemblies. Fracture mechanics is therefore a versatile tool, increasingly applied to evaluate a wide range of situations. There is a growing wealth of experience in the application of fracture mechanics to all such situations in polymers, adhesives and composites.

Arising from the experience now available, this book presents a collection of articles and case studies in the application of fracture mechanics to many different situations in the development of polymer materials and the design of polymer products. The collection is divided into three sections, with separate sections on polymers, adhesives and composites respectively. The papers in each section are further classified under headings of *Performance* and *Design*.

Under *Performance* are grouped papers that consider the use of fracture mechanics in association with materials comparison and performance, as well as the detection and monitoring of fracture mechanisms. Other papers consider the use of fracture mechanics in understanding and improving test methods, particularly in assessing current industry tests. Further papers review the use of fracture mechanics in the prediction and modelling of the performance of materials and classes of materials generally.

Under *Design* are grouped papers where fracture mechanics considerations have been applied in the design of the material in terms of the type of material, its composition and the system in which the material may exist, such as within a laminate structure. Further papers consider the use of fracture mechanics in the design of polymer components and products. In some cases the use in design moves towards the inclusion of fracture mechanics analyses in Codes of Practice.

REFERENCES

1. Fracture Mechanics Testing Methods for Polymers, Adhesives and Composites. Ed. D.R. Moore, A. Pavan and J.G.Williams. ESIS Publication 28. Elsevier (London) 2001

2. Williams, J.G. Fracture Mechanics of Polymers. Ellis Horwood. (1984)

Table 1
TC4 Activities leading towards fracture toughness test Standards

Activity	Status	Standard Designation
Determination of fracture toughness (G_{IC} & K_{IC}) for plastics. An LEFM approach	ISO standard published March 2000	ISO 13586-1
Mode I interlaminer fracture toughness, G_{IC}, of unidirectional fibre-reinforced polymer composites	ISO standard published in 2001	ISO 15024
Mode II interlaminer fracture toughness, G_{IIC}, of unidirectional fibre-reinforced polymer composites	TC4 protocol available	
Mixed-mode interlaminer fracture toughness, $G_{I/IIC}$, of unidirectional fibre-reinforced polymer composites	TC4 protocol available	
J-Crack growth resistance curve tests on plastics	TC4 protocol available	
Determination of fracture toughness (G_{IC} & K_{IC}) for short-fibre composites	TC4 protocol available Submitted to ISO	ISO DIS 13585-2
Determination of fracture toughness (G_{IC} & K_{IC}) of plastics at high loading rates (1m/s)	TC4 protocol published as final international standard	ISO 17281
Determination of fracture toughness (G_{IC} & K_{IC}) of plastics at loading rates greater than 1m/s	TC4 protocol available	
J-Crack growth resistance curve tests for plastics under impact conditions	TC4 protocol available	
Essential work of fracture	TC4 protocol available	
Peel testing of flexible laminates	TC4 protocol available Proposed to ISO	
Determination of the mode I adhesive fracture energy, G_{IC}, of structural adhesives	TC4 protocol available Published as BS standard	BS 7991
Test method for tension-tension fatigue crack propagation in plastics		ISO 15850

Note: Items listed under status as '*TC4 protocol available*' reflect on-going studies still under review and development within TC4.

Application of Fracture Mechanics to Polymers

The Application of Fracture Mechanics to Polymers, Adhesives and Composites
D.R. Moore (Ed)

DUCTILE-BRITTLE FRACTURE TRANSITIONS IN POLYETHERSULPHONE

D.R. MOORE

INTRODUCTION

A fascinating feature in polymers is their ability to exhibit both ductile and brittle fracture. This fascination deepens when it is apparent that more than one mechanism can lead to this behaviour. Polyethersulphone (PES) is one such polymer whose fracture characteristics encompass these phenomena.
PES is a so-called engineering polymer with a glass-rubber transition temperature (T_g) of 225 °C and therefore is used in load-bearing applications where elevated temperature and long times under load are potential service conditions. However, the scientific fascination of its fracture characteristics is replaced by some concern on the possibility of a brittle fracture in a service component. Moreover, since this possible event can arise from more than one source then understanding its performance in components becomes a key issue.
A ductile-brittle fracture transition can occur in fatigue for PES. In addition, such a transition can also occur in elevated temperature creep rupture. Both fracture processes involve breaking the molecular backbone of the polymer, but with different characteristics controlling the fracture. The fatigue fracture characteristics are governed by molecular fracture which happens to be independent of molecular weight for these particular polymers. The creep rupture fracture characteristics are governed by molecular disentanglements, resulting in a molecular weight dependence for the fracture. Fortunately, these events are well understood and need not become a problem in the application of PES in engineering applications. Application of linear elastic fracture mechanics (LEFM) can provide a foundation for understanding these events, particularly when reinforced with the polymer science of molecular entanglements. This article gives an account of these characteristics.

MATERIALS AND THEIR FRACTURE CHARACTERISTICS.

Four grades of PES were the subjects of this study. The polymers differed in terms of their molecular weight as determined by GPC. The materials are designated MW1....MW4, for simplicity, where the weight average molecular weights for these materials are 35900 (MW1), 41700 (MW2), 51700 (MW3) and 59500 (MW4). Injection moulded plaques, with a coat-hanger gate, were made with these materials, 150mm x 150mm x 3mm. Material MW1 was also moulded into plaques with three different thickness in the range 1.8 mm to 5.8 mm.
Three experimental measurements were made; the measurement of tensile modulus (E) by flexure of a beam, measurement of tensile yield stress (σ_y) and measurement of the linear elastic fracture mechanics parameters K_c and G_c. The details of these measurements are given elsewhere (1). Yield stress was either measured directly in tension, when a yield process could be detected, or in compression. A yield stress in compression was divided by 1.3 in order to obtain the tensile value (1). The modulus and yield stress measurements were conducted at 23 °C at 1mm/min, whilst the fracture mechanics measurements were conducted at –65 °C at 1m/sec in order to comply with the

requirements of achieving plane strain geometry independent values. However, this experimental programme was conducted prior to full development of the test method protocol for LEFM at impact speeds (3). As a consequence, the fracture toughness values were determined without an indentation correction as recommended in the protocol (3) and as such G_c may be too large by up to about 20%. Nevertheless, they have high comparative accuracy.

The purpose in obtaining some general fracture data was twofold. First, such data are valuable in terms of describing the fracture toughness of the samples. Second, since many of the creep rupture tests were conducted at non-ambient temperatures (as discussed in a later section), it was helpful to examine via these fracture characteristics whether the storage at these elevated temperatures had an influence on the properties of the sample. Results on the 3mm mouldings are summarised in Table 1 where the influence of molecular weight and conditioning can be examined.

Sample	**E (GPa)**	**σ_y (MPa)**	**K_c ($MPam^{1/2}$)**	**G_c (kJ/m^2)**
Unconditioned				
MW1	3.2	84	2.4	2.5
MW2	3.3	78	2.6	2.1
MW3	3.2	78	2.6	2.0
MW4	3.1	76	2.6	2.3
Conditioned (150 °C for 24hours)				
MW1	3.2	85	1.9	1.3
MW2	3.3	94	2.2	1.7
MW3	3.3	93	2.2	1.6
MW4	3.3	88	2.1	1.7
Conditioned (220 °C for 24hours)				
MW1	3.1	94	1.6	1.2

Table 1 Fracture Measurements for the PES Samples.

With reference to the results in Table 1, the influence of molecular weight is seen to be quite small for the four PES samples; none of the properties show a significant trend with molecular weight. However, conditioning at elevated temperatures has a pronounced influence on the fracture properties. PES has a glass-rubber transition temperature (T_g) measured by DMA at about 225 °C. Conditioning of an amorphous thermoplastic at or near to its T_g is known to reduce the free-volume in the material (4) with the consequence of making the material stiffer and less ductile. The moduli data provide no evidence of a stiffness increase, but there is a suggestion that the yield stress data support such a trend. Moreover the toughness data show a reduction in ductility since both fracture toughness properties are seen to be smaller with higher conditioning temperature. Fracture measurements were also conducted on the unconditioned MW 1 mouldings at the three different values of thickness (1.8mm, 3mm and 5.8mm). The average stiffness, strength and toughness values agreed well with those data summarised in Table 1.

These fracture results provide the foundation for an explanation of the fatigue and creep rupture behaviour of the PES samples.

FRACTURE IN FATIGUE FOR PES.

All the fatigue measurements were conducted on pneumatically controlled tensile fatigue stations as described by Gotham (5). The specimens were width waisted tensile bars of dimensions, length 75 mm and width 19 mm with a central waisted region of width 6.4 mm. Testing was conducted at 23 °C under load control using a square waveform at a frequency of 0.5Hz.

Samples of three different thicknesses of PES MW 1 were used in these tests (1.8 mm to 5.8 mm) and results are shown in Figure 1. Examination of the fractured specimens enables us to interpret the features in these fatigue plots. At relatively large levels of applied stress the fractured surfaces show large deformation and often "draw-down" of the polymer; such features are synonymous with ductile fracture where yielding is the dominant failure process. At relatively small levels of applied stress, the fracture surfaces are apparently little deformed and crack-like in nature; such features are synonymous with brittle fractures. Therefore between the application of high and low applied stresses a ductile-brittle transition in fracture has occurred.

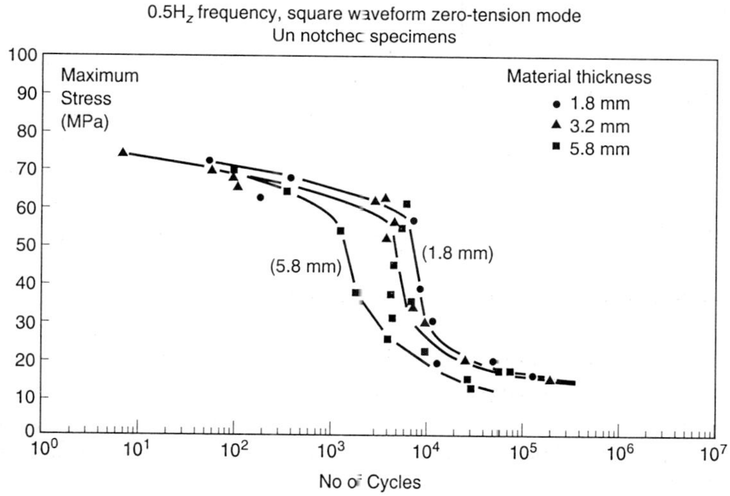

Figure 1 Fatigue for PES MW1 at 23 °C for different thickness specimens.

It is known that the level of constraint present in a fatigue test will ensure a ductile-brittle (d-b) fracture transition in PES (6). In addition, Davies et al (7) observed such a transition in 3 mm mouldings of PES and provided an explanation in terms of the size of the plastic zone relative to the size of the specimen. It would therefore be reasonable to expect that the fatigue measurements conducted on PES samples of different thickness as shown in

Figure 1 would also exhibit a ductile -brittle fracture transitions. The number of cycles required to generate this ductile-brittle fracture transition depends on the thickness of the sample. Therefore Table 2 shows the number of cycles to the onset of the d-b transition as a function of specimen thickness.

Specimen thickness (t) (mm)	Number of cycles to d-b transition. (k cycles)
5.8	1.5
3.2	5.8
1.8	8.5

Table 2 Relationship between thickness and cycles to d-b fracture transition in fatigue for PES MW1

If this ductile-brittle transition (d-b) is explicable in terms of the size of the plastic zone relative to sample thickness then the previously described LEFM fracture results should be able to account for the phenomenon. For example, ductile fracture will persist provided that the plastic zone size (r_P) is large compared with sample thickness (t). For brittle fracture to occur, then the sample thickness will be an order of magnitude larger than the plastic zone size. It is only necessary to have knowledge of the ratio $(\frac{t}{r_P})$ in order to describe the fracture behaviour. Consequently, the precise definition of plastic zone is not required and a general expression can be used, of the form:-

$$r_P \propto (\frac{K_c}{\sigma_y})^2 \qquad [1]$$

The property data in Table 1 will enable the calculation of r_P provided that allowance is made for the influence of time under load . Time dependence is derived from two experimental programmes (2). First, the dependence of yield stress on time under load at 23 °C for PES sample MW 1 (at 3 mm thickness). Second, the fatigue strength versus number of cycles to fracture for double-notched tensile specimens (tip radius 100 μm) tested at 23 °C where macroscopic cracking is observed for all specimens. These functions enable the results in Table 1 to be derated and the following values are then used in order to procedure with the analysis (full details are given in reference 2):

$K_c = 1.5$ MPam $^{1/2}$

$\sigma_y = 73$ Mpa

These two properties control fracture and in order to experience a transition from ductile to brittle fracture, the term $(\frac{t}{r_P})$ will change from less than 1 (ductile fracture) to a value in excess of 10 (for brittle fracture). In general terms, the larger the value of $(\frac{t}{r_p})$ then the more brittle the fracture.

In terms of the thickness part of this term, it is clear that as thickness increases so the term increases i.e. thicker material exhibit less ductile behaviour when everything else is equal. In terms of the d-b fracture transition in fatigue, we observe that thicker samples give rise to earlier transitions from ductile to brittle behaviour (i.e. at smaller number of cycles), therefore there should be some inverse correlation between the of the term $(\frac{t}{r_p})$ with the number of cycles to the d-b transition.

Consequently, there should be a simple trend between $(\frac{K_c^2}{\sigma_y^2 t})$ and the number of cycles to reach the d-b fracture transition. This is shown in Figure 2 albeit limited to just three data points.

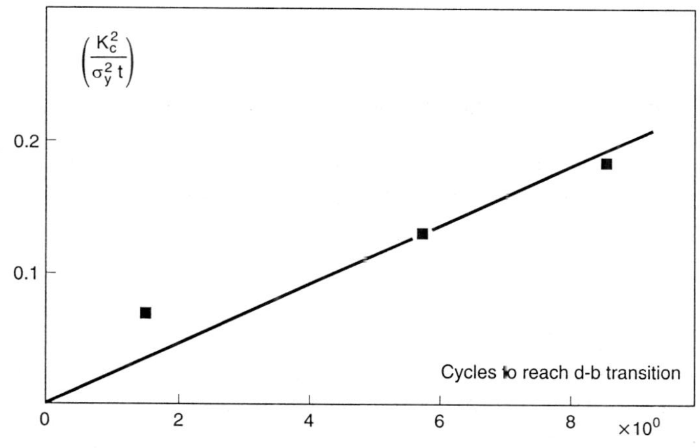

Figure 2 Comparative plastic zone to thickness ratio for PES MW1 at 23 °C plotted against the number of cycles to reach the d-b fracture transition in fatigue

This analysis of the factors that control the ductile-brittle transition at 23 °C in fatigue for PES confirms our understanding of the mechanism causing the process. That is, the transition is related to the intrinsic fracture toughness of the material and that the transition is a macroscopic event i.e. size related.

FRACTURE IN CREEP RUPTURE

Creep rupture testing involved another width waisted specimen that was subjected to a static tensile force and where the time to fracture was recorded as well as the nature and development of the failure/fracture process. The dimensions of the specimen were length 75 mm, width 12.7mm which reduced to a waist of 3.2mm. Full details of the lever loading creep rupture sites and other aspects of specimen preparation are discussed by Gotham (8). Creep rupture tests were conducted in the temperature range 23 °C to 200 °C.

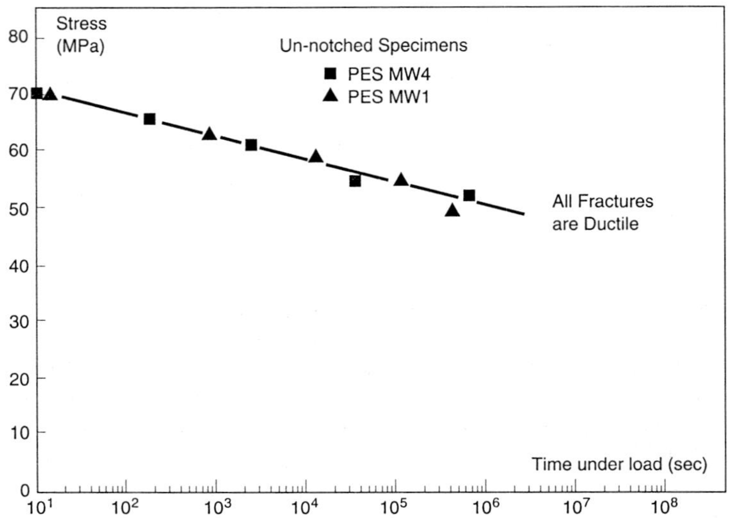

Figure 3 Creep rupture for PES samples at 100 °C

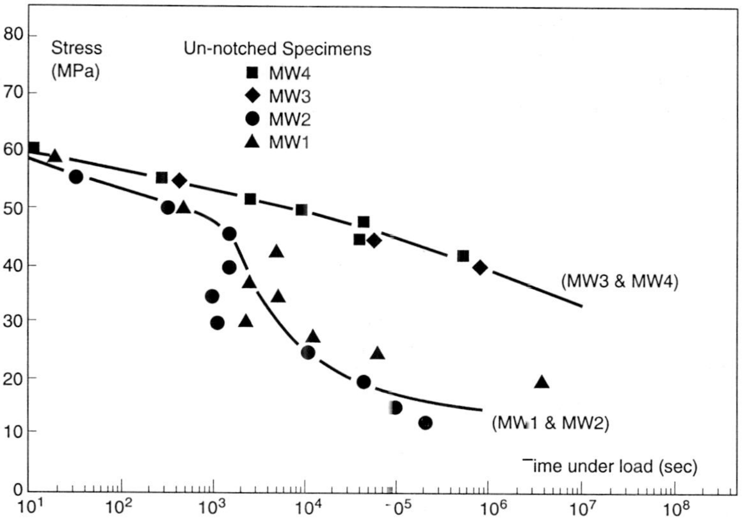

Figure 4 Creep rupture curves for PES samples at 150 °C.

Creep rupture provides a measure of the strength versus time under load for constant load conditions. The intention in our creep rupture study is to examine these functions in terms of different test temperatures and for different molecular weights. For example, Figure 3 shows data at 100 °C for PES MW 1 and MW 4, the lowest and highest molecular weight samples, respectively. All specimens relating to these data exhibited ductile fractures. Figure 4 shows further creep rupture data for all four of the samples. In this case MW 3 and MW 4 exhibited ductile fractures throughout the tests, but both MW 1 and MW 2 showed ductile-brittle fracture transitions at about 5000 seconds under load. Further measurements were conducted at 200 °C for PES MW 3 and 4 as shown in Figure 5. In this case all the MW 3 specimens exhibited brittle fracture whilst most of the MW 4 specimens were ductile, with the exception of those that followed the strength-time function for the MW 3 sample, which were brittle.

A primary question relating to these creep rupture curves and the observed d-b fracture transition is whether they are explicable in terms of the specimen size i.e. in terms of the fracture mechanics argument. It has already been seen that the term $\left(\frac{K_c^2}{\sigma_y^2 t}\right)$ has to be small for brittle fractures to occur. However, it is also necessary to take into account the changes in strength and toughness that occur as a consequence of storing the materials at the elevated temperature prior to creep rupture testing, as the fracture properties in Table 1 imply.

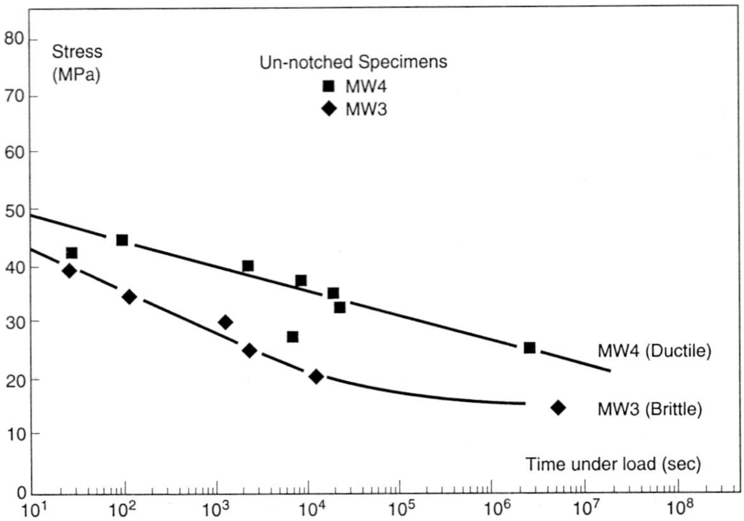

Figure 5 Creep rupture for PES at 200 °C.

The term $(\frac{t}{r_P})$ should be larger than 10 in order to observe brittle fracture. If we are to comment on the absolute performance of these PES materials in the creep rupture tests then we can no longer deal with comparative values for the plastic zone size. Therefore a specific definition has to be used. Several terms are possible, but the Dugdale line length (9) for the deformation ahead of a crack might be most suitable for fracture in PES because it accommodates transitions from yielding to crazing mechanisms which have been shown to be the most relevant for these samples (10). Hence the plastic zone length can be written

$$r_P = \frac{\pi}{8}\left(\frac{K_c}{\sigma_y}\right)^2 \qquad [2]$$

The results in Figures 3 to 5 enable us to determine values for yield stress at the appropriate times under load, where for the sake of simplicity we will assume that the operative time should be 10^5 seconds

We need to know the dependence of Kc on temperature. General experience suggests that plane strain values of K_c are independent of temperature and since in the context of these calculations this would be a conservative assumption, then it will be used. Rather than proceed with the calculations for all of the samples we can limit analysis to PES MW 1 (in a conditioned state, namely 24 hours at 150 °C) because that is the most critical material. Therefore Table 3 summarises the calculations of $(\frac{t}{r_P})$ at the three temperatures.

Temperature (°C)	$(\frac{t}{r_p})$ at 10^5 sec.
100	6.5
150	4.2
200	1.9

Table 3 Values of $(\frac{t}{r_p})$ as a function of temperatures for PES MW1. This ratio should exceed 10 for brittle fracture.

All the values of $(\frac{t}{r_p})$ in Table 3 are less than 10, therefore brittle fractures are inexplicable on the grounds of inherent material size and the fracture mechanics treatment of them. It remains blatantly the case that brittle fractures are observed in the temperature regime 150 °C to 200 °C and for every molecular weight tested. An alternative mechanistic cause of such fracture behaviour is therefore necessary and McLeish et al have provided one in terms of chain disentanglement (10), as mentioned in the introduction.

Returning to the creep rupture results depicted in Figures 3 to 5, it is clear that the fracture transitions in elevated temperature creep rupture are molecular weight dependent, whilst the fracture toughness data of Table 1 are molecular weight independent. For example, MW 1 is first to exhibit a d-b transition at 150 °C, MW 2 next, then MW 3 (although the severity of test temperature has to be raised to 200 °C) whilst results at the most severe temperature only just hint at the d-b transition for MW 4. Such molecular weight dependence is consistent with the molecular theory of chain disentanglement reported by others (10).

CONCLUDING COMMENTS.

We have established that ductile-brittle fracture transitions can occur in PES in both fatigue and creep rupture tests. The fracture behaviour in fatigue at 23 °C is governed by a fracture mechanics intrinsic flaw model where properties such as K_c and σ_y can be used to describe and hence predict performance. The fracture transitions at elevated temperature in the creep rupture tests are not explicable in terms of continuum fracture mechanics because a different mechanism is prevailing; one of molecular chain disentanglement where a transition from shear banding to crazing is causing the ductile brittle transition. However, application of fracture mechanics principles to these data suggests that an additional mechanism is the cause of the d-b fracture transitions.

Although the ductile-brittle fracture transitions in PES are caused by different mechanisms for different environments and loading configurations, it is quite apparent that the pattern of fracture behaviour is quite predictable on phenomenological grounds.

Consequently, engineering design with PES over a vast range of times, temperatures and types of load configuration is possible, where the designer can be confident that unexpected crack-like fractures should not occur.

REFERENCES

1 Davies M, Moore D. R, Composites Science & Technology **40** (1990), 1
2 Moore D.R. Slater B. Smith J.M. Plastics, Rubber and Composites Processing and Applications **17** (1992) 237-245
3 Moore D.R., Pavan A, Williams J.G. ed *Fracture Mechanics Testing Methods for Polymers Adhesives and Composites* ESIS Publication 28, Elsevier 2001, Ch 1 p27
4 Struik L C E *Physical Aging in Amorphous Polymers and other Materials* TNO Communication 565, Proefschrift Delft, 1977.
5 Gotham K V Plast & Polym August 1969
6 Gotham K V, Turner S SPE Antec, May 1972.
7 Davies M, Moore D R, Slater B Inst Phys Conf Ser 89 session 3 Sept 1987
8 Gotham K V Plastics & Polymers 40, April 1972, 59
9 Dugdale D.S. *J Mech Phys Solids* **8**, 100, 1960
10 McLeish T C B, Plummer C J G, Donald A M Polymer, 1989, 30 Sept 1651

The Application of Fracture Mechanics to Polymers, Adhesives and Composites
D.R. Moore (Ed)

THE APPLICATION OF FRACTURE MECHANICS TO THE FATIGUE CRACK PROPAGATION OF TOUGHENED THERMOSETS

H. SAUTEREAU

INTRODUCTION

During the last decade several papers have been devoted to the toughening of thermosets [1,2] and especially to epoxy networks. Generally the addition of a functionalized rubber is needed in order to form a separated phase with a low modulus and a low glass transition temperature. The rubbers used can be initially miscible and then will separate due to the increase of molecular weight during the polymerization of the system. It is the case of the well-known carboxyl terminated butadiene acrylonitrile copolymer (CTBN). In this case the final morphologies are strongly dependent on the curing conditions [3,4]. These rubbers are very efficient in toughening despite a significant loss of thermo-mechanical properties due to their miscibility in epoxy resins. Another approach uses preformed particles as core-shell particles [2,5] or Polydimethylsiloxane (PDMS) stable suspensions [6,7,8]. In this case the morphologies are strongly affected by the dispersion tool and shear rate [5]. Many studies have dealt with elastic plastic and fracture properties [1,2] but very few concerned fatigue crack propagation data on toughened epoxies [9,10] compared with the numerous works on thermoplastics [11]. Very often fatigue resistance of polymers is studied with a global approach of the durability. Generally Wolher curves are drawn, plotting the amplitude of stress versus the total number of cycles up to failure. This description, unfortunately, does not separate the crack initiation and crack propagation phenomena, and more, these curves depend on defects (and thus processing) and geometry of the specimens. Using Linear Elastic Fracture Mechanics it is possible to record the crack propagation rate which is really an intrinsic property and can be discussed with the polymer structure The aim of this paper is to measure and compare the mechanical properties (in both static and fatigue conditions) of rubber modified epoxies.

EXPERIMENTAL

Materials and preparation

The formulae of chemical products are listed on Table 1.The two epoxy systems are based on diglycidyl ether of bisphenol A (DGEBA) and the same hardener DDA (dicyandiamide). The only difference is due to the catalyst. Diuron for epoxy 1 and BDMA for epoxy 2 systems. All details are given in references [8] and [4,5] respectively. The neat epoxy 2 system is slighty more crosslinked than epoxy 1 and thus exhibits a higher glass transition temperature (see Table 2). The liquid reactive rubber (CTBN 8) was introduced as an epoxy terminated copolymer following an already described procedure [3]. The polydimethyl siloxane particles, (with a mean diameter of about 4 μm) were introduced at different amounts [8]. Two kinds of core-shell particles (CSR) were used, with various volume fractions, named CSR1 and CSR2, and are described in Table 1. The mean diameter for CSR1 is 220 nm and around 300 nm for CSR2.

The particles were mixed with the reactive epoxy system at room temperature with a high speed stirrer or a twin-screw extruder and then cast into a mould. The cure schedules were adapted in order to obtain completely cured samples with the maximum glass transition temperature.

Mechanical property characterizations at room temperature.
Static properties were measured using a 2/M MTS set-up. Tensile tests with ISO60 specimens at a strain rate of $3x10^{-3}$ s^{-1} lead us to determine the Young's modulus E_{RT}. The yield stress was measured in uniaxial compression at $\dot{\varepsilon} = 8x10^{-4}$ s^{-1}. The critical stress intensity factor, K_{Ic}, was determined according to the procedure of the European Group on Fracture(ESIS) [12] at a crosshead speed of 10 $mm.min^{-1}$ with SENB specimens (70 x 14 x 7 mm^3 – span 56 mm). At least five samples were fractured to obtain the average value of K_{Ic} from the maximum load at failure.

Fatigue crack propagation.
The fatigue crack propagation tests (FCP) were performed on compact-tension (CT) specimens using a hydraulic ZWICK machine (REL 1853) with sinusoidal cycling at a frequency of 5 Hz, following the main recommendation of the ESIS Technical Committee [13]. The tension-tension mode was applied, with load control, generally with a stress ratio $R_\sigma = \frac{F_{min}}{F_{max}}$ close to 0.1. The maximum load was calculated in order to start the FCP test with $\Delta K \sim K_{Ic}/2$. Crack growth was determined by using CPA1 Vishay-micromeasure gauges having 20 parallel conductors with 0.25 mm interval placed perpendiculary to the crack direction. Thus the crack length, a, was recorded without test interruption from the electrical resistance, versus of the number of cycles, N. As suggested [13] video recording, travelling microscope or compliance calculations [14] can also be used.

The FCP rate, $\frac{da}{dN}$, was determined graphically or numerically as the slope of the curve a = f(N).

The stress intensity factor range, ΔK, was calculated according to the following equation for CT samples [13].

$$\Delta K = \left(\frac{\Delta P}{b.w^{1/2}}\right)\left(\frac{2+A}{(1-A)^{3/2}}\right)\left(0.886 + 4.64\,A - 13.22\,A^2 + 14.72\,A^3 - 5.6\,A^4\right)$$

where ΔP is the load amplitude at each cycle (P_{max} - P_{min}). w and b are the specimen width and thickness of the sample respectively. A is the ratio of the crack length to the width of the specimen, a/w. For some comparison ΔK was calculated at a given crack growth rate, especially $7.5x10^{-4}$ mm/cycle as often done before [9,11] or at the crack instability at the end of the test (ΔK_{max}).

Table1. Chemical products used in the synthesis of materials

Name	Chemical Formula	Supplier and Tradename
* Epoxy prepolymer - DGEBA	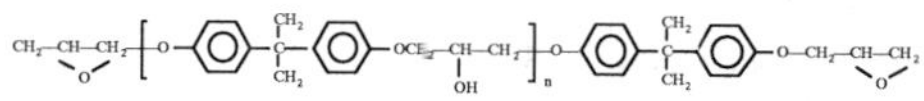 $\overline{n} = 0.15; \overline{M}_n = 380$ g	Bakelite 0164 or LY 556 Ciba Geigy
* Hardener - DDA (dicyandiamide)	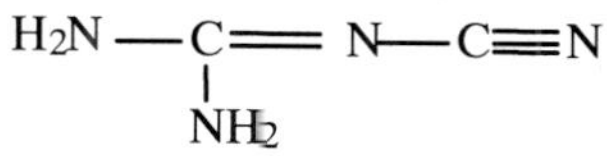	Bakelite VE 2560 or Dyhard 100S
* Catalysts or initiators - Diuron (epoxy 1)	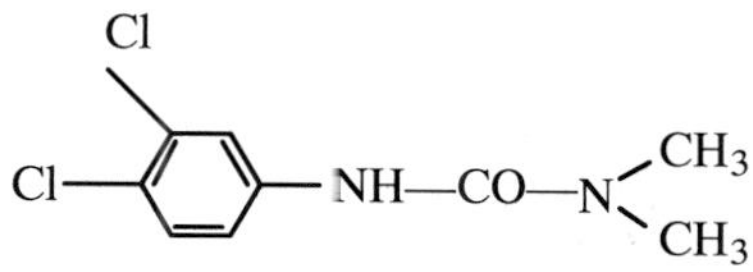	Dyhard UR200 SKW Trotsberg
- BDMA (epoxy 2)	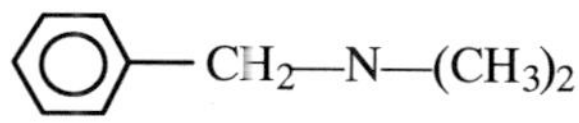	Aldrich
* Toughening agents - CTBN 8	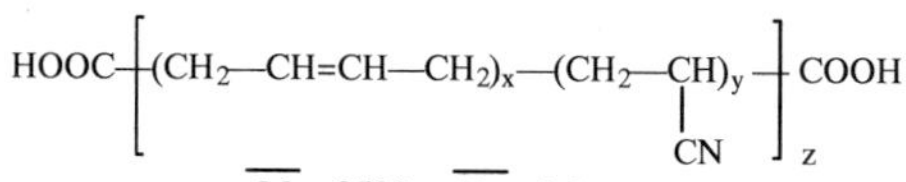 $\overline{M}_n = 3500$ g ; $\overline{F}_n = 1.8$	Hycar CTBN BF Goodrich
- CSR particles	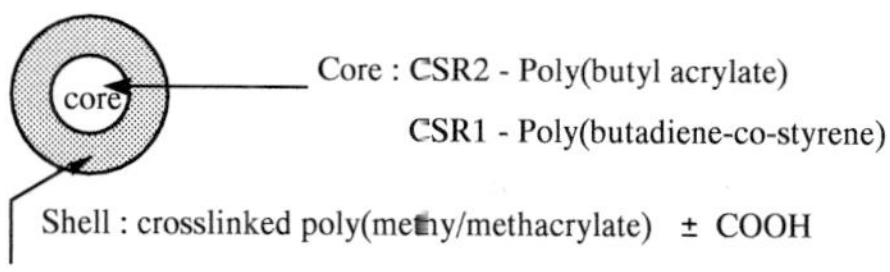	Rohm & Haas EXL 8866 EXL 2611
- Polydimethylsiloxane	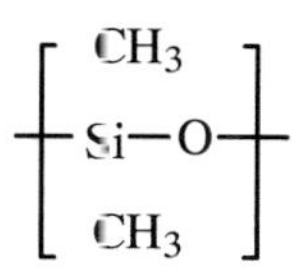	

RESULTS

All the results for static and fatigue tests are summarized in Table 2.

Static properties

Whatever the toughening agent introduced and the neat epoxy system used the same trends are found . They are detailed in previous papers [4-10].

As the volume fraction of rubber increases the Young's modulus and yield stress decrease while the toughness increases (from around 1 MPa.$\sqrt{m}$ for the neat system up to 1.7 MPa.$\sqrt{m}$ with CSR). Due to the partial miscibility of PDMS and CTBN, Tg slightly decreases although Tg remains constant with the introduction of CSR particles. This increase in toughness is explained by the deformation mechanisms of the rubber itself (cavitation and debonding) which promote the shear yielding of the epoxy matrix, crazing being not being found in such highly crosslinked networks [1,2]. It can be noticed as a general trend, that K_{Ic} increases when σ_y decreases, thus the presence of rubber particles favours the plastic deformation of the surrounding networks.

Fatigue crack propagation

Figures 1 and 2 show the FCP curves for PDMS and CSR toughened systems respectively. The experimental data can be linearly fitted in log-log diagrams in good agreement with the Paris law

$$\frac{da}{dN} = C\,(\Delta K)^{m}$$

where C and m are material dependent constants.

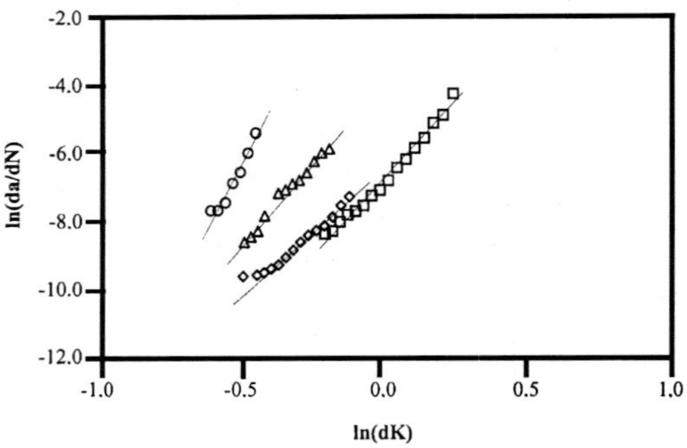

Fig. 1. FCP curves for the neat epoxy 1 (O) and PDMS modified networks with 4% vol. (Δ), 8 % vol. ( ) and 15 % vol. (◊).(Rey et al., 1999, with kind permission from Kluwer Academic Publisher).

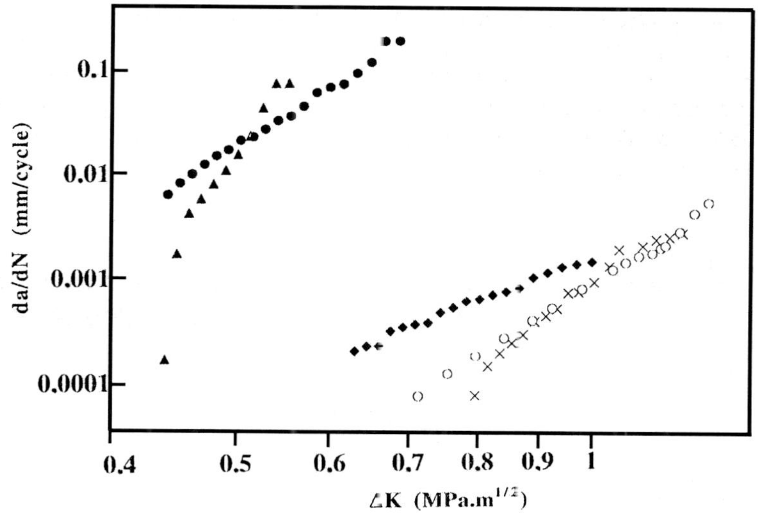

Fig. 2. FCP curves for the neat epoxy 2 (▲) and CSR1 epoxy 2 blends : (●) 9.5 % vol., (◆) 17.5 % vol., (x) 24 % vol., (o) 17.5 % vol. CTBN8. (Becu et al., 1997, Copyright 2002 – Reprinted by permission of John Wiley & Sons, Inc.).

Some similar results on epoxy toughened networks have already been obtained [5,6,8,9,10]. Other quantitative models were proposed in the literature and applied to toughened epoxies [9-10] but their application is out of the scope of this paper. Additional tests were performed showing that the effect of frequency is negligable in the range 1-10 Hz. The influence of the stress ratio, R_σ, studied with 15 % PDMS sample in the range 0-0.29 is also very limited [15].

DISCUSSION

Whatever the chemical nature of the rubber (PDMS, CTBN or CSR), the presence of a rubbery separated phase improves the resistance to FCP. In every case the FCP curves move toward higher ΔK values with a strong decrease of the exponent m.

Influence of the epoxy system

In Table 2, it is shown, that the epoxy 1 system is tougher than the epoxy 2 (K_{Ic} = 0.93 MPa.$\sqrt{m}$ compared with 0.8 MPa.$\sqrt{m}$). This is in agreement with well-known theories [1]. epoxy 1 being the less crosslinked material (with the higher molecular weight between crosslinks), has higher toughness and is also the more fatigue resistant with, for example, the higher ΔK value for $\frac{da}{dN}$ = 7.5x10^{-4} mm.min^{-1}. Some previous works [9,14] confirmed that the exponent m decreases when the fatigue resistance increases and a simple model based on crack opening displacement was proposed and checked by Fisher et al. [9].

Table 2.1. Static fracture properties of pure and toughened networks

Composition	Modifier content (% by vol.)	Tg [a)] (°C)	E_{RT} (GPa)	σ_y (MPa)	K_{Ic} (MPa. $\sqrt{m}$)
Neat epoxy system 1 (Diuron)	0	125	3.86	118	0.93
Epoxy 1 + PDMS	4	123	3.45	107	1
	8	122	2.82	98	1.36
	15	116	2.68	85	1.25
Neat epoxy system 2 (BDMA)	0	141	3.2	105	0.8
Epoxy 2 + CSR1	9.5	140	2.65	95	1.1
	17.5	142	2.67	85	1.1
	24	140	2.17	72	1.4
Epoxy 2 + CSR2	24	139	2.2	71	1.7
Epoxy 2 + CTBN8	17.5	135	1.8	76	1.3

a) DSC

Table 2.2. Fatigue crack propagation of pure and toughened networks

cComposition	Modifier content (% by vol.)	ΔK [b)] (MPa). $\sqrt{m}$	ΔK_{max} [d)] (MPa). $\sqrt{m}$	C [c)]	m [c)]
Neat epoxy system 1 (Diuron)	0	0.58	0.64	7.3	16.9
	4	0.72	1.01	0.017	9.5
Epoxy 1 + PDMS	8	0.97	1.28	0.001	8.8
	15	0.93	0.89	0.0012	7
Neat epoxy system 2 (BDMA)	0	0.43	0.6	0.437	8.9
	9.5	0.44	0.62	0.204	6.9
Epoxy 2 + CSR1	17.5	0.85	1.01	0.001	4.2
	24	0.86	1.14	0.0007	8.4
Epoxy 2 + CSR2	24	0.88	1.2	0.0007	8.2
Epoxy 2 + CTBN8	17.5	0.85	1.1	0.0008	6.9

b) for da/dN = 7.5×10^{-4} mm/cycle ; c) coefficients of the Paris law ; d) ΔK at the crack instability

Influence of rubber and volume fraction

On Figures 1 and 2 the introduction of increasing amounts of rubbers improves the FCP resistance of the materials. As a same trend C and m decrease in each series and ΔK_{max} (at the crack instability) increases regularly, except for 8 % PDMS where there is a maximum [8]. The mechanisms for rubber toughening both in fatigue and in static conditions were already described [1,2,6,9] and we can mention the pronounced ability to plastic deformation (which is confirmed by the decrease in σ_y), cavitation of rubber particles, dilatation bands in the case of CSR particles [2], crack front pinning and branching. We plotted on Figure 3, K_{Ic} versus ΔK_{max} and it appears that the correlation is rather good as shown previously in the literature [6,9] giving support to the similar toughening mechanisms appearing in both static and fatigue conditions.

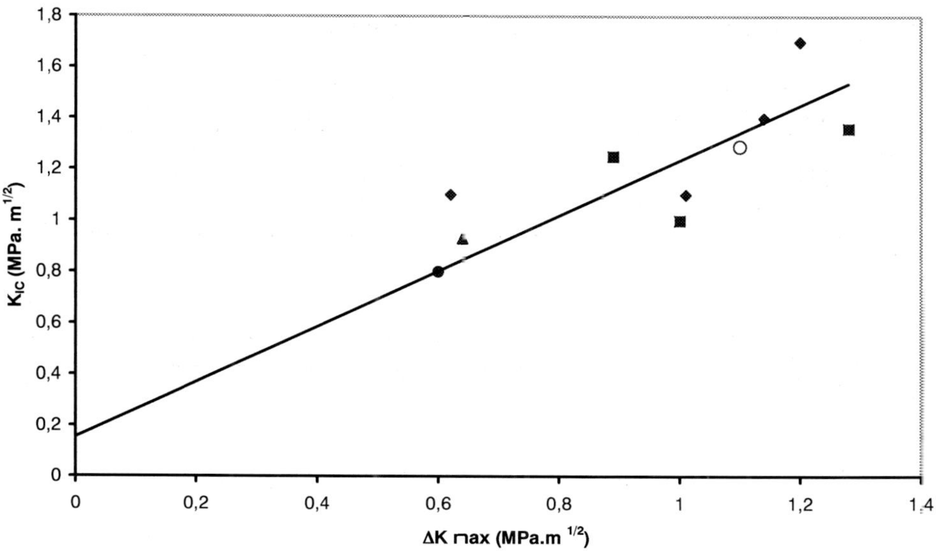

Fig. 3. Correlation between the static fracture toughness (K_{Ic}) and ΔK_{max} values from the FCP experiments for the various materials investigated. (▲) neat epoxy 1, (●) neat epoxy 2, (■) epoxy 1 + PDMS, (◆) epoxy 2 + CSR, (O) epoxy 2 + CTBN

CONCLUSIONS

Using linear elastic fracture mechanics, the toughening effect of various rubbers, on two different epoxy networks are proven. The fatigue crack propagation was recorded and the Paris law was well confirmed.

The toughening mechanisms in static and fatigue conditions are very similar as demonstrated by the correlation between K_{Ic} and ΔK_{max}. Some synergetic effects were found in FCP by mixing soft particles (rubbers) with glass beads [16,17] or glass bubbles [18] to form what is called hybrid composites. One of the main interests of fracture mechanics in fatigue experiments is that the FCP is well described, by simple models, and it is a very useful tool with no dependence on the initial crack tip radius. Furthermore with FCP tests a great deal of time can be saved compared with conventional Wölher curves. The knowledge of FCP is also useful to calculate the life duration of samples or structures submitted to fatigue loading.

REFERENCES

1. Pascault, J.P., Sautereau, H., Verdu, J. and Williams, R.J.J. (2002) *Thermosetting Polymers*, Marcel Dekker, New York.
2. Arends, C.B. (1996) *Polymer Toughening*, Marcel Dekker, New York.
3. Verchère, D., Pascault, J.P., Sautereau, H., Moshiar, S.M., Riccardi, C.C. and Williams, J.J. (1991) *J. Appl. Polym. Sci* , **42**, 701.
4. Maazouz, A., Sautereau, H. and Gérard, J.F. (1992) *Polymer Networks & Blends*, **2**, 65.

5. Becu, L., Maazouz, A., Sautereau, H. and Gérard, J.F. (1997) *J. Appl. Polym. Sci.*, **65**, 2419.
6. Karger-Kocsis, J., Friedrich, K (1992) *Colloïd Polym. Sci.*, **270**, 549.
7. Karger-Kocsis, J., Friedrich, K (1993) *Compos. Sci. Technol.*, **48**, 263.
8. Rey, L., Poisson, N., Maazouz, A. and Sautereau, H. (1999) *J. Mat. Sci.*, **34**, 1775.
9. Hwang, J.F., Manson, J.A., Hertzberg, R.W., Miller, G.A. and Sperling, L.H. (1989) *Polym. Eng. Sci.*, **29**, 1477.
10. Abou-Hamda, M., Mai, Y.W., Wu, D.X. and Cotterell, B. (1993) *Polymer*, **34** (20), 4221.
11. Hertzberg, R.W. and Manson J.A. (1980) *Fatigue of Engineering Plastics*, Academic Press, New-York
12. Williams, J.G. (2001) In: Fracture Mechanics Testing Methods for Polymers Adhesives and Composites. ESIS Publication, pp. 11-26, Moore, D.R., Pavan, A. and Williams, J.G. (Eds), Elsevier, Oxford.
13. Castellani, L. and Rink, M. (2001) In: Fracture Mechanics Testing Methods for Polymers Adhesives and Composites. ESIS Publication, pp. 91-116, Moore, D.R., Pavan, A. and Williams, J.G. (Eds), Elsevier, Oxford.
14. Fisher, M., Martin, D. and Pasquier M. (1995) *Macromol. Symp.*, **93**, 325.
15. Poisson, N. (1996). PhD Thesis, INSA Lyon, France.
16. Maazouz, A. Sautereau, H. and Gérard, J.F. (1993) *J. Appl. Polym. Sci.*, **50**, 615.
17. Azimi, H.R., Pearson, R.A. and Hertzberg, R.W. (1995) *J. Appl. Polym. Sci.*, **58**, 449.
18. Azimi, H.R., Pearson, R.A. and Hertzberg, R.W. (1995) *Polym. Eng. Sci..*, **36** (18), 2352.

The Application of Fracture Mechanics to Polymers, Adhesives and Composites
D.R. Moore (Ed)

FRACTURE MECHANICAL BEHAVIOR OF THERMOPLASTIC POLYMERS AS A FUNCTION OF MOLECULAR AND SUPERMOLECULAR VARIABLES

J. KARGER-KOCSIS

INTRODUCTION

Toughness is a key property for many applications of polymeric systems both of thermoplastic and thermoset nature. Assuming that the toughness - determined by a suitable approach of fracture mechanics - represents a material property, it should be obviously related to the molecular and supermolecular build-up of polymers. This aspect, however, has not been studied accordingly in the past albeit its clarification would contribute to a wider acceptance of fracture mechanics. The toughness controlling molecular parameters in thermoplastic amorphous polymers in the glassy state are far better studied than those affecting the toughness of semicrystalline polymers. This fact is likely due to the following reasons: a) some analogy between the physical network structure in amorphous polymers and chemical network in thermosets and rubbers. Note that in case of the latter the network parameters were proved to govern the toughness (see later), and b) strong interrelation between crystalline and morphological (supermolecular) entities due to which the toughness can hardly be considered as a function one structural parameter in semicrystalline polymers. This contribution is aimed to give a brief summary on the (super)molecular dependence of the toughness for both amorphous and semicrystalline thermoplastic polymers.

AMORPHOUS POLYMERS

In order to find correlations between molecular, network variables and toughness attention should be paid to the right selection of model materials in respect to the fracture mechanical approach used. It was shown by the author's group that amorphous copolyester (aCOP), which fail exclusively by shear yielding, is the best choice when the essential work of fracture (EWF) method is used for the toughness assessment ([1-2] and references therein). This is reasoned by two facts: a) the total work of fracture in aCOPs can be easy split for yielding- and necking+tearing-related terms as these polymers undergo full ligament yielding prior to the onset of crack growth, and b) necking+tearing occur via cold drawing which is controlled by the entanglement network (e.g. [3]). Further, the mean entanglement molecular weight (M_e) is a constant for a given polymer provided that its mean molecular weight (MW) is beyond a critical threshold (>2M_e). The MW of commercial polymer is always above this threshold value.

Effects of molecular weight and network variables
It was shown that the yielding-related specific work of fracture ($w_{e,y}$) does not depend on the MW in aCOPs. MW affects solely the necking+tearing process (e.g. [4-5] and references therein). Based on experimental data an interesting correspondence between mean MW and strain rate was found. Accordingly, the response of an amorphous polymer of high mean MW

at high strain rate is similar to that of a low mean MW version at low strain rate [5-6]. Note that based on some experimental evidence $w_{e,y}$ is considered by the author as an inherent toughness value, which is closely related or even equal to that of the plane strain value [5].
If MW does not influence the toughness, the decisive role should be assigned to characteristics of the physical network structure. This assumption is supported by the fact that the "plastic" zone is fully restored when heating the specimen beyond the T_g for a short time, i.e. the plastic zone was formed previously by cold drawing.
The entanglement network is usually characterized by the mean MW between entanglements (M_e) and by the entanglement network density (ν_e). ν_e is defined by:

$$\nu_e = \frac{\rho N_A}{M_e} \tag{1}$$

where ρ is the density and N_A is the Avogadro number (6.023×10^{23} mol^{-1}).

Toughness interpretation models
According to Lake and Thomas (e.g. [7]) the fracture energy (G_c) for crosslinked rubbers is given by:

$$G_c = K M_X^{1/2} \tag{2}$$

where K is a constant involving the polymer density and aspects of the bond strength, and M_X is the mean MW between crosslinks. How can be Equation 2 adopted for amorphous polymers? Assuming that $G_c = w_{e,y}$ and replacing M_X by M_e for polymers of physical network structure, a linear relationship between $w_{e,y}$ and the square root of M_e can be expected. Figure 1 seems to support this prediction. Needless to say that the M_e data underlay a large scatter due to problems of their experimental determination [5]. It is the right place to draw the attention that G_c of thermosets usually also obeys the Lake-Thomas Equation (e.g. [8]).

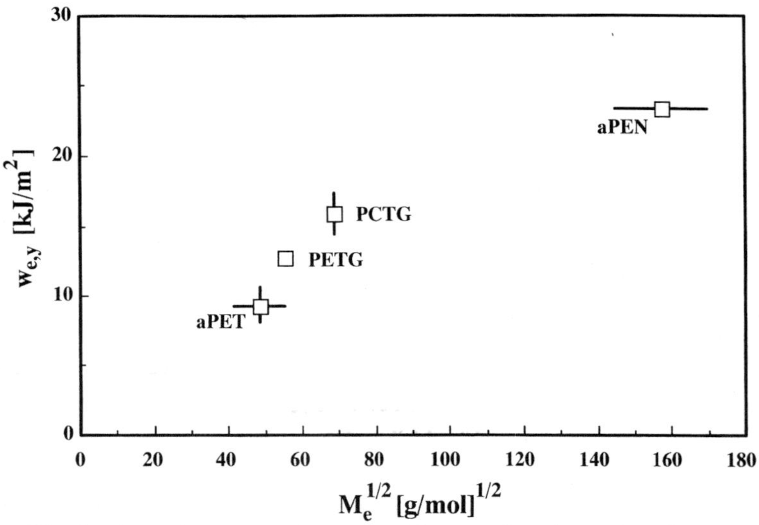

Fig.1. Specific yielding-related essential work of fracture ($w_{e,y}$) as a function of square root of M_e for amorphous copolyesters.
Designations: PET- polyethylene terephthalate, PETG- and PCTG- amorhous PET-type copolyesters with various amounts of cyclohexylenedimethylene units, PEN- polyethylene naphthalate [2,5].

According to the model proposed by Brown [9] for glassy polymers which fail by crazing, the fracture energy (G_c) increases linearly with the square of ν_e. This is opposed to results given in Figure 1. Recall, however, that the aCOPs studied by us never fail by crazing. Wu claimed [10] that the molecular aspects of the yielding/crazing behavior are controlled by ν_e and by the so called characteristic ratio. The latter is a measure of the intrinsic chain flexibility. In our case the characteristic ratio should increase from aPET towards aPEN due to the bulky rigid groups along the molecular chain. Nevertheless, there is a strong discrepancy between the above models. Its possible cause is that some of them consider the work-hardening (via shear yielding and crazing) which is excluded in the term $w_{e,y}$. The effects of work hardening should be, however, noticeable in the necking-related EWF terms. Figure 2 plots the specific necking-related essential ($w_{e,n}$) and non-essential ($\beta''w_{p,n}$) parameters as a function of ν_e. Based on the trend in Figure 2 one can claim that the necking+tearing related work of fracture parameters increase with the entanglement network density, indeed.

The above treatise already highlights the basic problem: effects of the initial structure should be separated properly from those affected by loading-induced structural changes (referred above as work-hardening) both in the experiments and theory. This statement holds also for semicrystalline polymers and represents a great challenge. Accordingly it is indispensable to perform the fracture mechanical tests on suitable specimens with in-situ inspection of the microstructural change. This could be done by using various techniques, like microbeam synchrotron X-ray, polarized IR and Raman spectroscopy, high-voltage electron microscopy, atomic force microscopy.

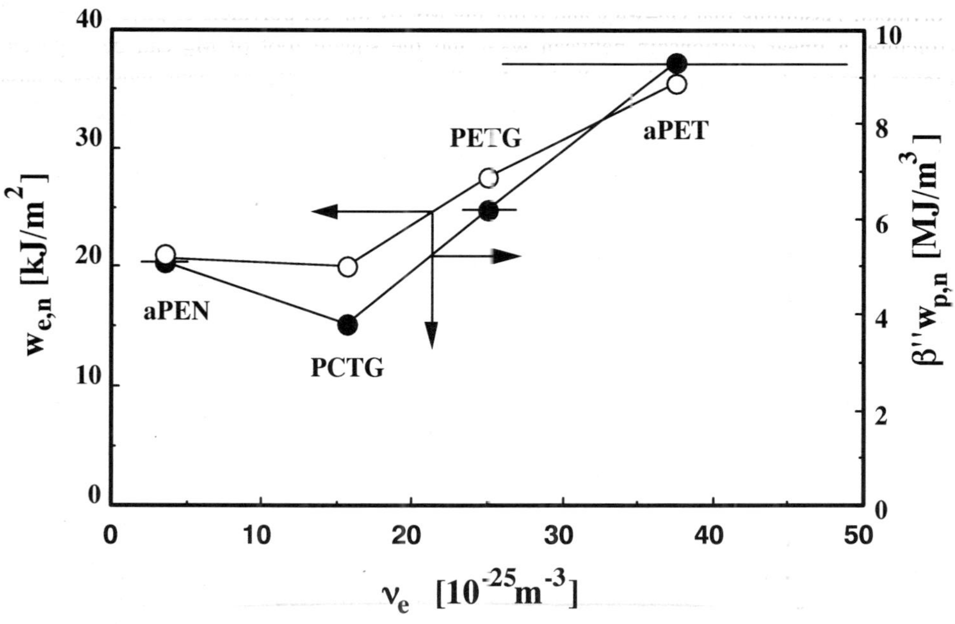

Fig.2. Necking-related specific work of fracture parameters ($w_{e,n}$ and $\beta''w_{p,n}$) as a function of ν_e for amorphous copolyesters.

Notes: data taken from Refs. [5,11]; for designations cf. Fig. 1; only the x-scatter range is indicated

SEMICRYSTALLINE POLYMERS

It was already shown that the presence of processing-induced higher order structures (e.g. skin-core morphology) may blur the use of fracture mechanics. The effect of morphology (spherulitic structure) on the fracture toughness proved to be also not negligible. Studies performed on β-phase PP exhibited a strong dependence of the work of fracture parameters as a function on both crystallinity and mean MW [1,12].

In the literature highly contradictory data can be traced on how the toughness is affected by molecular and supermolecular variables. This can be reasoned by the fact that one single parameter of the crystalline structure can hardly be varied without affecting some others. Variation in the degree of crystallinity for example is associated with manifold changes both in the spherulitic structure (type, size) and within its constituents (thickness and order of lamellae, amorphous layer thickness, tie molecules density). According to the model of the author both toughness and resistance to fatigue crack propagation (FCP) go through a maximum as a function of the degree of crystallinity (cf. Figure 3) [5,12]. The increasing left hand flank of the curve represents the resistance resulting from the crystalline structure, while the declining right hand part is owing to the lacking connection between the crystalline lamellae (especially the tie molecules which are capable to redistribute the stresses locally, are missing). The resulting curve in Figure 3 indicates that the toughness response is controlled by the "weaker" contribution. There are many indirect hints for the validity of the model depicted in Figure 3, however, an elegant experimental evidence is still lacking. One should also keep in mind that the tie molecules density is affected by the mean MW, viz. its value increases with the MW. Yeh and Runt (e.g. [13]) have demonstrated the role of MW and tie molecules density of the FCP behavior of semicrystalline polymers. The decisive role of the tie molecules (which can be estimated both experimentally [14] and theoretically), along with that of MW has been stated by Beerbaum et al. [15-16] recently. The authors feeling is that the abovementioned strong interaction between the structural parameters in semicrystalline polymers necessitates a complex statistical evaluation. That this is the right tool for data reduction and "mining" has been proposed by Egan and Delatycki [17].

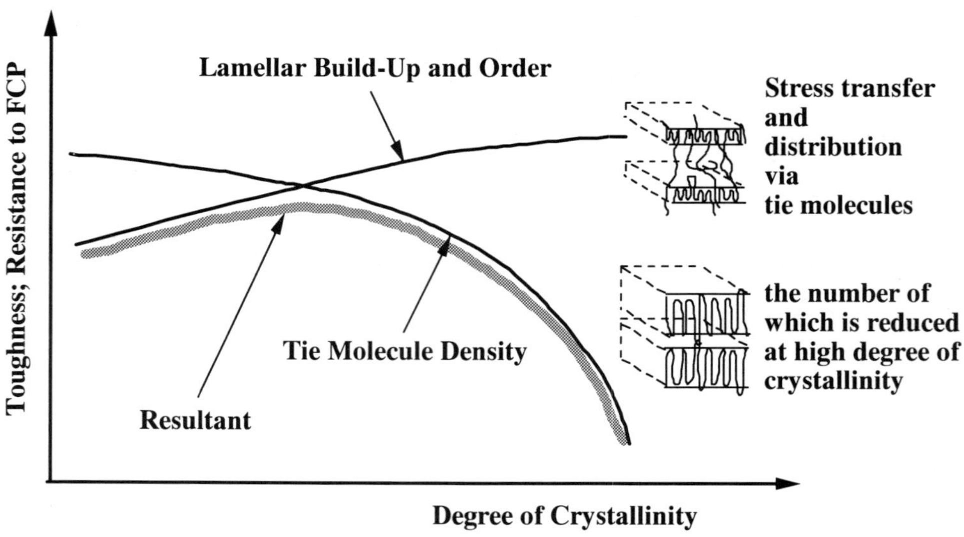

Fig.3. Predicted change of toughness and resistance to FCP as a function of degree of crystallinity [5,12]

CONCLUSIONS

The toughness of amorphous polymers depends unequivocally on characteristics of the entanglement network structure. Further work is needed, however, to separate effects of the initial physical network (likely related to the toughness at fracture initiation) from that of the loading-induced one (fracture propagation resistance). In this case special attention should be paid to distinguish between failure via shear yielding and crazing.
The microstructural dependence of fracture mechanical parameters is far less understood for semicrystalline polymers. Albeit some tendencies can be deduced, a reliable description is missing. This is due to the strong interrelation between parameters of the crystalline structure which does not allow us to study the effects of one or other structural parameter in a separate way. As a consequence, combined action with respect of specimen preparation, polymer characterization and data mining/reduction are required in order to close the gap between the toughness and structural parameters.

REFERENCES

1 Karger-Kocsis, J. (2000). In: *Fracture of Polymers, Composites and Adhesives*, ESIS Publ. 27, pp. 213-230, Williams, J.G. and Pavan, A. (Eds.), Elsevier Sci., Oxford
2 Karger-Kocsis, J. and Moskala, E.J. (2002). *SPE-ANTEC*, **60**, 1751.
3 Strobl, G. (1996) *The Physics of Polymers*, Springer, Berlin
4 Karger-Kocsis, J. and Moskala, E.J. (1997) *Polym. Bull.*, **39**, 503.
5 Karger-Kocsis, J. (2002). In: *Handbook of Thermoplastic Polyesters*, pp.717-753, Fakirov, S. (Ed.), Wiley-VCH, Weinheim
6 Karger-Kocsis, J. and Czigány, T. (2000) *Polym. Eng. Sci.*, **40**, 1809.
7 Lake, G. and Thomas, A.G. (1967) *Proc. Roy. Soc. London Ser. A*, **A300**, 108.
8 Karger-Kocsis, J. and Gremmels, J. (2000) *J. Appl. Polym. Sci.*, **78**, 1139.
9 Brown, H.R. (1991) *Macromolecules* **24**, 2752.
10 Wu, S. (1990) *Polym. Eng. Sci.*, **30**, 753.
11 Karger-Kocsis, J., Moskala, E.J. and Shang, P.P. (2001) *J. Thermal. Anal. Calorim.*, **63**, 671.
12 Karger-Kocsis, J. (2000). In: *Structure Development During Polymer Processing*, pp. 163-179, Cunha, A.M and Fakirov, S. (Eds.), Kluwer, Dordrecht
13 Yeh, J.T. and Runt, J. (1989) *J. Mater. Sci.*, **24**, 2637.
14 Brown, N. and Ward, I.M. (1983) *J. Mater. Sci.*, **18**, 1405.
15 Beerbaum, H. and Grellmann, W. (2000). In: *Fracture of Polymers, Composites and Adhesives*, ESIS Publ. 27, pp. 163-174, Williams, J.G. and Pavan, A. (Eds.), Elsevier Sci., Oxford
16 Beerbaum, H., Grellmann, W. and Seidler, S. (2001). In: *Deformation and Fracture Behavior of Polymers*, pp. 161-180, Grellmann, W. and Seidler, S. (Eds.), Springer, Berlin
17 Egan, B.J. and Delatycki, O. (1995) *J. Mater. Sci.*, **30**, 3307

The Application of Fracture Mechanics to Polymers, Adhesives and Composites
D.R. Moore (Ed)

FAILURE OF POLYMERS BY RUPTURE VERSUS DISENTANGLEMENT OF MOLECULES

F. RAMSTEINER

INTRODUCTION

Polymers fracture by rupture or disentanglement of the molecules [1] at the tip of the propagating crack. Rupture must prevail in crosslinked polymers and is likely to happen in thermoplastic materials at low temperatures and high deformation rates. If the fracture time and the test temperature are high enough for relaxation of the molecular structure, disentanglement is assumed to predominate as in polymer melts. Fracture mechanics allows the study of crack propagation within solid polymers and therefore it is an adequate tool to distinguish between disentanglement and rupture. This use of fracture mechanics to elucidate the details of failure is demonstrated in the following examples where the fracture mode is shown to depend on molecular weight, cross linking, temperature, and deformation rate.

EXPERIMENTAL METHODS

Mainly CT-specimens which had been milled from compression moulded sheets were used in this paper to study the behaviour of crack propagation independent of the initial flaw formation. In fatigue tests [2], crack propagation was measured by the compliance method [3]. Further experimental details or special arrangements are given in the individual paragraphs of this chapter.

EXPERIMENTAL RESULTS

Molecular weight and crystallinity

The dependence of the fracture time on the molecular weight at a given stress is one of the most important tests to study the transition from rupture to disentanglement of molecules. Disentanglement depends on the time the molecules need to slip through their long tube formed by the surrounding molecules, and this in turn depends on the molecular weight. Rupture of the molecules however is independent on the molecular weight, if the molecules are long enough to be fixes in the entanglement network.

Fig.1 shows the rate of fatigue crack in CT specimens of HDPE with different molecular weights as a function of the amplitude of the stress intensity factor.

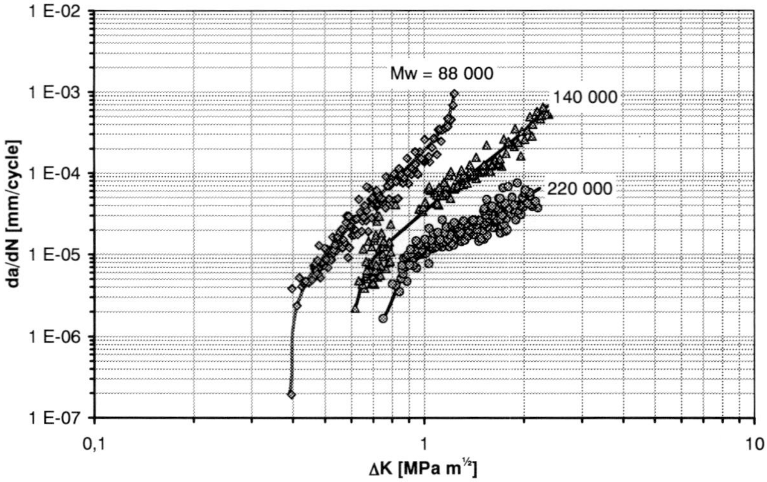

Fig.1 Rate of crack propagation in HDPE with different molecular weights as a function of the amplitude of the stress intensity factor in fatigue at 23°C and 10 Hz

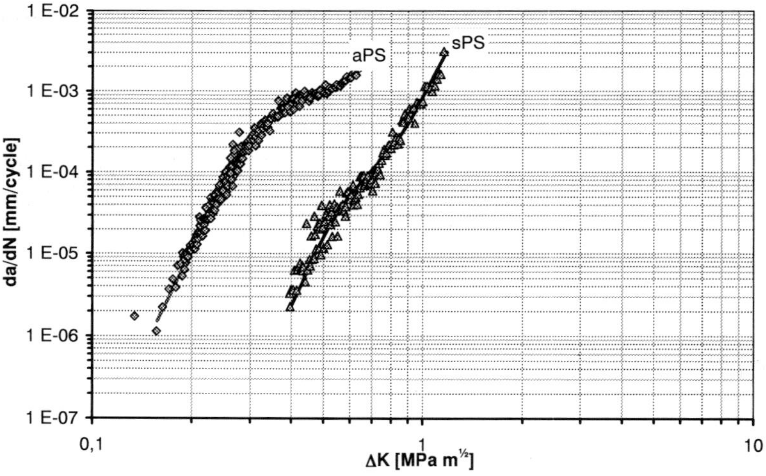

Fig.2 Crack propagation rate in fatigue of a syndiotactic and an atactic PS with the same molecular weight of 190 000. (frequency 10 Hz)

The higher the molecular weight the lower the crack propagation velocity. This observation confirms that crack propagation occurs by disentanglement and not by rupture. In semi-crystalline polymers like HDPE crystals can hinder the slippage of the molecules by their stronger bonding within the crystals than entanglements can do. Thus this effect from crystallinity is superimposed on the molecular weight effect. If the number of the tie

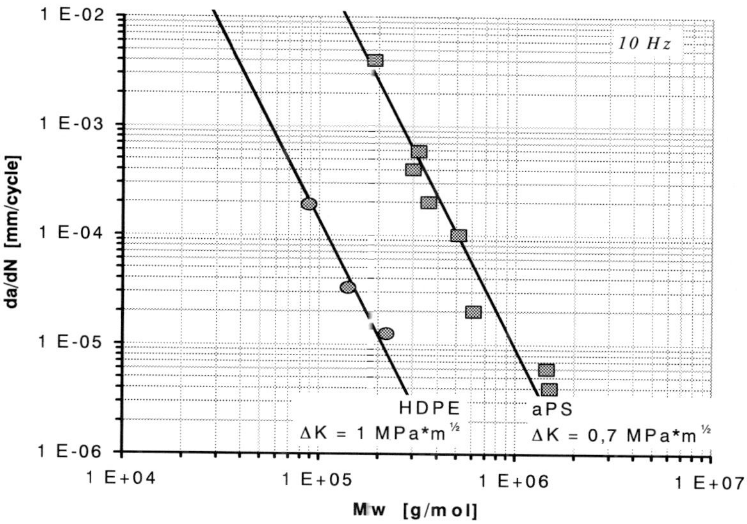

Fig.3 Crack propagation rates per cycle da/dN at 23°C and 10 Hz as a function of the molecular weight M_w of HDPE and a PS at a constant stress intensity factor.

molecules between the crystals and spherulites is sufficient enough for bonding then semi-crystalline polymers are expected to be stronger than amorphous ones.

To demonstrate the influence of crystallinity independent of the molecular weight, a syndiotactic and an atactic PS with the same molecular weight of 190 000 g/mol were tested in fatigue. The results are shown in fig. 2. In amorphous atactic PS (aPS) cracks propagate especially at low speeds, in the region of disentanglement, nearly two orders of magnitude faster than in the semi-crystalline syndiotactic PS.

The rates of crack propagation per cycle da/dN at a constant amplitude of stress intensity factor (ΔK) are plotted in fig.3 as a function of the molecular weight M_w for aPS (ΔK=0.7MPa $m^{1/2}$) and HDPE (ΔK=1.0 MPa $m^{1/2}$) respectively. The dependence follows roughly the relationship

$$da/dN \propto 1/M^{3.5}$$

which is very similar to the molecular weight dependence of the viscosity of polymer melts where reptation is the basis for understanding the flow behaviour.

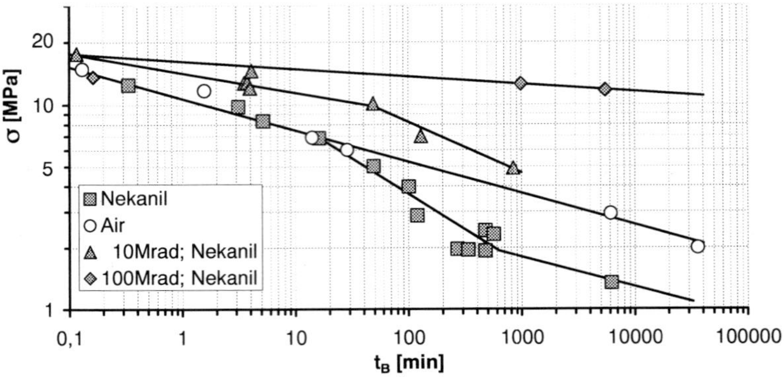

Fig. 4 Applied stress versus the corresponding time to failure of notched specimens of cross linked HDPE in a surface active agent (Nekanil®) at 50°C.

Cross-linking

If disentanglement prevails in failure of polymers cross linking of the molecules should increase life time enormously because slippage is practically impossible. Fig. 4 shows the plot of the failure stresses over the corresponding time to failure of the notched specimens of HDPE in a surface active agent (Nekanil®) at 50°C. For these experiments 3mm thick notched dumbbell bars with a 0.7mm deep notch were stressed at various constant loads in the

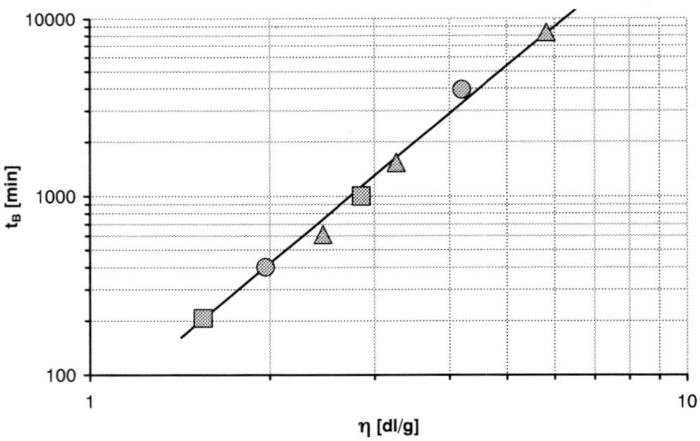

Fig. 5 Time of failure as a function of the intrinsic viscosity of HDPE at 2 MPa tensile stress in a surface active agent (Nekanil®) at 50°C.

liquid. The pure HDPE in air shows for comparison typical failure under creep, that is to say an increasing life time with decreasing stress. In the liquid surface active agent, behaviour of un-crossed or only slightly cross-linked polymer changes to the well known three-stage drop. At high stresses near the failure stress measured in air the failure time is not influenced by the agent because its diffusion time to the crack tip is too long compared to the failure time. In the middle region, failure time is reduced by the agent due to the relieved entanglement. At low stresses, relaxation caused by the agent determines the life time [4]. If the material is highly cross linked, failure time is enormously increased as expected for materials where

entanglement is suppressed. Therefore it is not surprising that in many applications e.g. for pipes in the heating industry the materials are cross-linked to reduce disentanglement at low stresses and high temperatures, thus extending their life times.
In the low stress region where disentanglement dominates the failure in uncrosslinked polymers life time becomes strongly dependent on the molecular weight of the sliding molecules. Fig.5 shows for notched specimens of HDPE the plot of lifetime in the surface active agent Nekalin® at 50°C as a function of the intrinsic viscosity [η] of the polymer. The measurements were performed with 3 mm thick specimens with a 0.7mm deep notch at a creep stress of 2 MPa. Experimental values of the life times t_B follow in this region of middle/low stress transition the relation

$$t_B \propto [\eta]^{2.5}$$

and with the dependence of $[\eta] \propto M^{0.7}$ we are very close to the simple gliding model, without deformation of the surrounding, in which the quadratic dependence is expected [4].

Temperature
On the basis of the disentanglement model it is to be expected, that crack propagation is dependent on the temperature at higher temperatures, because sliding molecules take time. At lower temperatures, in the brittle state, polymer scission predominates which only depends on the activation energy of the bonding in the main chain. In accordance with this model Hasegawa et al. [5] showed that the critical energy release rate G_C of PS increases with the molecular weight at higher temperatures. However at lower temperatures near 230K the critical energy release rate approaches a plateau at molecular weights where the molecules are sufficiently entangled to allow only scission.
In tab.1 [6] the stress intensity factors for two PS with different molecular weights are shown for different temperatures. The measurements were performed with SENB specimens and at a 5mm/min deformation speed. In the brittle state near 23°C the critical stress intensity factors are nearly independent of the molecular weight, whereas with increasing temperatures slippage also increases and the critical stress intensity factors increase more rapidly for the higher molecular weight PS.

Test temperature Molecular weight	23°C	50°C	70°C	85°C
185 000	1.75	1.30	0.95	0.50
333 000	2.0	1.55	1.45	1.05

Tab.1 Critical stress intensity factors [MPa√m] of PS at different temperatures

Deformation speed
According to the correspondence principle of time-temperature the mode of low temperature, and low deformation speed corresponds to high speed deformation at high temperatures. Therefore transitions in failure are observed with deformation speed as well as temperature. As an example, fatigue crack growth rates in HIPS are plotted in Fig.6 as a
function of the stress intensity factor amplitude [7]. The change of slope of this curve may indicate a transition in the resistance to crack propagation from the low to the high rates of crack growth. The fracture surface of the specimen are shown in fig.7. In the low speed region near $\Delta K \approx 0.6$ MPa $m^{1/2}$ the blend morphology consisting of rubber particles in a matrix is

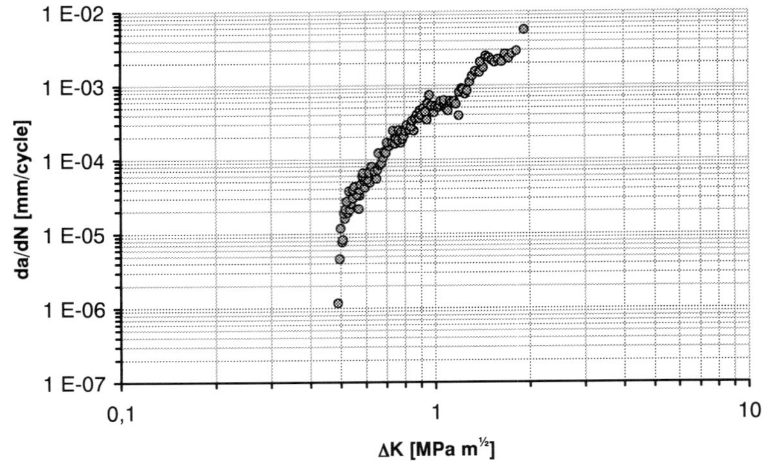

Fig.6 Crack growth rates in fatigue test as a function of the amplitude of the stress intensity factor for HIPS at 23°C and 10 Hz.

clearly seen without bigger deformations (fig.7a). At higher speeds, however, the plastic deformation in the rubber modified material prevails (fig.7b). In this region the molecules had had no time to disentangle simply, but where forced to deform their surrounding by stretching before rupture or final disentanglement

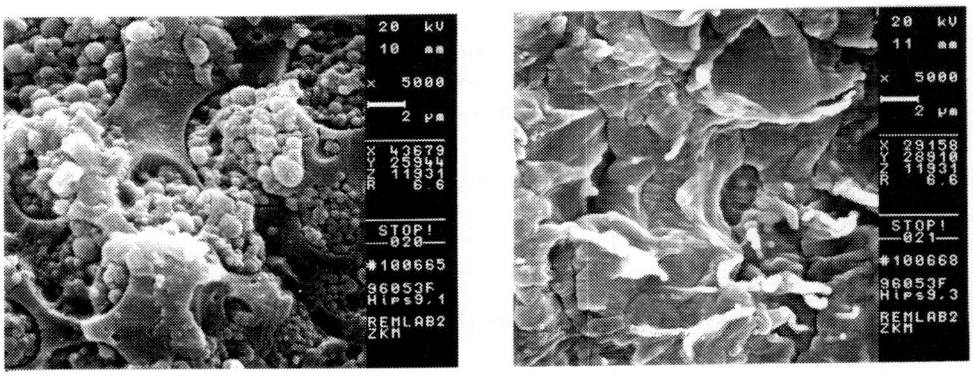

a b

Fig.7 Fracture surface of the specimen of fig. 6

ACKNOWLEDGEMENT

The experimental contributions by Mr Armbrust and Mr Forster are gratefully acknowledged.

REFERENCES

[1] Berger L.L., Kramer E.J. (1987) Macromolecules **20**, 1980
[2] ISO/TC 61/SC /WG 7 -] ASTM E 642

[3] Ramsteiner F., Armbrust T., (2001) Polymer Testing **20**, 321
[4] Ramsteiner F., Kunststoffe 80 (1990) 695
[5] Hasegawa S., Nakano N., (1991) Polymer Physics in Japan XXXIV
[6] Ramsteiner F., (1996) in Kunststoff Handbuch 4, Polystyrol, Hanser Verlag München Wien, p. 196
[7] Ramsteiner F., Schuster W., Forster S. (2001) Deformation and Fracture of Polymers, ed. by W. Grellmann and S. Seidel, Springer-Verlag Berlin-Heidelberg-New York, p. 27

The Application of Fracture Mechanics to Polymers, Adhesives and Composites
D.R. Moore (Ed)

RATE DEPENDENT FRACTURE OF POLYMERS - APPLICABILITY AND LIMITATIONS OF FORCE BASED FRACTURE MECHANICS APPROACHES

Z. MAJOR and R.W. LANG

INTRODUCTION

For many structural engineering applications of polymeric materials, fracture behavior under monotonic and impact loading conditions is of prime practical importance. Due to the viscoelastic nature of polymers, fracture properties are significantly affected by the loading rate, the test temperature and the local (if notches or cracks are present) and global stress state. As a result of the complex combination of the influence of these parameters, fracture values determined by conventional test methods (e.g., monotonic tensile test, standardized bending and penetration type impact tests, etc.) are only of limited use for advanced engineering design purposes based on finite element methods and numerical simulations as well as for a detailed and reliable material characterization to aid material ranking and selection for a given application. This is especially true for situations where parts or component are exposed to high loading rates.

Due to the wide range of fracture behavior generally observed in plastics from brittle (i.e., quasi-brittle failure) to highly ductile failure, numerous fracture mechanics concepts have been proposed and applied to characterize the failure behavior of plastics in terms of specific fracture parameters [1]. The different concepts reflecting the various degrees of crack tip yielding generally were applied to specific engineering polymers at pre-defined single testing rates or in a rather limited testing rate regime, so that hardly any investigations were reported using various fracture mechanics methods for a given polymer type over a wide loading rate range.

Considering the above, the overall objectives of this paper is to describe and discuss the rate dependent fracture behavior of polymers in terms of material specific property functions and to determine the applicability range of various force based fracture mechanics concepts as characterization tool for the generation of fracture parameters for various failure modes and regimes. The paper will be primarily concerned with the ductile and the brittle regimes (i.e., quasi-brittle) of rate dependent fracture along with the definition of limitations for the applicability of the associated fracture mechanics concepts. Furthermore, based on experimental results proposals will be made to define characteristic values for the lower and upper end of the rate dependent ductile-brittle transition. Having thus quantitatively established the upper and lower plateau curves along with the relevant transition points, rate dependent fracture behavior of polymers may then be characterized over a wide loading rate range using material specific, specimen configuration independent fracture concepts.

BACKGROUND AND SCOPE

There are two main categories of fracture mechanics analysis. The force based analysis (FBA) requires only critical loads to determine a particular fracture parameter, while the energy based analysis (EBA) requires either the entire load-displacement record of a fracture experiment or directly measured critical energy values.

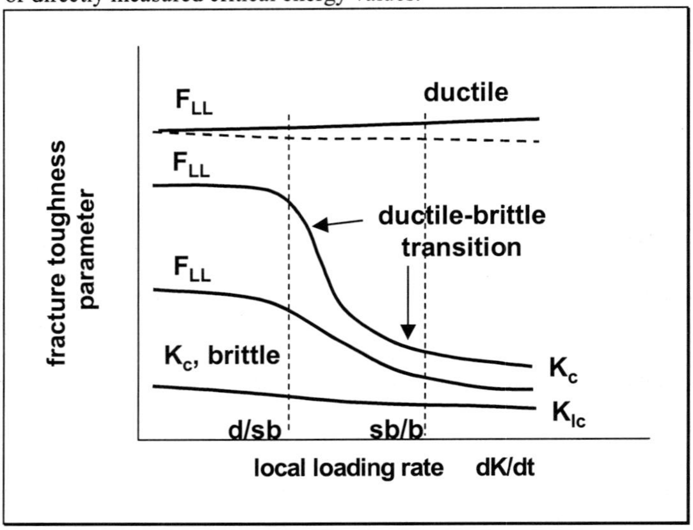

Fig. 1. Schematic illustration of rate dependent fracture property functions of polymers for various types of fracture behavior also indicating the appropriate fracture parameters and the lower (d/sb) and upper (sb/b) embrittlement transition points; FBA fracture mechanics concepts.

For an FBA approach, based on the experimental findings [2], rate dependent fracture property functions in terms of an arbitrary FBA fracture toughness parameter may be expected as depicted in Fig. 1. The relevant fracture parameters in a force based analysis (FBA) are the limit load, F_{LL}, for highly ductile fracture behavior in the sense of cross-sectional yielding (PYFM concept) [3], and the critical stress intensity factor, K_c, or K_{Ic}, for brittle fracture behavior (LEFM concept).

In terms of test procedure and measurement technique, the essential pre-requisites of an FBA approach is to determine material specific fracture toughness values, which requires a sufficient good quality of the load signal and the correspondence of the true specimen load with the experimentally determined external load. The relevant loading rate parameter controlling the failure mode is the local crack tip loading rate, *dK/dt* [4]. Also indicated in Fig. 1 are the limitations of the applicability of the various concepts on the loading rate scale in terms of relevant embrittlement transition points (ductile/semi-brittle transition point (d/sb) for the lower end of the embrittlement transition on the loading rate scale; semi-brittle/brittle transition point (sb/b) for the upper end of the embrittlement transition on the loading rate scale).

EXPERIMENTAL

To examine the applicability of various fracture mechanics concepts to rate dependent fracture of polymers, monotonic fracture experiments were performed with various polymers over a loading rate range of up to 7 orders of magnitude from 10^{-5} to 8 m/s (corresponding to 10^{-3} to 10^{4} $MPam^{1/2}s^{-1}$). The polymers selected for the investigations were the amorphous thermoplastics poly(carbonate) (PC) and poly(vynilchloride) (PVC), and the semi-crystalline thermoplastics poly(ethylene) (PE), poly(oxymethylene) (POM) and various grades of poly(propylene) (PP).

RESULTS AND DISCUSSION

The applicability limits of force based fracture mechanics concepts on the loading rate scale along with a definition of the lower and upper embrittlement transition points are shown in Fig. 2 for β^{+}-PP(H) as an example. First it must be noted that the FBA methods for ductile and brittle failure do not allow for a continuous function of fracture parameters over a wide loading rate scale due to the different mechanical concepts both fracture parameters are based upon. Hence, the diagram in Fig. 2 exhibits two ordinate descriptions, one being related to the LLA approach which is based on overall specimen loads (peak loads and limit loads), the other corresponding to the stress intensity factor concept which describes the local crack tip stress field.

The lower transition point on the loading rate scale, termed ductile/semi-brittle transition (d/sb) point, is defined as the loading rate where significant deviations occur between the peak loads and the calculated, specimen specific limit loads using rate dependent yield stress values. In other words, the lower transition point refers to a loading rate above which crack growth commences prior to achieving full cross-sectional yielding of the remaining ligament. The upper transition point, termed semi-brittle/brittle (sb/b) transition is defined as the loading rate above which specimen configuration and geometry independent fracture toughness values K_c (plane stress) and K_{Ic} (plane strain) are obtained. Both of these transition points simultaneously define the limits of applicability of the respective fracture mechanics concepts.

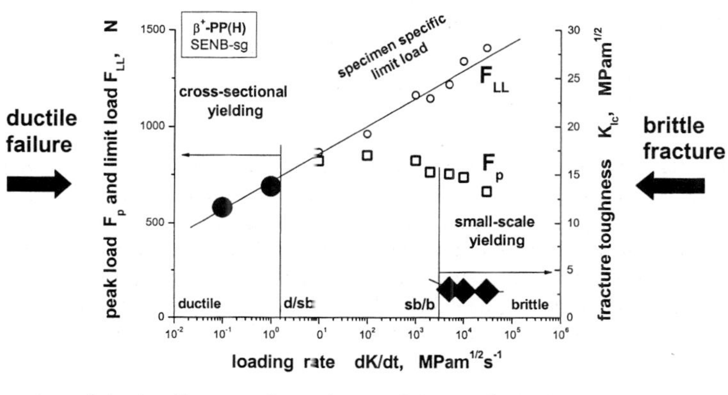

Fig. 2. Illustration of the loading rate dependence of the applicability of force based fracture mechanical concepts.

Limit Load Analysis Concept

In the following, first effects of stress state - plane stress (pss) vs. plane strain (psn) - on the ductile/semi-brittle transition rates (i.e., $dK/dt_{d/sb}^{pss}$ and $dK/dt_{d/sb}^{psn}$) will be discussed. Subsequently, various polymeric materials will be compared with regard to their ductile/semi-brittle transition rates.

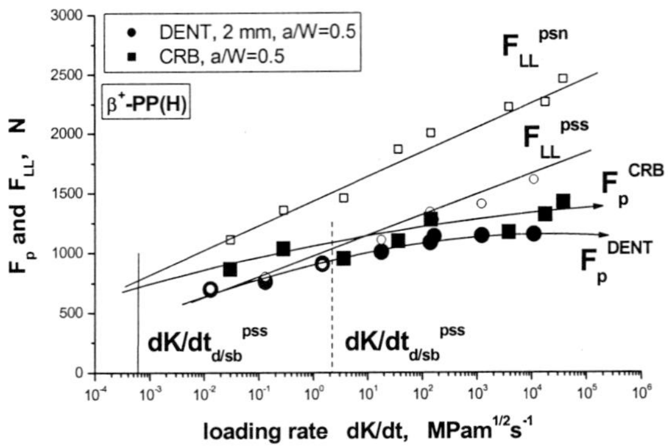

Fig. 3. Influence of plane stress (pss) vs. plane strain (psn) conditions on the d/sb-transition over a wide loading rate for β^{+}-PP(H); 2 mm thick DENT (near plane stress; $dK/dt_{d/sb}^{pss}$) and CRB specimens (near plane strain; $dK/dt_{d/sb}^{psn}$).

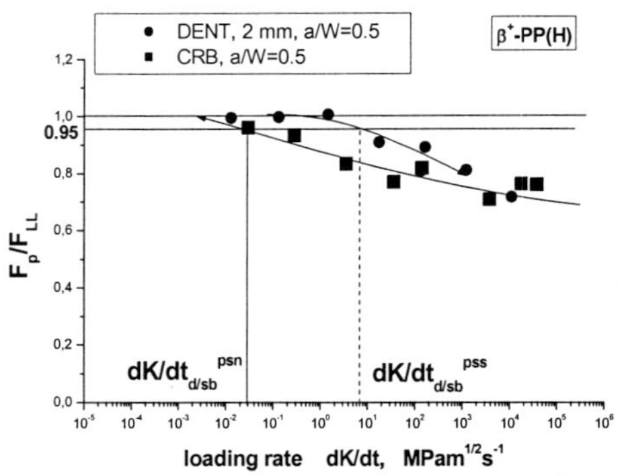

Fig. 4. Comparison of the constraint dependence of the ductile/semi-brittle transition for 2 mm thick DENT (near plane stress; $dK/dt_{d/sb}^{pss}$) and CRB specimens (near plane strain; $dK/dt_{d/sb}^{psn}$) for β^{+}-PP(H).

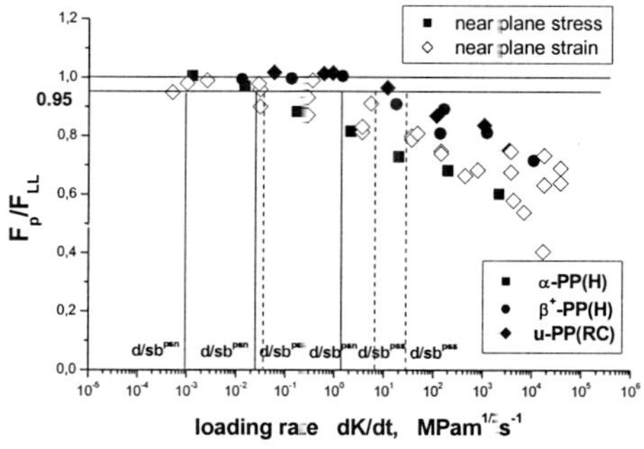

Fig. 5. Loading rate dependence of plastic limit load-to-peak load ratio F_p/F_{LL}, of 2 mm thick DENT specimen (near plane stress) and of CRB specimen (near plane strain) for α-PP(H), β^+-PP(H), and u-PP(RC).

The ductile/semi-brittle transition point.
The influence of plane stress (pss) vs. plane strain (psn) conditions on the d/sb-transition on the loading rate scale is shown in Fig. 3 for β^+-PP(H). For near plane stress conditions a 2 mm thick double edge notched tensile (DENT) specimen and for near plane strain conditions a cracked round bar (CRB) specimen was used. As expected, for near plane strain conditions peak loads and limit loads are higher, and the d/sb-transition occurs at a lower loading rate. To be able to define the d/sb-transition region more precisely, the data of Fig. 3 are replotted in Fig. 4 in terms of the F_p/F_{LL}-ratio (relative limit load), and the d/sb-transition is now defined as the loading rate at which the F_p/F_{LL}-ratio is reduced by 5% and thus intersects with an F_p/F_{LL} value of 0.95. This definition of the d/sb-transition yields values of 10^{-3} MPam$^{1/2}$s^{-1} for plane strain conditions and of $7*10^0$ MPam$^{1/2}$s^{-1} for plane stress conditions.

Material comparison.
In Fig. 5 various poly(propylene) (PP) materials (α-PP(H), β^+-PP(H) and u-PP(RC)) are compared in terms of the loading rate dependence of their respective F_p/F_{LL}-ratios. Clearly, the materials investigated differ in terms of both the quantitative positions of the d/sb-transition points on the loading rate scale and the loading rate difference between plane stress and plane strain d/sb-transition points. More detailed numerical comparisons for all materials investigated see elsewhere [2].

Stress Intensity Factor Concept
Analogous to the limit load analysis concept above, in the following first some examples and aspects related to the definition of the sb/b-transition point will be discussed. Furthermore, various polymeric materials will be compared with regard to their sb/b-transition rates and the loading rate sensitivity of crack tip stress field based fracture toughness values.

The semi-brittle/brittle transition point.
As to the stress intensity factor concept, the upper transition point, termed semi-brittle/brittle transition rate ($dK/dt_{sb/b}$) is defined as the loading rate above which specimen configuration and geometry independent fracture toughness values, K_{Ic} are obtained. Examples of results of fracture tests over a wide loading rate range in terms of K_c^{app} vs. dK/dt and for various specimen configurations and geometries are shown in Figs. 6a to 6c for POM, PC and β^+-PP(H). These examples were selected because they illustrate specimen configuration independent K_{Ic} values over the entire loading rate range investigated (i.e., POM in Fig. 6a), specimen configuration independent K_{Ic} values only above $2*10^0$ MPam$^{1/2}$s^{-1} (i.e., PC in Fig.

6b), and strong specimen configuration dependent behavior over the entire loading rate range investigated (i.e., β^+-PP(H) in Fig. 6c). Further examples of the influence of specimen configuration and loading rates on fracture toughness values are presented elsewhere [Major, 2002].Values for $dK/dt_{b/sb}$ may be defined as local loading rates for which the scatter of K_c^{app} values for various specimen configurations is less than ±20 % relative to the respective mean value. Above the so defined semi-brittle/brittle transition rate, K_c^{app} values may be interpreted as true K_{Ic} values.

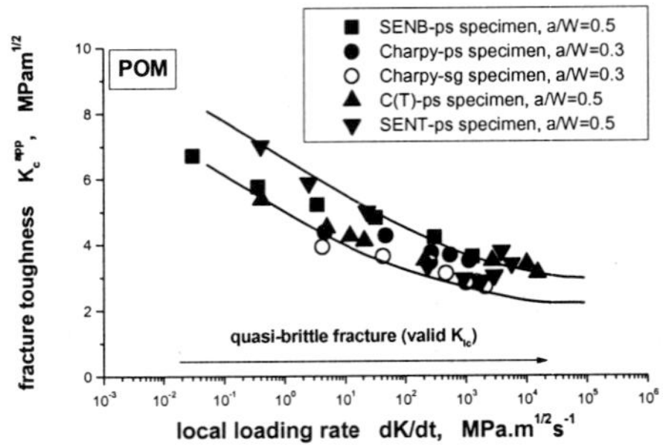

(a)

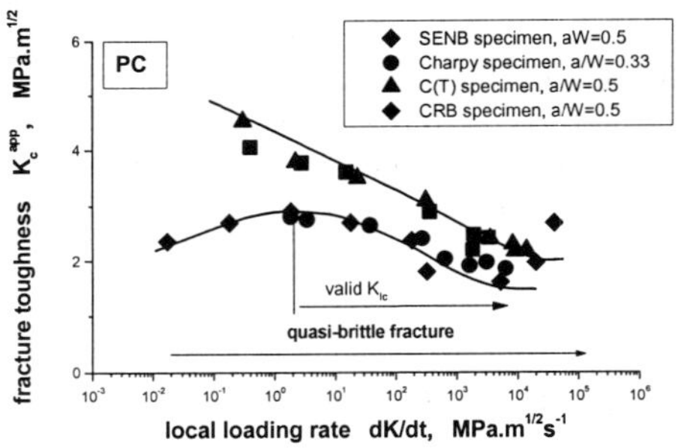

(b)

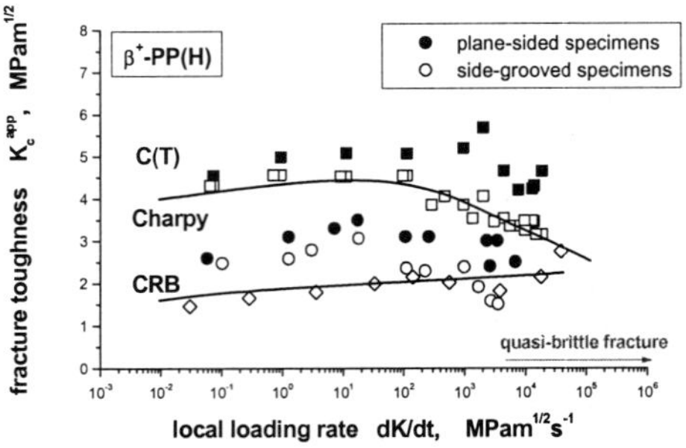

(c)

Fig. 6: Effect of local loading rate on fracture toughness for β^+-PP(H) using various specimen configurations and data reduction schemes; (a) POM, (b) PC and (c) β^+-PP(H).

The good agreement of force based analysis K_{Ic} values and K_{Id} values obtained from dynamic key curve (DKC) experiments in the high loading rate regime is shown in Fig. 7 for PC and POM. While in the DKC experiments only bending type specimens were used, tensile and bending type specimens were used in the FBA experiments.

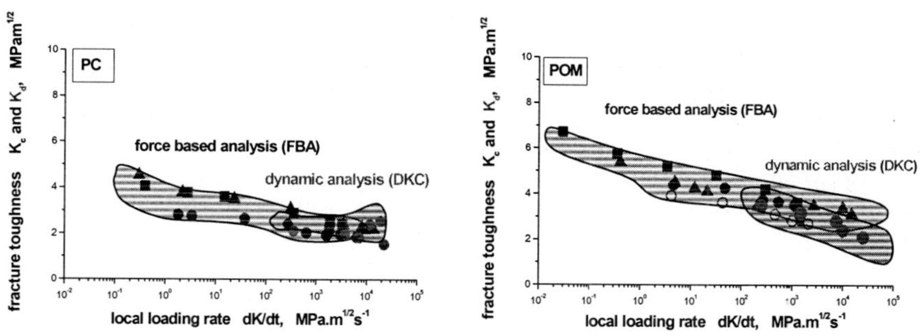

(a) (b)

Fig. 7. Comparison of loading rate dependence of fracture toughness values (FBA method) and dynamic fracture toughness values (DKC method) for, (a) PC, and (b) POM.

Material comparison.

Fracture toughness values, in terms of true K_{Ic} values according to the above definition are plotted vs. loading rate in Fig. 8 comparing various materials. Each of the data points for a given material represents a mean value of all tests performed with different specimen configurations and geometries, including data of both the FBA and the DKC experiments (the latter of course only at high loading rates). Also indicated in Fig.8a by large full circles are the semi-brittle/brittle-transition points for each of the materials. With a value of about $5*10^{-2}$ $MPam^{1/2}s^{-1}$, POM exhibits the lowest sb/b-transition point, whereas β^{+}-PP(H) reveals the highest sb/b-transition point with approximately 10^{4} $MPam^{1/2}s^{-1}$. In other words, the sb/b-transition rates of the materials investigated differ by more than 5 orders of magnitude.

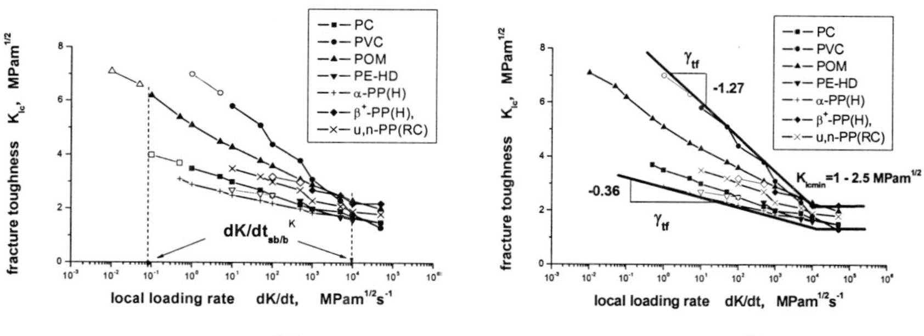

(a) (b)

Fig. 8: Fracture toughness, K_{Ic}, as a function of local loading rate on for all engineering polymers investigated in this study (average curves of all specimen configurations for a specific material type); (a) indication of sb/b-transition rates ($dK/dt_{sb/b}^{K}$), and (b) indication of slopes and lower bound K_{Ic} values (K_{Icmin}) values.

Moreover, the rate dependence of the K_{Ic} values in the brittle failure regime of the various materials investigated also varies to a significant degree. As is shown in Fig. 8b, with values of -1.27 and -0.36 for PVC and PC, respectively, these materials exhibit the highest and lowest slopes in the semi-logarithmic diagram. Finally, also worthwhile noting is that rate dependent K_{Ic} values of all materials investigated converge at about 10^{4} $MPam^{1/2}s^{-1}$, reaching a lower bound fracture toughness (K_{Icmin}) in the range of 1.0 to 2.5 $MPam^{1/2}$.

CONCLUSIONS

To examine the applicability of various fracture mechanics concepts to rate dependent fracture of polymers, monotonic fracture experiments were performed with various polymers over a loading rate range of up to 7 orders of magnitude from 10^{-5} to 8 m/s (corresponding to 10^{-3} to 10^{4} $MPam^{1/2}s^{-1}$). In the brittle failure regime at high loading rates the critical stress intensity factor concept of linear elastic fracture mechanics (LEFM) may be applied. Depending again on the stress state, the relevant fracture parameters are the plane stress fracture toughness, K_c, and the plane strain fracture toughness, K_{Ic}, respectively. In contrast to the rate dependence of the limit loads in the ductile failure regime, the fracture toughness values decrease with loading rate before reaching a lower bound level.

REFERENCES

1. Williams, J.G. (1995) Fracture mechanics, in Physics of Glassy Polymers; Haward, R.N. and Young, R.J. (Eds.), Chapman&Hall, London.

2. Major, Z. (2002) PhD Thesis, University of Leoben

3. Anderson, T.L. (1991) Fracture Mechanics, CRC Press, Boca Raton

4. Béguelin, Ph. (1996) PhD Thesis (1572), University of Lausanne

The Application of Fracture Mechanics to Polymers, Adhesives and Composites
D.R. Moore (Ed)

CREEP CRACK GROWTH IN HIGH DENSITY POLYETHYLENE

G. PINTER and R.W. LANG

INTRODUCTION

Since high density polyethylene (PE-HD) is increasingly used in long-term structural applications (e.g., gas and water pipes, geo membranes or other mechanically, thermally and chemically loaded components), in recent years considerable attention was paid to the effects of material parameters on failure behaviour under static loads. Thus it is well documented that at least two different types of time dependent failure modes are typical for PE-HD. While at high stress levels and correspondingly short failure times ductile failure occurs, controlled by shear yielding mechanisms on a larger scale, quasi-brittle failure involving the initiation and growth of creep cracks is frequently observed at lower stresses and correspondingly longer failure times, particularly when the materials are exposed to multiaxial stress fields at elevated temperatures. These creep cracks generally originate from pre-existing material defects (impurities, voids, etc.) or in areas of high stress concentrations (notches, etc.). Differences in failure times of various PE-HD types in the ductile failure regime are generally attributed to different yield stress values, which primarily depend on the material density and thus on the degree of crystallinity. In other words, for a given degree of crystallinity the molecular structure, molecular mass distribution and supramolecular structure have at most only minor direct effects on failure behaviour at high stress levels (ductile failure regime). On the other hand, in the quasi-brittle failure regime at lower stress levels and longer lifetimes, all of these latter factors are known to play a significant role [1-7]. Whereas in recent years considerable attention was paid to the effects of various material parameters (average molecular mass and molecular mass distribution, concentration and length of short chain branches, crystalline morphology, etc.) on the creep crack growth (CCG) behaviour of PE-HD, no detailed study exists on the influence of stabilisation on CCG kinetics. This is somewhat surprising, as it is well known that stabiliser type and concentration may significantly affect the lifetime of pressurised pipes in the quasi-brittle and brittle failure regime [8, 9].

It is the main objective of this paper to provide more insight into some of the details of long-term mechanical failure of PE-HD at elevated temperatures under static loads. Particular attention is paid to effects of molecular and supramolecular parameters as well as the influence of stabilisation on the kinetics of CCG initiation and propagation in PE-HD. For this purpose 5 grades of PE-HD with different molecular and morphological characteristics were selected and included in the investigations. One of the PE-HD types was stabilised in 3 different ways. Linear elastic fracture mechanics (LEFM) methods were applied to obtain more comprehensive information on the micromechanisms and kinetics of CCG.

EXPERIMENTAL

Materials

Five grades of PE-HD (linear copolymers designated PE-HD 1 to PE-HD 5) were supplied by Borealis AG (Linz, Austria). Except for PE-HD 4, which contained approximately 2 mass % carbon black, all other materials were provided in natural colours. PE-HD 1 was stabilised in three different ways. Stabiliser S is a sulfur containing secondary antioxidant (Santonox R;

Monsanto, Belgium) and was added at 0.1 mass % to the base polymer. The stabiliser system K is a mixture of a primary and secondary antioxidant (Irganox B225, Ciba Specialty Chemicals Inc., Switzerland) and was added at 0.1 mass % (formulation code K1) and 0.2 mass % (formulation code K2) It is specifically the primary stabiliser that protects the polymer against thermo-oxidative degradation during the application stage. All materials were compression moulded into plaques with a nominal thickness of 12 mm. Some of the relevant material properties of the compression moulded plaques are summarised in Table I.

Creep crack growth tests

The crack growth experiments under static loads were performed at 60 and 80 °C with specimens of the compact-type (CT) in distilled water, in a test apparatus designed and constructed at the Institute of Materials Science and Testing of Plastics (University of Leoben, Austria) [10]. The specimens were loaded with dead weights, and values of the crack length at the specimen side surface and the crack opening displacement at the specimen front edge were measured with a travelling microscope at a magnification of 30X.

Table I. Material properties of the investigated PE-HDs (M_w: weight average molecular mass, SCB: number of short chain branches, X_C: degree of crystallinity, L_C: lamella thickness, E: tensile modulus, σ_y: yield stress).

material code	comonomer	M_w (kg/mol)	SCB (1/1000C)	X_C (%)	L_C (nm)	E* (N/mm^2)	σ_y* (N/mm^2)
PE-HD 1	Hexene	80	1	83	24	1650	32.2
PE-HD 2	Hexene	106	2	68	18.4	1140	26.5
PE-HD 3	Butene	213	8.2	68	18.4	1100	25.5
PE-HD 4	Hexene	290	4	60	13	950	24
PE-HD 5	Butene	320	3.2	77	21.8	1400	30

* tested at 23 °C and 50 % r. h.

Based on concepts of LEFM the stress intensity factor K_I was calculated according to the following expression [1]:

$$K_I = \frac{F}{B \cdot \sqrt{W}} \cdot f\left(\frac{a}{W}\right) \tag{1}$$

where F is the applied load, B the specimen thickness, W the specimen width and a the crack length; f(a/W) is a non-dimensional correction function that accounts for the specimen configuration and geometry and the loading conditions.

To study CCG initiation times t_{in}, individual specimens were loaded to different initial stress intensity factor values K_{Iin}. Values for t_{in} were then determined from measurements of the crack opening displacement at the specimen front face (COD_{ff}) as a function of loading time. By plotting COD_{ff} data as a function of loading time in a double logarithmic diagram, the point of crack initiation corresponds to the first deviation of these data from the linear base line. The CCG initiation times so determined were found to be in good agreement with travelling microscope observations of the immediate crack tip region.

Following the CCG initiation period, crack growth rates da/dt were calculated by a method analogous to that proposed in ASTM E647-93 for the determination of fatigue crack growth. Further details as to the test apparatus and the test and data reduction procedure are described elsewhere [10, 11].

RESULTS AND DISCUSSION

Applicability of the LEFM approach

In order to verify the applicability of LEFM, constant K_I experiments were performed with all PE-HD types included in this investigation. The data depicted show remarkably constant crack growth rates with very little scatter over the entire a/W-range, thus providing good support for K_I as the controlling factor of CCG rates in this material. Most importantly they indicate that CCG rates are indeed material specific properties uniquely related to the applied stress intensity factor and independent of specimen geometry and configuration.

Effect of material parameters on creep crack growth kinetics

Creep crack growth initiation behaviour. CCG initiation times t_{in} as a function of the initial stress intensity factor K_{Iin} are depicted in Fig. 1a for the five PE-HD types investigated at a test temperature of 80 °C (data for PE-HD 4 from [12]). In good agreement with previous findings [12, 13], the behaviour of all materials reflects a power law dependence of t_{in} on K_{Iin}.

To gain more insight into the relevant material parameters controlling CCG initiation, t_{in} values of the various PE-HD types are plotted for given values of K_{Iin} as a function of the average molecular mass M_w in Fig. 1b. Clearly, CCG initiation values are seen to increase with M_w in the range from 80 to 290 kg/mol, which corresponds to the materials PE-HD 1 to PE-HD 4. On the other hand, for PE-HD 5 the material with the highest M_w value of 320 kg/mol, a sharp drop in CCG initiation time to a value corresponding to the behaviour of a material with about half the M_w value equivalent to PE-HD 2 to PE-HD 3 is observed.

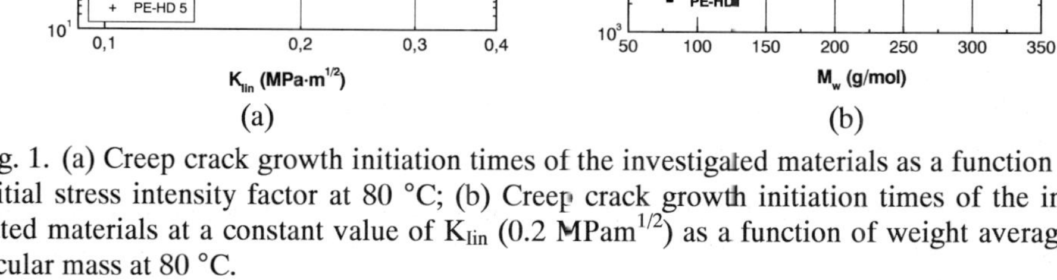

(a) (b)

Fig. 1. (a) Creep crack growth initiation times of the investigated materials as a function of the initial stress intensity factor at 80 °C; (b) Creep crack growth initiation times of the investigated materials at a constant value of K_{Iin} (0.2 $MPam^{1/2}$) as a function of weight average molecular mass at 80 °C.

When comparing the relevant material characteristics of Table I, the increase in CCG initiation times from PE-HD 1 to PE-HD 4 may be explained as follows. The degree of crystallinity continuously decreases from PE-HD 1 to PE-HD 4 (from 83 to 60 %) which results in a concomitant decrease of yield stress from 32 to 24 MPa. This in turn leads to the development of larger plastic zone sizes which more effectively blunt the crack. Moreover, it may also be deduced from Table I that the lamella thickness L_c decreases from PE-HD 1 to PE-HD 4 (from 24 to 13 nm). With the simultaneous increase in the average molecular mass M_w in this order, this implies that the density of tie molecules and interlamellar entanglements also increases from PE-HD 1 to PE-HD 4. Particularly PE-HD 4 contains the highest fraction of high molecular mass polymer chains which are most favourable in developing effective tie molecules and interlamellar entanglements. All of these factors combined are believed to be the cause for the significant enhancement in CCG initiation times from PE-HD 1 to PE-HD 4.

On the other hand, PE-HD 5 exhibits rather high values for the degree of crystallinity (77 %) and correspondingly for σ_y (30 MPa). Simultaneously it also contains lamellae with higher values of L_c (21.8 nm) than in materials PE-HD 2 to PE-HD 4. In other words, the beneficial effect of increased molecular mass is apparently offset by a diminished ability to develop larger plastic zones and blunting and by a lower density of tie molecules and interlamellar entanglements. The high degree of crystallinity of PE-HD 5 - compared to, for example, PE-HD 4, a material with a roughly equivalent molecular mass and an approximately equal concentration of short chain branches (see Table I) - is undoubtedly related to the short chain branch length. PE-HD 5 with butene as comonomer and thus shorter chain length values crystallises to a significantly higher degree than PE-HD 4 with hexene as comonomer.

Of particular relevance to any effects related to the concentration of short chain branches is a comparison of the materials PE-HD 3 (8.2/1000 C; M_w= 213 kg/mol) and PE-HD 5 (3.2/1000 C; M_w= 320 kg/mol), for both of which butene was used as comonomer (see Table I). Due to the higher concentration of short chain branches in PE-HD 3, this material exhibits a lower degree of crystallinity with lower values in the lamella thickness and consequently a lower yield stress. This in turn favours crack tip plastic zone development and hence increases the CCG initiation time (compare PE-HD 3 and PE-HD 5 in Figure 1a). An increase in the concentration of short chain branches, as is the case for PE-HD 3, is apparently more effective in enhancing the CCG initiation resistance than increasing the molecular mass of a polymer with a low concentration of short chain branches, as is the case for PE-HD 5. This correlation is of great practical importance, at least within the range of materials investigated here.

Interestingly, no clear differences in creep crack growth initiation times could be detected for the 3 stabiliser formulations of PE-HD 1. This may at least in part be due to difficulties in precisely determining the point of CCG initiation, and further research is needed on examining the role of stabilisers in the creep crack growth initiation stage.

Creep crack growth behaviour. Equilibrium CCG data for all 5 PE-HD types generated at 80 °C are shown in Fig. 2a (data for PE-HD 4 from [12]). Again, significant differences are observed in the behaviour of the various materials. The fact that crack growth rates in all materials are higher at 80 °C than at 60 °C [11] indicates that creep crack growth may be considered as thermally activated process.

Analogous to Fig. 1b for CCG initiation times, CCG rates da/dt are shown in Fig. 2b as a function of the average molecular mass M_w for constant K_I values of 0.22 and 0.33 MPam$^{1/2}$, respectively. In terms of material ranking similar tendencies to those discussed above for CCG initiation can be inferred when considering that higher t_{in} values should correspond to lower da/dt values for predefined stress intensity levels. Apparently all material parameters (degree of crystallinity controlling the yield stress and the plastic zone size; molecular mass distribution and lamella thickness controlling the tie molecule and interlamellar entanglement density) affect stable crack growth properties and crack growth initiation in similar ways.

To support the argument on the role of the crack tip plastic deformation zone as one controlling factor for CCG resistance, microscope pictures of crack tip regions are shown in Fig. 3 for the various materials (except PE-HD 4) at a testing temperature of 60 °C. All crack tip plastic zone photographs were taken from the stable CCG region at a predefined K_I value of 0.4 $MPam^{1/2}$. The smallest crack tip plastic zones for a given K_I value apparently occur in the materials with the highest yield stress, PE-HD 1 (32 MPa) and PE-HD 5 (30 MPa). Conversely, material PE-HD 3 with the lowest yield stress value of 25.5 MPa develops the largest plastic zone size. It should be recognised, however, that in a strict sense plastic zone sizes at the tip of slow growing creep cracks should reflect the relevant time dependent yield stress values rather than those determined in a monotonic tensile test [14]. In other words, some part of the increased plastic zone sizes of PE-HD 3 is certainly also due to the slower crack growth rates developed in this material at given values of K_I.

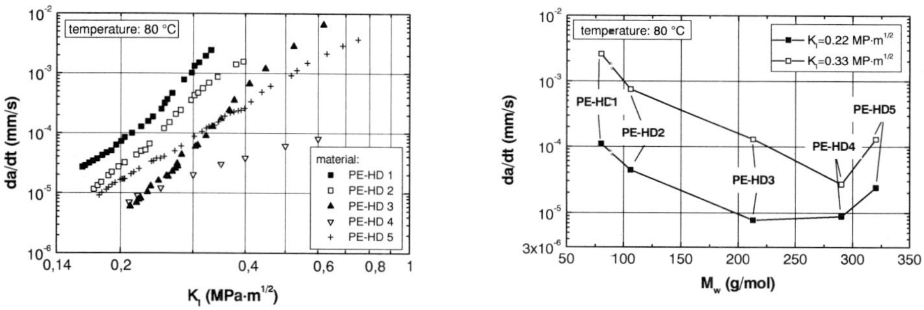

(a) (b)

Fig. 2. (a) Crack speed versus stress intensity factor for the investigated materials at 80 °C; (b) Creep crack growth rates of the investigated materials at constant values of K_I (0.22 and 0.33 $MPam^{1/2}$, respectively) as a function of weight average molecular mass at 80 °.

When directly comparing PE-HD 3 and PE-HD 4, it is of interest to note that PE-HD 3 exhibits slower crack growth rates in the low crack growth rate regime (i.e., at small K_I values), while PE-HD 4 reveals slower crack growth rates in the higher crack growth rate regime (i.e., at higher K_I values). Although an unambiguous explanation requires further investigations, it seems that the superior behaviour of PE-HD 3 at low K_I values may be related to the existence of a higher concentration of short chain branches in PE-HD 3, despite its lower molecular mass, higher degree of crystallinity and larger lamella thickness (compare Table I). Thus short chain branches may act to delay the disentanglement process of the tie molecules within both the amorphous and crystalline domains and to enhance the energy absorption during crack growth. The higher slope of the crack growth curves of PE-HD 3 seems to indicate that the short chain branches are effective especially at low crack speeds, whereas at higher crack growth rates molecular mass and the density of tie molecules and interlamellar entanglements dominates CCG.

(a) (c)

(b) (d)

Fig. 3. Dimensions of surface near craze zones of the investigated PE-HDs (60 °C; K_I= 0.4 MPam$^{1/2}$); (a) PE-HD 1 (da/dt= 2×10^{-4} mm/s), (b) PE-HD 2 (da/dt= 7×10^{-5} mm/s), (c) PE-HD 3 (da/dt= 7×10^{-6} mm/s), (d) PE-HD 5 (da/dt= 2×10^{-5} mm/s).

Creep crack growth data for the differently stabilised PE-HD 1 formulations are shown in Fig. 4 again for a test temperature of 80 °C. In the low creep crack speed region formulation S results in considerably higher crack growth rates than formulations K1 and K2. As the characterisation methods used so far (i.e., average molecular mass and molecular mass distribution, degree of crystallinity, lamella thickness) could not detect any differences between the stabiliser formulations, global ageing can be excluded in these fracture mechanical crack growth tests lasting only for a few days to a few weeks. Hence, the differences observed for the formulations at low creep crack growth rates are believed to be a result of the influence of the different stabiliser systems on local ageing around the crack tip related to the combined influence of time, the elevated temperature, the presence of oxygen and water, and the high mechanical stresses in the immediate crack tip region [15, 16]. As the time scale for local ageing is reduced at high crack speeds, the creep crack growth curves of the formulations in Fig. 4 converge at high K_I values.

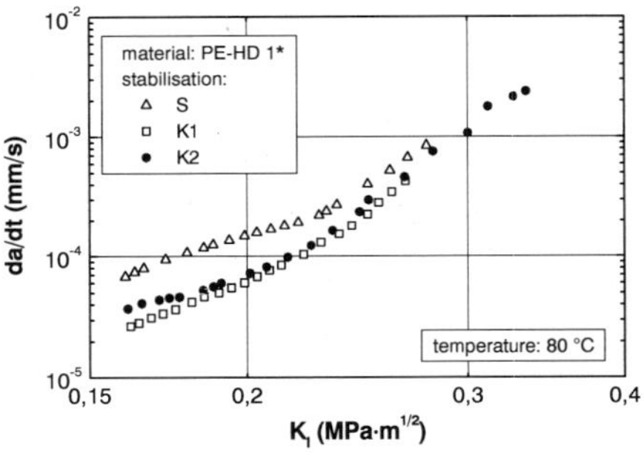

Fig. 4. Creep crack growth behaviour of the formulations of PE-HD 1* under equilibrium conditions at (a) 60 °C and (b) 80 °C.

Surprisingly, stabiliser K1, containing the same antioxidants as K2, but in a lower concentration, showed the highest crack growth resistance. Changes in the molecular and morphological structure (i.e. branching, crosslinking, lamella thickness) during processing on an extreme local or molecular scale beyond the sensitivity of the characterisation methods used (i.e. high performance liquid chromatography, rheological analyses, thermo-analytical analyses, small angle X-ray scattering) may serve as possible explanations for the anomalous behaviour of compounds with K1 versus K2. In other words, it may also well be that the lower concentration of stabiliser in K1 compounds resulted in a somewhat higher density of effective tie molecules and interlamellar entanglements after exposure to the various processing conditions (compression moulding plaques), leading to an enhanced crack growth resistance [4, 5].

SUMMARY AND CONCLUSIONS

Comparing the various PE-HD materials investigated in terms of CCG initiation times and CCG rates, the same material ranking was found especially in the slow CCG rate regime (below 10^{-4} mm/s), which implies that these two basic failure mechanisms are largely controlled by the same molecular and morphological material parameters. On the other hand some deviation in this ranking was observed for fast CCG rates (above 10^{-4} mm/s), indicating that a direct correlation of long-term properties controlled by CCG and short-term failure associated with rapid crack growth may not be granted in all cases.

As to the molecular and morphological parameters, the average molecular mass and the molecular mass distribution were found to be important factors controlling CCG, as expected. However, on the other hand, the experimental results also revealed that molecular mass effects may be outweighed by the influence of short chain branching for which both short chain branch density and chain length are of importance. The various effects of molecular structure on CCG initiation and CCG rates were first correlated with the resulting morphology in terms of degree of crystallinity and lamella dimensions and then interpreted in terms of the effects of these morphological parameters on yield stress, which controls the crack tip plastic zone size (i.e. plastic zone development), and on the density of tie molecules and interlamellar entanglements, which control the plastic zone breakdown process. Both of these mechanisms (plastic zone development and plastic zone breakdown) are, of course, of prime importance in controlling the overall CCG rates.

Crack growth rates were found to depend on the stabiliser system, particularly in the low CCG regime, with diminishing stabiliser effects at higher CCG rates. No molecular and morphological differences between the formulations containing various stabilisers could be detected. Hence, the phenomena observed are explained in terms of local ageing processes in the immediate vicinity of the crack tip, which are controlled by the presence and content of various stabilisers. The concept of local crack tip ageing assumes that the high stresses approaching the materials yield and craze stress in the immediate vicinity of the crack tip assist and accelerate the thermo-oxidative degradation of the material.

REFERENCES

1. Kinloch, A.J. and Young, R.J. (1983). *Fracture Behaviour of Polymers*. Applied Science Publishers Ltd., Barking.
2. Michler, G.H. (1992). *Kunststoff–Mikromechanik*. Carl Hanser Verlag, Müchen Wien.
3. Brown, N., Donofrio, J. and Lu, X. (1987) *Polymer* **28**, 1326.
4. Brown, N., Lu, X., Huang, Y.L., Harrison, I.P. and Ishikawa, N. (1992) *Plastics Rubber and Composites Processing and Applications* **17**, 255.
5. Egan, B.J. and Delatycki, O. (1995) *J. Mater. Sci.* **30**, 3307.
6. Lang, R.W., Stern, A. and Dörner, G. (1997) *Die angew. Makrom. Chemie* **247**, 131.
7. Kausch, H.H. (1986). In: *Failure of Plastics*, pp. 84-97, Brostow, W. and Corneliussen, R.D. (Eds). SPE Hanser, Munich.
8. Gebler, H. (1989) *Kunststoffe* **79(9)**, 823.
9. Dörner, G. and Lang, R.W. (1997) *3R international* **11(36)**, 672.
10. Stern, A., Novotny, M. and Lang, R.W. (1998) *Polym. Test.* **17**, 403.
11. Pinter, G. (1999). Ph.D. Thesis, University of Leoben, Austria.
12. Stern, A. (1995). Ph.D. Thesis, University of Leoben, Austria,.

13. Tonyali, K., Rogers, C.H.E. and Brown, H.R. (1989) *J. Macromol. Sci.–Phys.* **B28(2)**,. 235.
14. Lang, R.W. (1984). Ph.D. Thesis, Lehigh University Bethlehem, USA.
15. Popov, A., Rapoport N. and Zaikov G. (1991). *Oxidation of Stressed Polymers*. Gordon & Breach, New York.
16. Terselius, B., Gedde, U.W. and Jansson, J.F. (1986). In: *Failure of Plastics*, pp. 273-286, Brostow, W. and Corneliussen, R.D. (Eds). SPE Hanser, Munich.

The Application of Fracture Mechanics to Polymers, Adhesives and Composites
D.R. Moore (Ed)

PLASTIC ZONE CORRECTED LEFM: BENEFITS AND LIMITS

C. GREIN, H.-H. KAUSCH, Ph. BÉGUELIN

INTRODUCTION : WHY USE PLASTIC ZONE CORRECTED LEFM?

Compared to industrial standards (Charpy, Izod, Elmendorf), the application of fracture mechanics (FM) constitutes a notable progress in the assessment of mechanical properties of polymers because it provides intrinsic parameters. These allow the unequivocal ranking of materials and provide material properties for modeling work. However, the choice of a unique FM method to describe material properties over a range of test temperatures, test speeds and stress states is rather challenging.

The information provided by conventional FM methods is sufficient when the material exhibits the same macroscopic mode of failure. Linear elastic fracture mechanics, LEFM, is suited for brittle behaviours, whilst elastic-palstic fracture mechanics accommodates ductile fractures in the form of the J-Integral for polymers exhibiting small scale yielding and Essential Work of Fracture, EWF, for ductile mode of failures. Therefore, when the material exhibits multiple modes of fracture then more than one FM approach is required. However, a strategy can be even more difficult if a ductile-brittle fracture transition is to be measured.

To take into account these issues, an alternative approach that provides geometry independent materials parameters for both ductile and brittle fracture is proposed. It consists of determining experimentally the size of the plastic zone developed at a crack tip using specimens of different ligament lengths and LEFM as a tool for data reduction. The analysis will be restricted to the stress intensity factor, K, although the proposed plastic zone correction works also with the energy release rate, G. Our approach will be illustrated with two ethylene-propylene rubber toughened isotactic polypropylene blends (iPP/EPR) which differ only in the molecular weight of their matrix (M_w(iPP/EPR-1) = 0.8 M_w(iPP/EPR-2)).

DETERMINATION OF INTRINSIC TOUGHNESS VALUES USING A PLASTIC ZONE CORRECTED LEFM ORIGINAL APPROACH

Materials under investigation were tested at between 0.001 and 10 m/s using a servo-hydraulic Schenk apparatus, an optical sensor for displacement assessment and Compact Tension (CT) specimens (24 x 24 x 4 m^3). Further detail about the mechanical set-up can be found in references1-3.

LEFM was chosen for data reduction. Using the classical approach, the location of a ductile-brittle fracture transitions, T_{db}, were assessed. A unique specimen geometry was utilized (a/W = 0.5 where a is the crack length and W the specimen width), toughness was calculated from equation 1 and T_{db} was taken as the maximum of K (determined from equation 1) over the speed range under consideration (as in ref. [1, 4-5]). With this method, it was found that iPP/EPR-1 exhibited its ductile-brittle transition at 0.6 m/s, iPP/EPR-2 at 0.8 m/s. A classic

LEFM approach was, however, not suitable for providing intrinsic toughness values in the ductile range of material behaviour since (i) the K-values deduced from equation 1 were not independent of the crack length of the specimen (table 1) and (ii) the plane strain conditions, described by equation 2, were not satisfied up to 3-10 m/s as can be seen by the data in Table 2.

$$K_I = f(a/W)\frac{F_{\max}}{B\sqrt{W}} \tag{1}$$

$$B, a, W-a \geq 2.5\left(\frac{K_I}{\sigma_y}\right)^2 \tag{2}$$

with $K_I = K_{IC}$, the critical stress intensity factor in mode I, when the LEFM fracture criteria are satisfied and $K = K_{Imax}$ when the global linearity criteria of LEFM is not satisfied. F_{max} is the maximun of force on the force-displacement curves; B, the thickness of the specimen; W, its width; (W-a), the ligament length; σ_y, the yield stress obtained at loading times commensurate with those for determing fracture toughness; f(a/W), a function depending on the specimen characteristics and available for all standard geometries.

Table 1. Evolution of the toughness, K_I, versus a geometry factor, a/W, for iPP/EPR-2 tested at room temperature. Tested at 0.1 m/s, K_I was an apparent toughness, K_{Imax}; tested at 6 m/s K_I was an intrinsic toughness, K_{IC}.

a/W	**0.3**	**0.4**	**0.5**	**0.6**	**0.7**	[-]
$K_I = K_{Imax}$ - 0.1 m/s	4.1	3.85	3.55	3.05	2.7	[MPa.m$^{1/2}$]
$K_I = K_{IC}$ - 6m/s	2.7	2.7	2.75	2.7	2.65	[MPa.m$^{1/2}$]

Table 2. Evaluation of the plane strain required thickness, B, calculated after equation 2. If B > 4 mm (the thickness of the tested samples), the plane strain criterion is not fulfiled. Material iPP/EPR-2 tested at 23 ^{0}C.

Test speed	**0.001**	**0.01**	**0.1**	**1**	**3**	**10**	[m/s]
K_{Imax}	3.10	3.25	3.42	3.60	3.15	2.40	[MPa.m$^{1/2}$]
σ_y	27.2	29.5	33.5	42.1	45.9	51.2	[MPa]
B	42.2	36.8	27.2	13.4	5.5	1.2	[mm]

According to equation 1, K can be determined graphically by plotting F_{max} versus $(B\sqrt{W})/f$ for different crack lengths.

In the case of linear elastic brittle behaviour (for which the distribution of the stress field in front of a crack tip is sketched in Fig. 2(i)), the straight line described by the experimental points runs through the origin (see Fig. 1(i)) and its slope gives K_{IC} [6-8].

In the case of ductile behaviour, microplasticity occurs at the crack tip, but F_{max} and $(B\sqrt{W})/f$ exhibit also a linear correlation (Fig. 1(ii)). However, a line through the data does not pass through (0,0) because of the development of a plastic zone in front of the crack tip (see Fig. 2(ii)). We propose to deduce numerically the size of this plastic zone, r_p, in such a way that the straight line describing all data is forced through the origin (according to an Irwin-like model for small-scale yielding [9]). The related slope gives the effective toughness, K_{eff}, and is a geometry-independent quantity. Practically the crack length, a, is replaced by a + r_p in the proportionnality factor f; f(a/W) becomes f((a+r_p)/W) and the value of r_p is then adjusted iteratively until the straight line passes through (0,0). Furthermore, the values of r_p found experimentally defines the radius of the equivalent Irwin damaged zone.

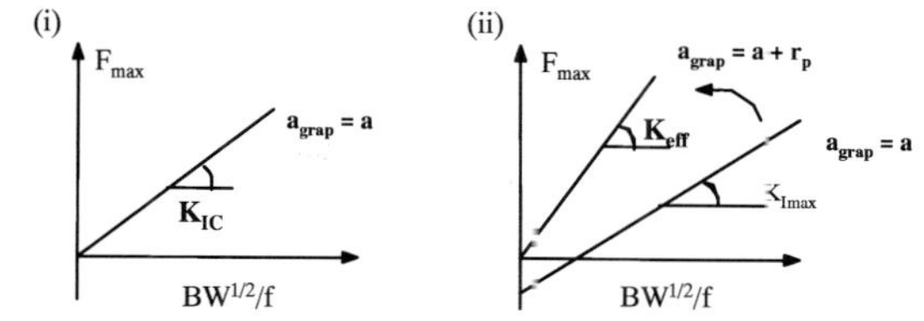

Fig. 1. Experimental determination (i) of K_{IC} in case of linear elastic material behaviour; (ii) of K_{eff} by correction with the radius of the plastic zone, present at the vicinity of the crack tip; a_{grap} refers to the crack length used for graphical determination of K.

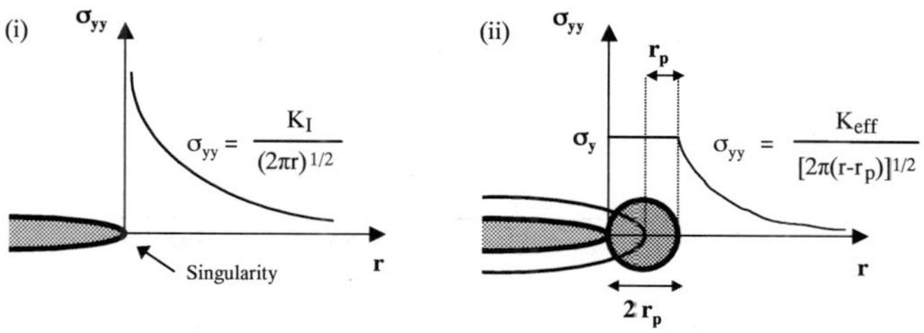

Fig. 2. (i) Normal stress towards the fracture plane in mode I for an elastic material; (ii) redistribution of stress by development of a plastic zone in front of a crack tip according to Irwin.

An example of a plastic zone correction is given in Fig. 3(i) for iPP/EPR-2 tested at 0. 1 m/s and 23 ^{0}C. The excellent coefficient of linear regression (r^2 > 0.99) for both sets of data is an indication of the reliability of the applied Irwin-like proposed correction. The effective toughness, K_{eff}, is higher than both the toughness measured by varying a/W without applying the plastic zone correction (Fig. 3(i), K_{Imax} = 4.66 MPa.m$^{1/2}$) and the mean value of K_{Imax} (Fig. 3(ii), $K_{Imax} \approx$ 3.45 MPa.m$^{1/2}$) directly calculated from equation 1 for a/W = 0.5. The fracture toughness of the material was therefore underestimated with the uncorrected values.

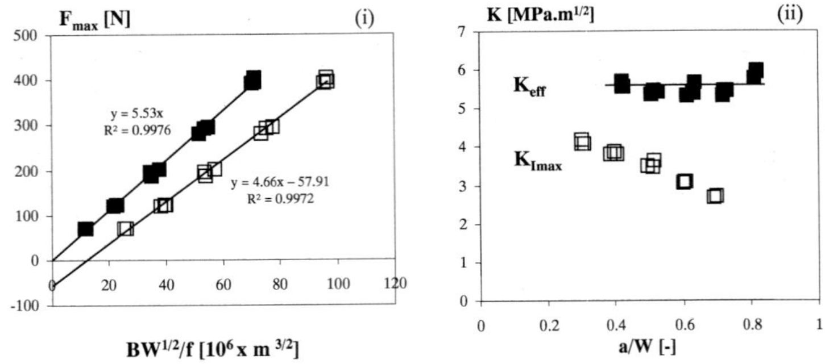

Fig. 3. (i) Graphical determination of the size of the plastic zone. (ii) Evolution of the toughness, K, with a geometry factor a/W before correction (K_{Imax}) and after correction (K_{eff}) with the size of the plastic zone.
Material iPP/EPR-2 tested at 0.1 m/s and 23 ^{0}C.

BENEFITS AND LIMITS OF THE PLASTIC ZONE CORRECTED LEFM

The graphical determination of geometry independent parameters in both ductile and brittle ranges using the same data reduction tool constitutes an improvement for proper characterisation of these polymers.
The calculation of toughness from the plastic zone corrected LEFM is presented now for iPP/EPR-1 and iPP/EPR-2. Table 4 shows that the size of the plastic zone, r_p, is independent of the test speed as long as they remained ductile. Whereas for unstable crack growth, r_p decreased and tended progressively to 0. , r_p had t values of 2.15 mm for iPP/EPR-1 and 2.40 mm for iPP/EPR-2 for stable crack propagation. The higher r_p of iPP/EPR-2 up to 1 m/s seemed to reflect the higher intrinsic ductility of its matrix compared to that to iPP/EPR-1. Moreover, the r_p of both materials near their respective ductile-brittle transitions (0.6 m/s for iPP/EPR-1, 0.8 m/s for iPP/EPR-2) were consistent with their ranking (grade 2 tougher than grade 1).

Table 4. Size of the plastic zone, r_p, as a function of test speed at room temperature for iPP/EPR-1 and iPP/EPR-2. Unstable crack propagation is marked in grey background..

Test speed	**0.001**	**0.01**	**0.1**	**1**	**6**	[m/s]
r_p-iPP/EPR-1	2.28 ± 0.40	2.1 ± 0.32	2.07 ± 0.29	0.74 ± 0.27	0.58 ± 0.47	[mm]
r_p-iPP/EPR-2	2.63 ± 0.29	2.43 ± 0.25	2.34 ± 0.24	1.56 ± 0.58	0.31 ± 0.34	[mm]

The evolution of the effective toughness, K_{eff} (deduced graphically), towards the apparent toughness, K_{Imax} (calculated for a/W = 0.5 using equation 1), over five decades of test speed at 23 ^{0}C can be summarised as follows according to Table 5:

- a decrease of both K_{Imax} and K_{eff} occurred in the ductile-brittle transition region;

- K_{eff} was always higher than K_{Imax}. In the ductile range, $K_{Imax}/K_{eff} \approx 0.68 \pm 0.02$ for iPP/EPR-1 and $\approx 0.61 \pm 0.01$ for iPP/EPR-2. In case of unstable crack propagation, $K_{Imax} \approx K_{eff}$;
- the values of K_{eff}(iPP/EPR-1)/K_{eff}(iPP/EPR-2) for given test conditions were close to those of K_{Imax}(iPP/EPR-1)/K_{Imax}(iPP/EPR-2) when both grades exhibit the same macroscopic behaviour. In other words, K_{Imax} is a semi-quantitative toughness parameter, whereas K_{eff} provides a quantitative description of the fracture resistance.

Table 5. Evolution of the apparent toughness, K_{Imax}, and the effective toughness, K_{eff}, for iPP/EPR-1 and iPP/EPR-2 tested over a wide range of test speeds at room temperature. In grey, unstable crack propagation.

Test speed	**0.001**	**0.01**	**0.1**	**1**	**6**	[m/s]
K_{Imax} - iPP/EPR-1	3.25 ± 0.08	3.33 ± 0.07	3.48 ± 0.06	3.38 ± 0.07	2.55 ± 0.10	[mm]
K_{eff} -iPP/EPR-1	4.88 ± 0.33	4.95 ± 0.23	4.96 ± 0.24	3.88 ± 0.15	2.82 ± 0.12	[mm]
K_{Imax} - iPP/EPR-2	3.10 ± 0.09	3.25 ± 0.04	3.42 ± 0.08	3.50 ± 0.08	2.60 ± 0.2	[mm]
K_{eff} -iPP/EPR-2	5.17 ± 0.41	5.36 ± 0.28	5.53 ± 0.25	4.80 ± 0.35	2.84 ± 0.09	[mm]

For conditions well removed from fracture in the ductile-brittle transition zone, it was possible to readily determine K_{eff}. However, at fracture conditions near to the transition, the determination of K_{eff} was more challenging.. Figure 4 shows the difficulty in determining a unique ductile-brittle transition test speed with the LEFM corrected parameters: the maximum value of K_{eff} does obviously not correspond to any change in the macroscopic mode of failure and the inflexion of r_p is difficult to localize precisely. Moreover, the testing of additional specimens (for r_p and K_{eff}) in this range would not be of any help. Indeed around 0.8 m/s (± 0.4 m/s), the crack grew in an unstable way for the biggest ligaments and in a stable way for the smallest ones. As a consequence, r_p and K_{eff} can not be assessed unambiguously there, especially because it has been remarked that the F_{max} recorded for brittle behaviour were about 7%-10% higher than the F_{max} for ductile behaviours when both modes of failure occurred at given ligament length. Moreover, suppression of part of the experimental data would not be acceptable because r_p is not totally independent of the ligament length as highlighed in Table 6 for iPP/EPR-2 tested at 0.1 m/s. Consequently, the speed at which the ductile-brittle transition occurs can not be determined with precision over 0.5 decades of test speed with the plastic zone corrected LEFM approach.

Table 6. Evolution of the size of the plastic zone, r_p, and the effective toughness, K_{eff}, as a function of the considered interval. Material iPP/EPR-2 tested at 0.1 m/s and 23 ^{0}C.

	$0.3 \leq a/W \leq 0.7$	**$0.3 \leq a/W \leq 0.6$**	**$0.3 \leq a/W \leq 0.5$**	**$0.4 \leq a/W \leq 0.7$**	**$0.5 \leq a/W \leq 0.7$**	
r_p	2.34 ± 0.24	2.59 ± 0.16	2.59 ± 0.16	2.09 ± 0.16	1.96 ± 0.16	[mm]
K_{eff}	5.53 ± 0.25	5.75 ± 0.13	5.88 ± 0.14	5.18 ± 0.15	5 ± 0.15	[MPa.m$^{1/2}$]

Speculation about the r_p values near the transition is not acceptable for an accurate characterisation of viscoelastic materials. An alternative approach is therefore recommended, consisting of (i) definition of the speed at which the ductile-brittle transitions occur using the

conventional LEFM approach with one ligament length (i.e one a/W) and (ii) determinination of intrinsic toughness values in order to conduct some associated modelling work.
It is apparent from both examples given in Fig. 4, that the transition (defined as the maximum of K_{Imax} as a function of the velocity) could easily be determined to within 0.1-0.2 decades of test speed. Moreover, as the apparent values (K_{Imax}) are always lower than the effective parameters (K_{eff}), none of the material paremeters would be overestimated. In addition, since K_{Imax}-values have been shown to provide a semi-quantitative evaluation of fracture resistance parameters, a coherent material comparison would be possible over the whole investigated range. This remark remains true as long as the materials have comparable r_p. For iPP grades, it should be checked (and considered with more caution) when materials exhibit different matrix and/or particles features. It should also be investigated in detail when different polymer families are compared.

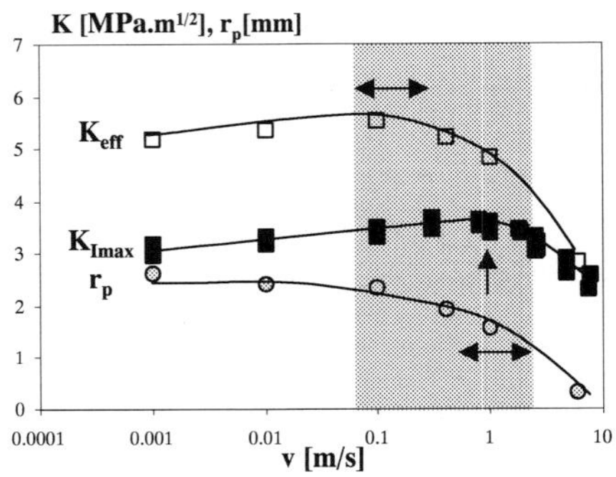

Fig. 4. Size of the plastic zone (r_p), of the effective fracture resistance (K_{eff}) and of the apparent fracture resistance (K_{Imax}) plotted against the logarithm of the test speed. Grey zone indicates a possible transition zone using either r_p or K_{eff} as material descriptor. The single arrow represents the speed at which the ductile-brittle transition occurs. Material iPP/EPR-2 tested at 23 °C.

CONCLUSION

A new approach has been proposed for assessing for fracture resistance. It takes into account the high amount of plasticity of tough polymers and consists of determining experimentally the size of the plastic zone developed at a crack tip using specimens of different crack lengths. The effective fracture toughness K_{eff} was determined from the slope of the plot F_{max} over $BW^{1/2}/f(a_{eff}/W)$ with $a_{eff} = a + r_p$ in such a way that the plastic zone radii, r_p, were obtained numerically using an iteration procedure in such a manner all data fall on a line through the origin. This approach has been shown to provide geometry independent values for ductile and brittle fracture behaviour, except near the ductile-brittle transition. Since this transition is a key-parameter for toughness characterisation of polymers, we have combined the plastic zone corrected LEFM with a more conventional approach using the speed at which the ductile-brittle transition is determinated at given ligament length (typically a/W = 0.5). Since, both of the approaches provide complementary information about toughness, they are thought to give a complete picture of the fracture behaviour of any given material. Further details about this method and its extension to G can be found in ref. [10]

REFERENCES

[1] C. Grein, P. Béguelin, C.J.G. Plummer, H.-H. Kausch, L. Tézé, Y. Germain, *in Fracture of Polymers, Composites and Adhesives*, J.G. Williams and A. Pavan Editors, ESIS Pulication 27, Elsevier Science, UK, 319-333, (2000).
[2] P. Béguelin and H.-H. Kausch, *in Impact and Dynamic Fracture of Polymers and Composites*, J.G. Williams and A. Pavan Editors, Mechanical Engineering Publications, UK, 3-19, (1995).
[3] P. Béguelin, C. Fond and H.-H. Kausch, International Journal of Fracture (1998), **89**, 85.
[4] C. Grein, P. Béguelin, H.-H. Kausch, *in Fracture of Polymers, Composites and Adhesives*, J.G. Williams and A. Pavan Editors, ESIS Pulication, Elsevier Science, to be published (2003).
[5] C. Grein, C.J.G. Plummer, H.-H. Kausch, Y. Germain, Ph. Béguelin, Polymer (2002), **43**, 3279.
[6] P. L. Fernando and J. G. Williams, Polymer Engineering and Science (1980), **20(3)**, 215.
[7] J. M. Hodkinson, A. Savadori and J. G. Williams, Journal of Materials Science (1983), **18**, 2319.
[8] J. I. Velasco, J. A. De Saja and A. B. Martinez, Fatigue and Fracture of Engineering Materials and Structure (1997), **20**, 659
[9] G. R. Irwin, Journal of Applied Mechanics (1957), **24**, 361.
[10] C. Grein, , H.-H. Kausch, Ph. Béguelin, accepted by Polymer Testing (2003).

The Application of Fracture Mechanics to Polymers, Adhesives and Composites
D.R. Moore (Ed)

WHY DUCTILE THERMOPLASTICS SOMETIMES FRACTURE UNDER IMPACT

P.S. LEEVERS

INTRODUCTION

Many crystalline thermoplastics such as polyethylene can be relied on for exceptional toughness and ductility when stressed or strained at moderate rates. Under impact (i.e. when loaded fast enough to challenge their strength in less than one second) the same materials can be vulnerable to the initiation and rapid propagation of brittle cracks.

Impact fracture test methods developed by ESIS register a significant reduction in G_c or K_c as loading rate increases up to 1 m/s (using ISO 17281) and beyond [1]. More specialised test methods [2] also indicate a relatively low dynamic fracture resistance G_D or K_D at crack speeds of 100-300 m/s [3], a finding which is important for the tough thermoplastics used to extrude pressure pipe. The effects of impact and of rapid crack propagation (RCP) seem to be related, but to bring the relationship to light one needs a physical model of the underlying mechanisms. If the model is correct, impact fracture resistance data can then be interpreted as RCP resistance data, and obtained more easily using more familiar methods. This article will show that Linear Elastic Fracture Mechanics is able not only to translate material property data into performance predictions but also to provide the framework for such a model.

LEFM can be approached from '*K*' or '*G*' directions. *K* defines the applied crack-opening stress field around a crack for comparison with some critical value (K_{Ic} etc.) for the material. The *K* approach presupposes an infinitely sharp crack and infinitely high stresses around it — i.e. it is rather 'unphysical'. However, it is more useful for design since it uses concepts which correspond to applied stress σ and strength. It adapts these quantities to a particular shape of cracked component through a geometry correction factor *Y*:

$$K_I = \sigma Y\sqrt{a} = \sigma Y(\alpha)\sqrt{\alpha W} \qquad (1)$$

where *a* is the crack size. *W* is a component size parameter; the crack breaks out to a free surface when $a = W$, i.e. $\alpha = 1$.

G defines the virtual force (per unit width) which a loaded component develops to drive a crack front in a specified direction. The effect of *G* depends on its magnitude relative to some critical value G_c defining fracture resistance. *G* can be calculated directly from the increase in the elastic compliance *C* at the load point as the crack length *a* increases:

$$G = \frac{P^2}{2WB_c}\frac{1}{E'B}\frac{dC^*}{d\alpha} = \frac{v^2}{2WB_c}E'B\frac{1}{C^{*2}(\alpha)}\frac{dC^*}{d\alpha}. \qquad (2)$$

Here B_c is the crack path width, *P* the applied external load and $v = CP$ its displacement. We have defined a dimensionless compliance $C^*(\alpha) = E'BC$, where *B* is the component thickness; it depends only on α and the component or specimen shape.

The *K* and *G* approaches are interchangeable through the equation $K^2 = E'G$, where the plane stress tensile modulus $E' = E$ is replaced by $E' = E/(1-\upsilon^2)$ for plane strain conditions. If $C(a)$ or $C^*(\alpha)$ is not already known, it can therefore be derived from $Y(\alpha)$:

$$C^*(\alpha) = C^*(0) + 2\frac{B_c}{B}\left(\frac{BW\sigma}{P}\right)^2 \int \alpha Y^2(\alpha)\mathrm{d}\alpha.$$

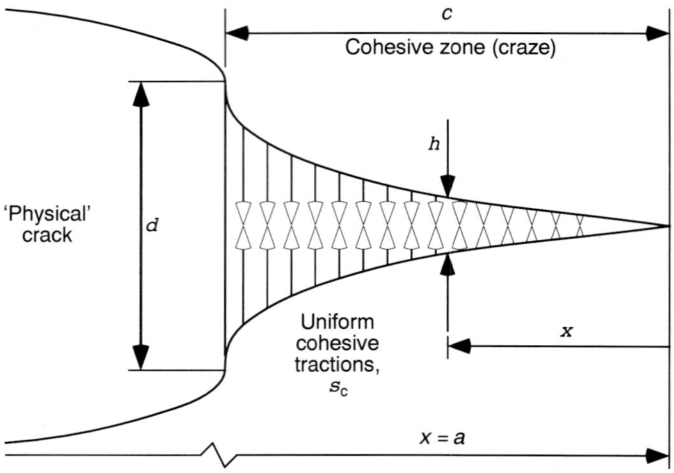

Fig. 1. The Dugdale-Barenblatt cohesive zone model, which provides a good representation of the crack-tip process zone in craze-forming polymers.

Historically, the G approach preceded the K approach. Because it was based on the physics of *how* materials resist fracture, it remains a more useful approach for physical modelling. In this article we start from Dugdale's model of a cohesive zone (Fig. 1) which describes the crack tip region in a craze-forming polymer [4] rather better than it did in the steels for which it was originally devised. Opening of the crack surfaces is assumed to be resisted by a *cohesive stress*, σ_c, which acts uniformly within a distance c ahead of the stress-free 'physical' crack. The cohesive zone supports all of the load which would otherwise be concentrated at the physically unrealistic crack-tip singularity of the K representation. On this basis its length c and thickness can be determined in terms of the applied crack driving force G using the two equations

$$c = \frac{\pi}{8}\frac{E'}{\sigma_c}\delta \tag{3}$$

and

$$G = \sigma_c\delta, \tag{4}$$

where δ is the *crack opening displacement,* the separation of the cohesive surfaces at the crack tip (Fig. 1). As G increases σ_c remains constant but both c and δ increase. When δ reaches some critical value δ_c cohesion fails at the tip of the crack, which extends. If we can predict or measure σ_c and δ_c then we can predict $G_c = \sigma_c\delta_c$ and hence K_c. This will now be attempted, for the two cases referred to in the introduction:

1 *impact*: a driving force G is rapidly applied to an existing, stationary physical crack, and
2 *rapid crack propagation*: a crack extends at high speed under a constant driving force G_D.

ADIABATIC HEATING AND DECOHESION

As a craze thickens under load, the cohesive stress draws polymer chains from the surrounding bulk material into the craze fibrils. The picture of craze thickening built up by Lauterwasser and Kramer [5] and others visualises all of the necessary deformation and chain disentanglement taking place within a thin *active layer*. The fibrils which emerge into the craze are highly orientated and relatively stable.

Deformation in the active layer generates frictional heat on the cohesive surface. Since symmetrical drawing at both surfaces seems unlikely we can assume that only one cohesive surface is active. At any point in the material on this active surface, heat will be generated at a rate per unit area which is the product of $\beta\sigma_c$ (β being the proportion of work converted to heat, 0.8-0.9 for most polymers) and the speed $\dot{\eta}$ at which the craze thickens. At a point fixed in the polymer and lying on the cohesive zone, this heating occurs either as the load on a static craze increases, or as the craze carried at the tip of a rapidly propagating crack sweeps past.

Polymers have very low thermal diffusivity, and if the craze thickens rapidly the active layer gets hot. The model proposes that in crystalline polymers the decohesion mechanism, in both impact and rapid crack propagation, is the disentanglement of polymer chains within a melt layer formed by this adiabatic heating. The melt layer has been assumed to separate catastrophically when it grows too thick to be spanned by an extended weight-average polymer chain, of length s_w (1-2 μm). However, we will treat the critical layer thickness as a generalised structural dimension s_c: in principle, the model could just as well be applied using any other temperature-time or temperature-space criterion more appropriate to another class of polymers.

IMPACT FRACTURE

Impact is usually better represented as a load-point displacement rate $\dot{v}$ (e.g. the pendulum speed in a Charpy test) than as a loading rate. Thus the displacement is $v = \dot{v}t$ and Eqns. (2) and (4) can be used to show that

$$\delta = \frac{E'}{\sigma_c}\left[\frac{1}{C^{*2}(\alpha)}\frac{dC^*}{d\alpha}\right]\frac{B}{2WB_c}\dot{v}^2t^2 \tag{5}$$

so that the craze thickening rate

$$\dot{\delta} = \frac{E'}{\sigma_c}\left[\frac{1}{C^{*2}(\alpha)}\frac{dC^*}{d\alpha}\right]\frac{B}{WB_c}\dot{v}^2t \tag{6}$$

increases linearly with time (unless the craze grows long enough to affect the compliance [6]).

Thermal conduction analysis can now be applied to determine the time taken for the active layer to heat adjacent polymer to the melt temperature $T = T_m$ throughout a layer s_c thick. Assuming constant density ρ, specific heat C_p and thermal diffusivity κ, there is an algebraically awkward but apparently general solution. G_c has a minimum value

$$G_{c,min} = 3.064 s_c \rho C_p (T_m - T_0), \tag{7}$$

where T_0 is the initial bulk temperature. Since ρC_p is far from being temperature independent in the materials of interest, the term $\rho C_p(T_m - T_0)$ can be approximated as the total heat to melt unit volume of polymer.

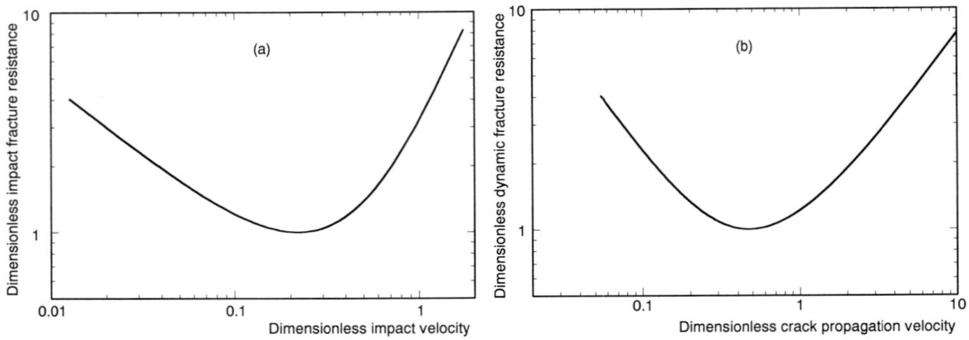

Fig. 2. The predicted dependence of the fracture resistance of a crystalline thermoplastic on (a) impact speed and (b) crack propagation speed (normalisations are defined in the text).

The dependence of G_c on $\dot{v}$ is plotted in Fig. 2(a) as a general function $G_c/G_{c,\min}$ *vs.* $\dot{v}/\dot{v}_0$, where the reference impact speed is given by

$$\dot{v}_0 = 6.522\kappa\left[\rho C_p(T_m - T_0)\frac{W}{E' s_c^3 \Gamma_v(\alpha)}\right]^{1/2}. \tag{8}$$

Figure 2 predicts an impact fracture resistance minimum at an impact speed of ~$0.2\dot{v}_0$, which for a Charpy-like impact bend specimen is a few metres per second — i.e. a typical pendulum speed. At lower impact speeds adiabatic heating becomes more diffused, the dependence of G_c on the structural dimension s_c becomes insignificant, and

$$G_c = E'^{-1/3}\left[\tfrac{3}{2}\rho C_p(T_m - T_0)\right]^{4/3}(\pi\kappa)^{2/3}\Psi(\alpha)W^{1/3}\dot{v}^{-2/3} \tag{9}$$

where $\Psi(\alpha) \equiv \left[\tfrac{1}{2}\Gamma_v(\alpha)\right]^{-1/3}$ is a geometry function (about 2.5 to 3 for a typical impact-bend specimen).

Equation (9) suggests that G_c decreases with the –2/3 power of impact speed, which had already been observed for pipe grade polyethylenes [7] and would be for other thermoplastics including even those for which a structural 'melting' temperature cannot be defined [8]. At a given impact speed it decreases with increasing tensile modulus, increases with the heat to melt per unit volume and increases weakly with specimen size. However, at some speed G_c must fall to the minimum value given by Eqn. (7). Although the exact magnitude of the predicted G_c value is sensitive to the detail of assumptions made about non-linear properties, it is often correct to within a factor of two.

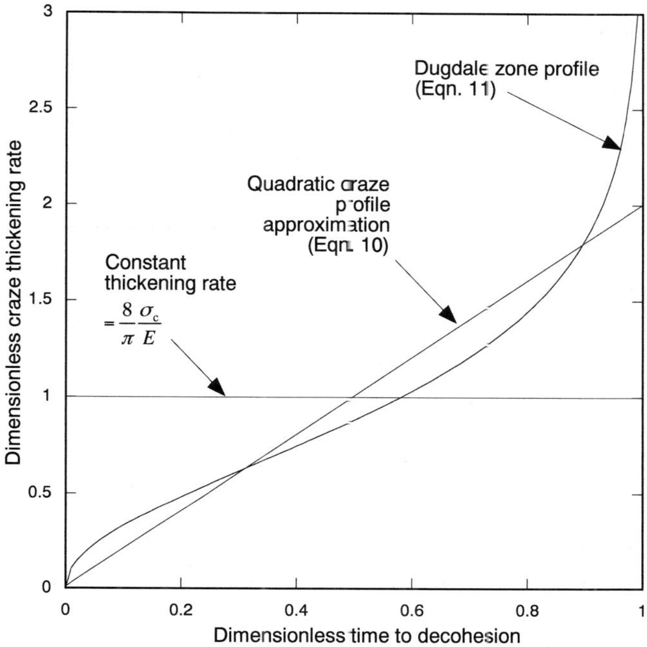

Fig. 3: The rate of craze thickening due to passage of a moving Dugdale cohesive zone

RAPID CRACK PROPAGATION

In the case of rapid crack propagation, the rate of craze thickening (which determines the intensity of the active surface heat source) is determined by the thickness profile of the passing cohesive zone. For a Dugdale zone this profile has the cusp shape suggested in Fig. 1, meeting the crack tip at a tangent which implies infinite rate. The thickening rate at any point $0 < \xi < c$ from the craze tip is plotted in Fig. (3):

$$\dot{\eta} = \frac{8}{\pi} \frac{\sigma_c}{E'} \left[\tfrac{1}{2} ln \frac{\left(1 + (\xi / c)^{1/2}\right)}{\left(1 - (\xi / c)^{1/2}\right)} \right] \dot{a} \, . \tag{10}$$

It can be seen that an adequate and very useful approximation is

$$\dot{\eta} \approx \frac{8}{\pi} \frac{\sigma_c}{E'} \left[2 \frac{\xi}{c} \right] \dot{a} \, . \tag{11}$$

If this quadratically-profiled craze sweeps past a fixed point in the material at a constant speed $\dot{a}$, $\xi = \dot{a}t$ and we obtain, using Eqns. (3) and (4).

$$\dot{\eta} \approx 2 \left(\frac{8}{\pi} \frac{\sigma_c}{E'} \right)^2 \frac{\sigma_c}{G} \dot{a}^2 t \, . \tag{12}$$

Equation (12) gives $\dot{\eta}$ the same linear dependence on time as Eqn. (6) gives $\dot{\delta}$. A point fixed in the material at the tip of a crack loaded in impact will, therefore, experience a similar thermal history as a fixed point passed by a craze driven at force G_D if

$$2 \left(\frac{8}{\pi} \frac{\sigma_c}{E'} \right)^2 \frac{\sigma_c}{G_D} \dot{a}^2 \approx \frac{E'}{\sigma_c} \left[\frac{1}{C^{*2}(\alpha)} \frac{dC^*}{d\alpha} \right] \frac{B}{WE_c} \dot{v}^2 \, . \tag{13}$$

Since (according to this theory) the crack tip will then suffer decohesion at the same time, we can write $G_D = G_c$ and simplify the expression to

$$\dot{a} = \frac{\pi}{16}\left(\frac{E'}{\sigma_c}\right)^2 \left\{\left[\frac{2}{C^{*2}(\alpha)}\frac{dC^*}{d\alpha}\right]\frac{B}{WB_c}\frac{G_c}{E'}\right\}^{1/2} \dot{v} \tag{14}$$

in which all of the geometrical parameters refer to the impact configuration.

If the underlying idea is correct Eqn (14) achieves the objective outlined in the introduction: it allows data from an impact G_c test to be interpreted in terms of an equivalent resistance to rapid crack propagation. Impact G_c tests, even at the high speeds required, are less complicated than dynamic fracture toughness tests, and much more work has been put into refining experimental methods for them [2].

The broader picture of how rapid crack propagation resistance might depend on crack speed can again be presented as a general function $G_D/G_{D,min}$ *vs.* $\dot{a}/\dot{a}_0$, where the reference crack speed is given by

$$\dot{a}_0 \equiv 4.43\frac{E'\rho C_p \kappa}{\sigma_c^2 s_c}(T_m - T_0), \tag{15}$$

and is of the order of 100 m/s. This function is plotted in Fig. 2(b). The most inaccessible parameter in Eqns. (14) and (15) is the cohesive stress σ_c, although Clutton [9] and Leevers *et al.* [10] have used craze size measurements to measure it for polyethylene. However, the most important result is the minimum dynamic crack resistance $G_{D,min}$ and this is expected to be essentially equal to $G_{c,min}$.

CONCLUSION

Impact tests are widely used for the mechanical characterisation of polymers. The results seem to represent some basic 'brittleness' property, and the refinement of instrumented sharp-notch impact fracture tests has already shown how fracture mechanics can be used to express this property quantitatively. However, without some physically based understanding of the fracture mechanism, there is a danger of applying impact data to fracture by entirely different modes, e.g. slow crack growth. Even the application of impact fracture data to rapid crack propagation, which is phenomenologically similar, is still under discussion. This article has shown how the physical models which underlie fracture mechanics can be used to guide such discussion, in this case arriving at a specific relationship between impact fracture resistance and rapid crack propagation resistance.

REFERENCES

1. Böhme (2001). In: *Fracture Mechanics Testing Methods for Polymers Adhesives and Composites*, ESIS Publication 28, pp. 73-90, Moore, D.R., Pavan, A. and Williams, J.G. (Eds). Elsevier, Oxford.
2. Ritchie, S.J.K. and Leevers, P.S. (1995). *Impact and Dynamic Fracture of Polymers and Composites*, ESIS Publication 19, pp.) 137-146, Williams, J.G. and Pavan, A (Eds), Mechanical Engineering Publications, Bury St. Edmunds
3. Wheel, M.A. and Leevers, P.S. (1993) *Int. J. Fracture* **61,** 349.
4. Williams, J.G. (1987) *Fracture Mechanics of Polymers*. Ellis Horwood, Chichester.
5. Lauterwasser, B.D. and Kramer, E.J. (1979) *Phil. Mag.* **A39**, 469.
6. Leevers, P.S. and Morgan, R.E. (1995) *Engng. Fracture Mech.* **5**, 999.

7. Leevers, P.S. (1995) *Int. J. Fracture* **73**, 109.
8. Rager, A. (2003). PhD Thesis, University of London.
9. Channell, A.D. and Clutton, E.Q. (1996) *Plastics, Rubber and Composites* **25**, 272.
10. Leevers, P.S. (2000) *Plastics, Rubber and Composites* **29**, 453.

The Application of Fracture Mechanics to Polymers, Adhesives and Composites
D.R. Moore (Ed)

RATE AND TEMPERATURE EFFECTS ON FRACTURE TOUGHNESS OF POLYSTYRENE AT VARYING MOLECULAR MASS

L. CASTELLANI, S. COLOMBARINI and R. FRASSINE

INTRODUCTION

Fracture behaviour of polymers may differ significantly from that of more conventional structural materials, mainly because of their viscoelastic nature. This may imply, for example, that the strain energy dissipation mechanisms taking place during fracture at different positions inside the body may vary at varying the loading time scale. Very complex interaction may develop between the deformation of the linear viscoelastic continuum surrounding the crack and the highly non-linear and/or yielding phenomena occurring at the crack tip, which are more likely to be influenced by the molecular structure of the material.

The fracture process in polymeric materials is almost invariably characterised by two stages: the initiation stage, in which a damage zone ("process" zone) progressively develops ahead of the crack tip until the crack starts to propagate, and the propagation stage, in which the material inside the process zone breaks creating new surfaces.

The polystyrene (PS) here examined is an amorphous polymer well below its glass transition at room temperature, and shows an intrinsically brittle fracture behaviour when tested at normal loading and temperature conditions. Its fracture toughness depends on molecular mass, although the effect is hardly apparent from test results obtained with conventional tests such as tension, bending or impact (either Charpy or Izod).

Testing using Double Torsion (DT) geometry clearly evidenced such dependence on the crack propagation stage. Scanning electron microscopy of the fracture surfaces has shown that different deformation micromechanisms develops at the crack tip at varying crack speed and temperature.

EXPERIMENTAL DETAILS

Two PS homopolymers having different molecular mass (Mw) and, to a lesser extent, different molecular mass distribution (Fig. 1 and Table 1) were kindly supplied by Polimeri Europa (Italy). In the remainder of the text, the material with lower M_w will be referred to as PS-B, while the other as PS-A.

DT specimen (Fig. 2a) is a rectangular plate having in-plane dimensions of 200 x 60 mm. The thickness, B, was varied between 2 and 10 mm. Specimens were edge-notched at 60 mm depth by saw-cutting with a razor blade. In some cases, the specimens were also been V-grooved to a depth of about 0.2 B in order to drive the crack to propagate along the symmetry axis of the sample. For testing, specimen is supported on two cylindrical rollers of radius 3 mm and spaced 50 mm apart. Two hemispheres of radius 3 mm and spaced 13.5 mm apart are placed across the notch and forced to move at constant displacement rate (Fig. 2b). This loading configuration results in a torsion of the two arms separated by the notch and drives the crack to propagate at a constant speed. Testing temperature was varied between 0° and 80°C with displacement rates ranging from 0.01 and 100 mm/min.

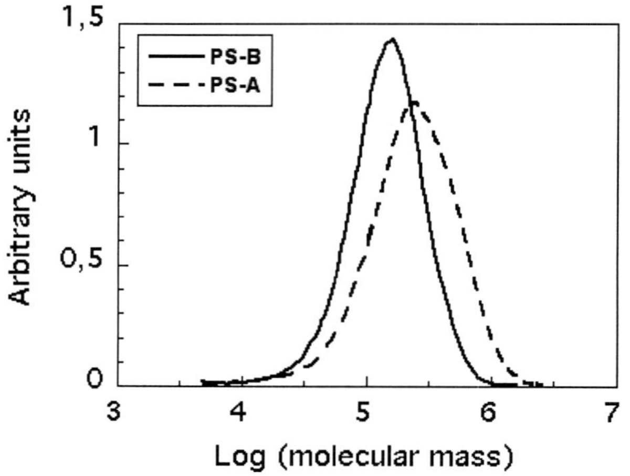

Fig. 1: Molecular mass distribution of the two materials

Table 1: Molecular parameters for the two materials examined

Property	PS-A	PS-B
M_w	312000	177000
M_n	134000	104000
M_w / M_n	2.33	1.70

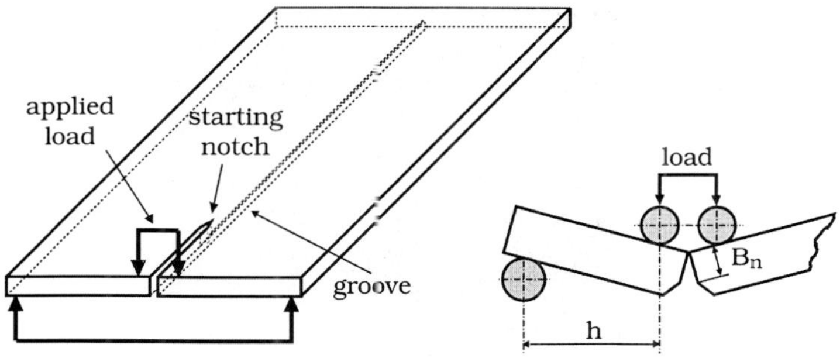

Fig. 2: Schematic of Double Torsion testing geometry

DOUBLE TORSION ANALYSIS

The strain energy release rate during crack propagation can be calculated using:

$$G_{Ic} = \frac{P_c^2}{2B_n}\frac{dC}{da} \tag{1}$$

in which P_c is the load during crack propagation, C is the specimen compliance and a is the crack length. The crack speed is proportional to the applied displacement rate as follows:

$$\dot{a} = \frac{\dot{x}}{P_c dC/da} \tag{2}$$

Theoretical analysis predicts that, for linear elastic materials, the specimen compliance, C, should be linearly dependent on the crack length, a. Therefore, if the test is carried out at a constant displacement rate, and G_{Ic} is a monotonically increasing function of crack speed, the theory also predicts that fracture will take place at constant load, P_c, and constant crack speed.

Due to experimental difficulties encountered in measuring the shear modulus of the materials at varying strain rate and temperature, direct experimental compliance calibration was adopted to calculate the value of dC/da for each testing condition. Corrections for crack front curvature (Fig. 3) and load moment arm reduction at large displacements were also applied for data analysis [1, 2].

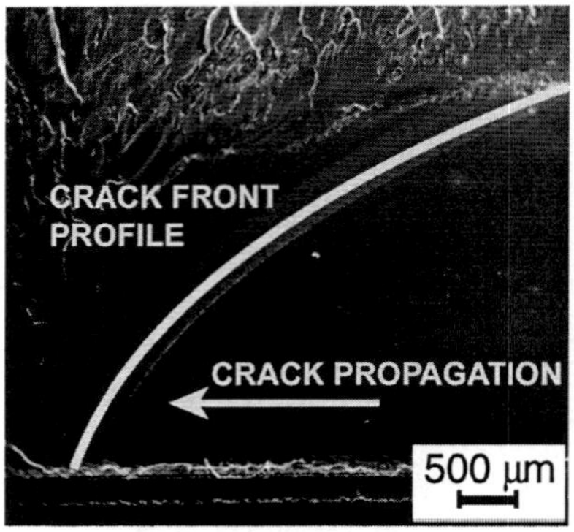

Fig. 3: Crack front in Double Torsion

Fig. 4 shows a typical load-displacement diagram obtained in DT. After an initial transient, the load stabilises at a nearly constant value for most of the test. Equation 2 therefore predicts that the crack will propagate at constant speed. Since the material is highly transparent, such prediction could be experimentally verified by direct optical observation of the crack front during the test [3]. The experimental steps were as follows: one side of the specimen was metallographically polished, the videorecording was synchronised with the load recording and the propagating crack was shooted perpendicularly to its plane through the polished face.

Results indicate that, after an initial transient corresponding to the maximum observed in the load curve of Fig. 4, a constant crack speed is reached and then maintained for most of the propagation. The videorecordings were also used to derive experimental calibration curves of specimen compliance vs. crack length, such as that shown in Fig. 5. Data indicate that the linear dependence is approximately verified for most of the crack growth, in agreement with the above observations and previous findings [3]. Moreover, the crack lenght range in which the crack was observed to propagate at constant speed is always contained inside the compliance linearity range.

It may therefore be concluded that the simplyfied theoretical analysis briefly recalled above is suitable for data reduction, at least for the two materials examined.

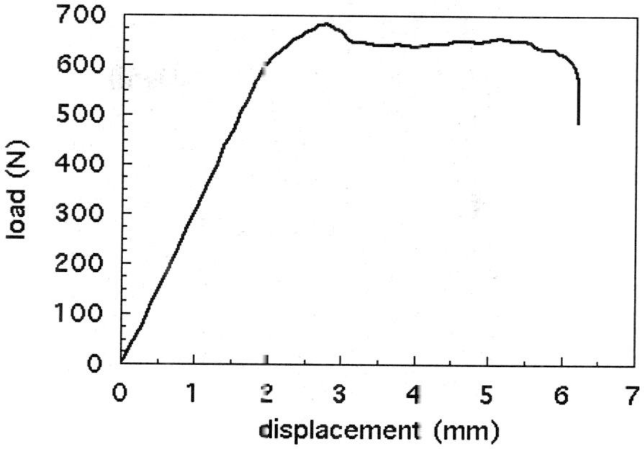

Fig. 4: Load diagram obtained in DT for PS-A at room temperature and 2 mm/mm displacement rate

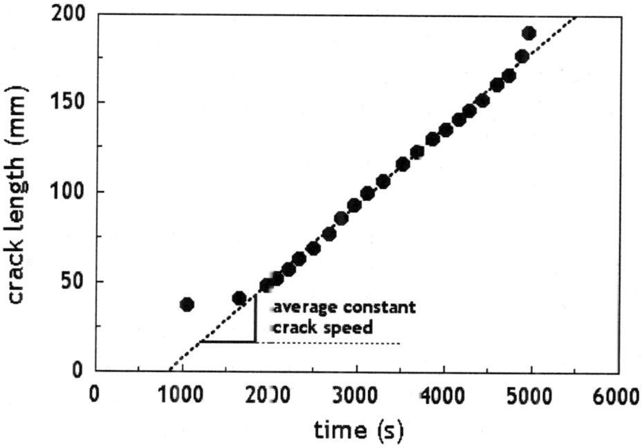

Fig. 5: Specimen compliance vs. crack length obtained on a 9.8 mm thick DT specimen

RESULTS

Figure 6 shows the master curves of fracture toughness vs. crack speed for PS-A and PS-B, obtained by shifting G_{IC} data obtained at different temperatures along the crack speed axis until superposition. This procedure has been previously successfully applied by one of the Authors to either polymers and composites [4, 5]. The relevant shift factors derived empirically from the above procedure are shown in Fig. 7.

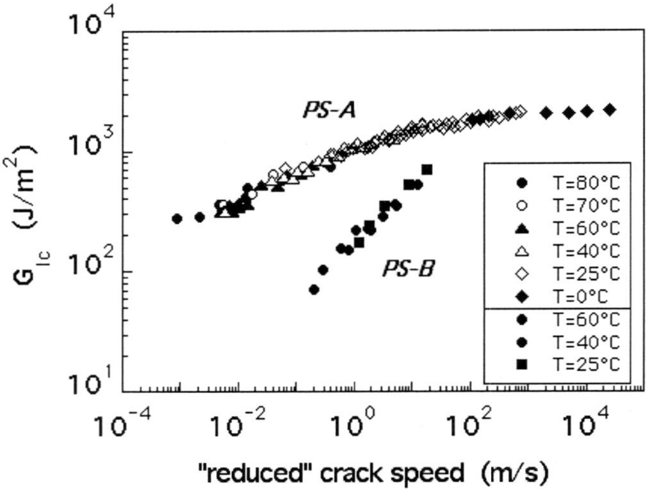

Fig. 6: Fracture toughness vs. crack speed master curves for the two materials examined (reference temperature: 23°C).

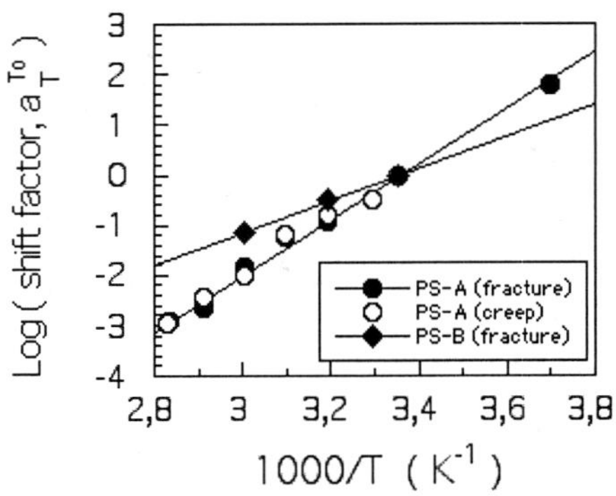

Fig. 7: Temperature shift factors relevant to Fig. 6

A good agreement is observed with shift factors obtained from small-strain tensile creep testing on PS-A, suggesting that fracture and deformation processes are controlled by the same kinetics. Data reported in Fig. 6 indicate that fracture toughness is very sensitive to M_w, especially at low crack speeds The values obtained on the two materials differ by more than one order of magnitude for crack speeds less than 0.2 mm/min. At much higher crack speeds, however, the curves appear to converge at the same asymptotic value (approx. 2.5 kJ/m^2), and this is in agreement with the observation that conventional impact testing cannot discriminate between the two materials. The different slope of the toughness vs. crack speed master curves in the lower crack speed range could arise from different dissipative mechanisms occurring either in the bulk (due to the viscoelastic behaviour of the material under load) or in the neighbourhood of the crack tip (in the so-called "process" zone, where a highly stressed, relatively small volume of material progressively accumulates damage until failure). Dynamical-mechanical testing conducted on the two materials, however, has shown that the loss tangent, tan δ, is roughly the same for the two materials (0.016 for PS-A and 0.018 for PS-B at 1 Hz and room temperature), indicating that molecular relaxation mechanisms associated with the β-transition are not affected by M_w. Therefore, the observed difference in slope cannot be attributed to the viscoelastic behaviour of the material in the bulk and should be interpreted in terms of fracture micromechanisms.

SEM microscopy

The failure micromechanisms have been investigated using scanning electron microscopy (SEM) of the fracture surfaces.

PS-A shows two distinct features at varying crack speed: for lower speeds, where the toughness is also lower, the fracture surface appears very smooth and perfectly transparent. Only at very high magnification (10000x) some characteristic features could be evidenced (Fig. 8): the fracture surface shows very thin marks, approximately perpendicular to the crack front, which probably indicates that microfailures occurs inside the single craze preceding the crack front.

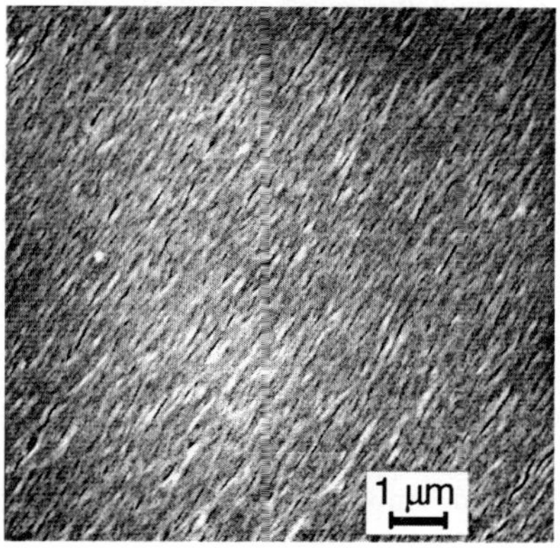

Fig. 8: Fracture surface obtained on PS-A at low speed

For higher speeds, the surface appears considerably rougher (Fig. 9) and reflects the incident light assuming the characteristic "silvered" appearance of the multiple craze process zone.

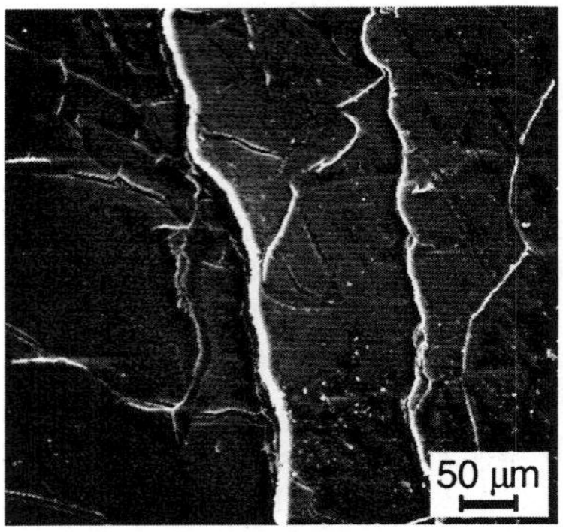

Fig. 9: Fracture surface obtained on PS-A at high crack speed

For material PS-B, however, fracture surfaces are significantly different, and no fundamentally distinctive features could clearly be identified. Tests conducted at high crack speeds again produced "silvered" fracture surfaces (Fig. 10), which, even under lower magnification, are considerably rougher than those observed on PS-A. The crack is propagating on different planes, which indicates a "multiple craze" crack growth mechanism.

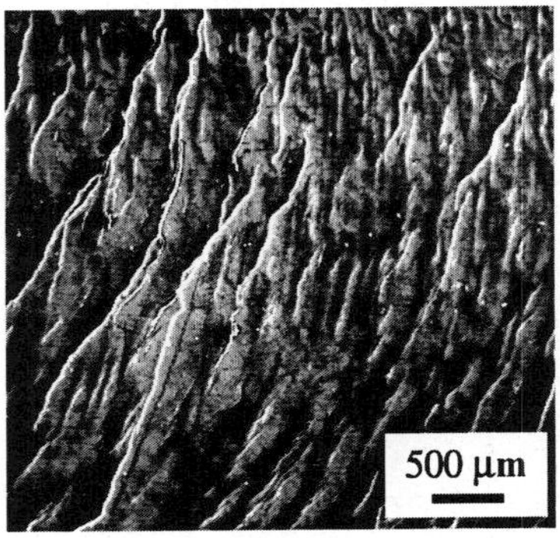

Fig. 10: Fracture surface of PS-B at high crack speed

Lower crack speeds produced smooth and transparent surfaces, which at high magnification (Fig. 11) again appear considerably rougher than those observed on PS-A (Fig. 8) with no marks to indicate the crack growth direction.

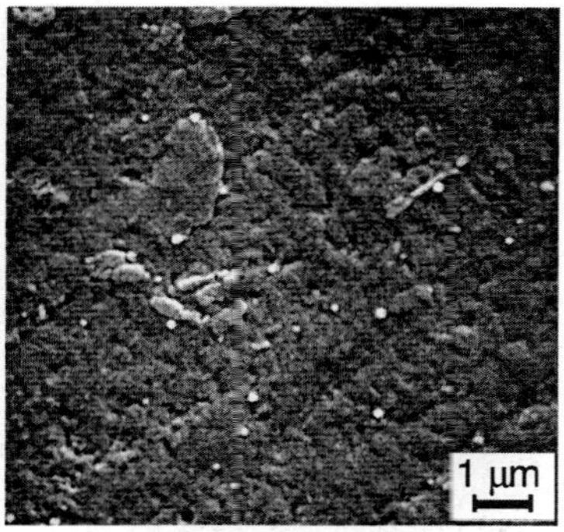

Fig. 11: Fracture surface of PS-B at intermediate crack speed

For much lower crack speeds, multiple craze "silvered" features appear starting from the bottom side of the specimen, which are considerably rough (Fig. 12a), while the upper side surface remains essentially unchanged (Fig. 12b).

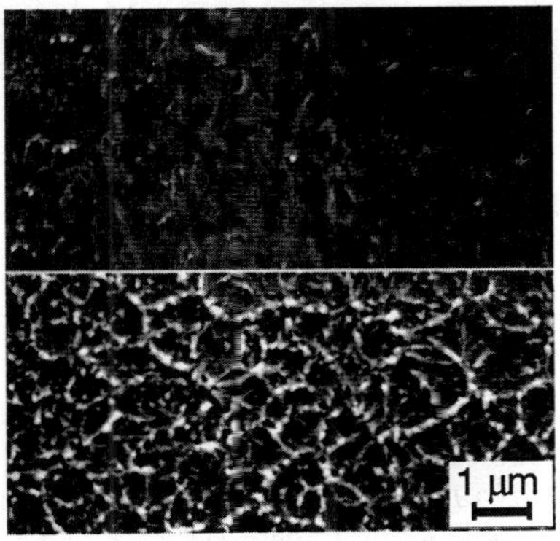

Fig. 12: Fracture surface obtained on PS-B at low crack speed on the upper (a) and lower (b) portion of the DT specimen

For very low crack speeds, most of the fracture surface is again smooth and transparent, but some bands appear having the characteristic shape of the curved crack front (Fig. 13). Closer observation of a single band shows three distinct zones: a first one, in which fracture occurs on different planes (Zone A), a second one, which reflects the incident light and is significantly rougher (Zone B), an a final one, which appears dark and smooth (Zone C).
Under higher magnification (Fig 14) the morphology of the three zones appears significantly different.

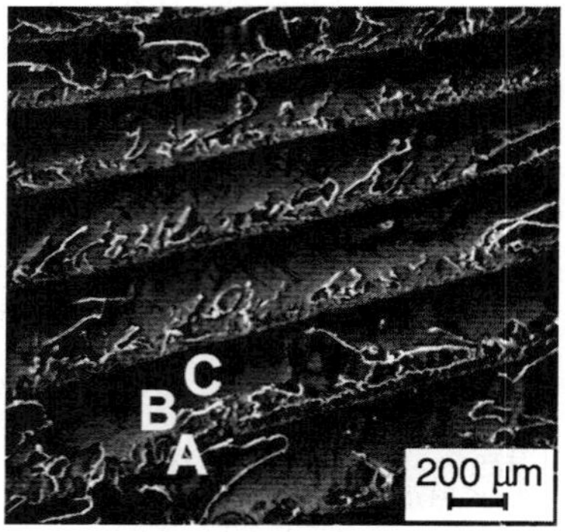

Fig. 13: Fracture surface of PS-B at very low crack speed

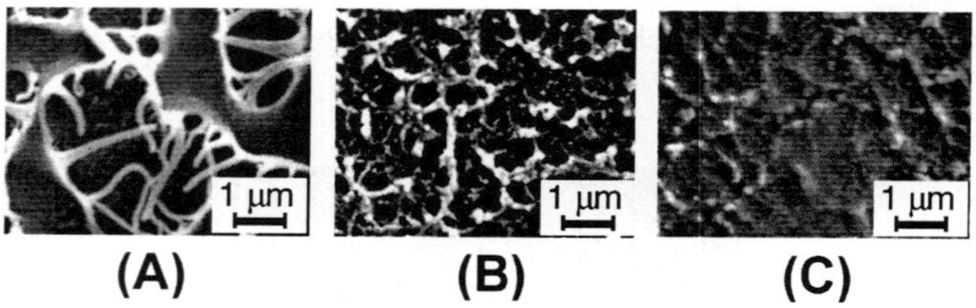

Fig. 14: Detail of the three zones in Fig. 13

DISCUSSION

Craze growth in PS has been shown to occur by drawing new material into the craze fibrils from an "active zone" at the craze-bulk interface [6]. This process requires a loss of entanglement density [7], and evidence has been given by Kramer and coll. [8] of two mechanisms being involved in it: chain scission and chain disentanglement. The entanglement network has been shown to play a key role in determining the stability of crazes. Only craze fibrils having significant entanglement density can survive to allow craze widening and extension. Extensive work [8] on monodisperse PS samples has shown that craze stability sharply increases from M_w = 50000 up to 200000, then levelling off for further increase of M_w. Greater stability of disentanglement crazes is also to be expected as a consequence of the lower stress they require to grow with respect to scission crazes.

The transition observed in PS-A from smooth (Fig. 7) to rough (Fig. 8) fracture surface for increasing crack rate well corresponds to the transition from disentanglement to chain scission crazing. At low crack speeds a single, large and highly stable disentanglement craze forms at the crack tip, due to small amounts of chains with $M<M_e$, giving rise to a smooth failure surface. At high propagation rates, however,

chain scission crazes are favoured, and they break into micro-cracks soon after formation. New crazes are then generated at the crack tip and the whole process is repeated. The great number of small crazes on different planes gives rise to the observed rough fracture surface.

In PS-B the entanglement network is "diluted" by a significantly larger fraction of molecules having $M<M_e$ with respect to PS-A. Transition from scission to disentanglement will therefore occur at higher propagation rates and craze stability will be lower because of the lower molecular weight. It is therefore not surprising that, even at intermediate crack speeds, smooth portions of the fracture surfaces are obtained. The gradual transition from scission to disentanglement and the repeated occurrence of craze breakdown give rise to the various different zones shown in Figs. 9 to 11. At very low crack rates, the disentanglement regime is presumably well established and the fibril breakdown probability is constant during the entire test. A phenomenon similar to the "retarded crack growth" observed in fatigue experiments [9] occurs, with the development of the growth bands shown in Fig. 12. It may be assumed that, when crazes reach a critical size, fibril breakdown occurs and the crack jumps through the growing craze. When the crack finally comes to stop, a new craze forms and the process repeats giving rise to the observed bands, the size of which reflects the critical size for craze breakdown.

REFERENCES

[1] P. S. Leevers (1982), J. Mat. Sci. 17: p. 2469.

[2] P. S. Leevers (1986), J. Mat. Sci. Letters 5: p. 191-192.

[3] R. Frassine, T. Riccò, M. Rink and A. Pavan (1988), J. Mat. Sci. 23: p. 4027.

[4] R. Frassine and A. Pavan (1995) Comp. Sci. and Techn. 54: p. 193.

[5] R. Frassine, M. Rink and A. Pavan (1996) Comp. Sci. and Techn. 56: p. 1253.

[6] B.D. Lauterwasser, E.J. Kramer (1977) Phil. Mag.A 36: p. 1217.

[7] E.J. Kramer (1983) Adv. Polym. Sci. 52/53, H.H. Kausch Ed., Springer-Verlag.

[8] E.J. Kramer and L.L. Berger (1990) Adv. Polym. Sci. 91/92, H.H. Kausch Ed., Springer-Verlag.

[9] W. Döll and L. Könczöl (1990) Adv. Polym. Sci. 91/92, H.H. Kausch Ed., Springer-Verlag.

The Application of Fracture Mechanics to Polymers, Adhesives and Composites
D.R. Moore (Ed)

APPLICABILITY OF LINEAR ELASTIC FRACTURE MECHANICS TO FATIGUE IN AMORPHOUS AND SEMI-CRYSTALLINE POLYMERS

R.W. LANG, W. BALIKA and G. PINTER

INTRODUCTION

Due to the increasing interest in engineering plastics for structural components exposed to cyclic loads, many research efforts have focused on the investigations of their fatigue behaviour [1-11]. Of particular importance, among all mechanical loading modes (monotonic, static, cyclic), fatigue loading is the most detrimental to the material´s performance. Moreover, many structural failures of load bearing plastics components are related to fatigue.

Depending on the viscoelastic state of the polymeric material and on the test conditions, fatigue failure may occur either by hysteretic-heating-induced thermal failure or via a mechanically-induced crack propagation process. Regarding the latter mechanism it is generally accepted that the fatigue process, as for any other engineering material, may consist of at least three different stages. These involve the initiation of a micro-crack, stable crack growth at a measurable growth rate, and unstable crack growth (catastrophic fracture) or ductile tearing in the last loading cycle.
Unfortunately very little evidence exists that would indicate how much of the fatigue life of polymeric components and test specimens that fail by a fatigue crack propagation (FCP) mechanism may be attributed to crack initiation and crack propagation, respectively. However, since many components contain stress raising defects or inhomogeneities that are characteristic of the material (e.g., composites), or are introduced during the fabrication process (e.g., voids, inclusions, flaws, microcracks) or subsequently by mechanical handling (e.g., scratches, cracks), cracks may be easily initiated and fatigue crack growth may become the service life determining factor.

GENERAL BACKGROUND AND OBJECTIVES

Kinetics of Fatigue Crack Growth
From the several different approaches that have been suggested to describe fatigue crack propagation in polymeric solids, linear elastic fracture mechanics (LEFM) has received the widest application [1, 2, 5, 7, 9, 10, 11]. Thus, it is increasingly recognised that fatigue crack growth experiments based on LEFM provide a powerful and effective tool for material characterization in terms of basic data generation for material ranking and material selection for specific applications, and for the development of novel polymeric materials via a better understanding of structure-property-relationships. Compared to other fatigue approaches (e.g., S-N-approach as originally proposed by Wöhler (compare [1])), potential advantages of the LEFM approach include a reduction in testing time and the number of specimens to be tested, a reduced data scatter and a high sensitivity to changes in the material micro-structure. Most importantly, however, if applicable LEFM provides a material specific, specimen configuration and geometry independent fatigue failure criterion in terms of the kinetics of fatigue crack growth.

According to LEFM principles the growth rate of a fatigue crack, da/dN, under a given set of test conditions is governed by the variations in the elastic stress intensity factor, ΔK, independent of specimen geometry and configuration. However, the unique dependence of crack growth rates on the stress intensity factor range may not necessarily hold for viscoelastic materials. Two of the factors that could cause deviations and which are the subject of this paper are the path or history dependence of mechanical properties even in the linear viscoelastic range, and the occurrence of pronounced nonlinearity in the viscoelastic stress-strain response.

For example, consider a center-cracked viscoelastic body experiencing a constant-ΔK loading situation in the tension-tension range. Assuming conditions for crack advance at a stable growth rate it is readily apparent that the time under load, or the number of loading cycles, experienced by each material element in the crack plane of the uncracked ligament increases with the distance from the original crack tip. As a result the crack essentially moves continuously into a region of material that has been exposed to a different loading history before separation by the propagating crack. Such changes in loading history originating in the so-called far-field region may at least in principle not only introduce continuous changes in the viscoelastic state of the material away from the crack tip, but may subsequently also affect the local yielding behavior of each material element as it is approached by the crack. Accordingly, if the far-field history component influences the mechanical response to any significant degree, macroscopic fatigue crack growth rates even under constant ΔK conditions should change.

Crack tip plastic deformation

For elastic-plastic materials it is generally acknowledged that plastic deformation takes place in a region around the crack tip wherever the elastic stresses meet the yield or flow criterion of the material. In this context a number of models have been proposed to describe the size or the shape of crack tip plastic zones or both [2, 12-15].

For a first estimate it is frequently assumed that the plastic zone, due to stress redistribution, should extend over roughly twice the distance at which the y-component in the elastic stress field equals the effective yield stress of the material. The effective yield stress is generally taken as the product ($m_p \cdot \sigma_{ys}$), where σ_{ys} is the yield stress under uniaxial tension, and m_p is a measure for the amount of constraint on the plastic zone; m_p essentially accounts for the increase in the stress necessary for yield to occur under the crack tip stress conditions in comparison to the case of uniaxial tension. The size of the plastic zone, r_p, is thus given by:

$$r_p = \frac{1}{\pi} \cdot \left(\frac{K_I}{m_p \cdot \sigma_{ys}} \right)^2 \quad (1)$$

where K_I is the stress intensity factor under pure tensile loading. While m_p equals 1 in the case of plane stress, a value of $\sqrt{3}$ has been estimated for plane strain [16].

In polymeric solids two principal mechanisms of crack tip plastic deformation - crazing and shear yielding - may occur. Although the actual shape of the plastically deformed region at the crack tip varies depending on the plastic deformation mechanism, plastic zone dimensions are also frequently assumed to be proportional to $(K_I/\sigma_0)^2$, where σ_0 is either the yield or craze stress, whichever is relevant. Of course, history related phenomena before yielding or crazing, pronounced viscoelastic nonlinearity, and hysteretic crack tip heating may also influence crack tip plastic deformation processes by affecting the materials craze or yield stress.

Objectives

The major objective of this paper is to discuss the applicability and potential limitations of

LEFM concepts to fatigue in engineering plastics. A first sub-objective is to evaluate the potential influence of the path or history dependence of mechanical properties on the macroscopic FCP behavior of polymeric materials. The role of far-field history effects will be analyzed based on fatigue crack growth data generated under constant-ΔK conditions and based on FCP results obtained from various specimen types with vastly different distributions in their far-field stresses. A second sub-objective of this study is to present results of plastic zone size measurements over a wide range of fatigue test conditions and to examine these results in terms of their agreement with fracture mechanical predictions.

EXPERIMENTAL

Materials and Specimen Preparation

The polymeric materals investigated included amorphous plastics such as polycarbonate (PC), poly(methyl methacrylate) (PMMA) and unplasticized poly(vinyl chloride) (PVC-U), and semi-crystalline materials such as various grades of high density poly(ethylene) (PE-HD), and poly(1-butene) (PB). Four types of specimen configurations including center-cracked tension (CCT), single-edge-notched (SEN), compact-type (CT) and wedge-open-loading (WOL) specimens were used to perform the FCP experiments (Fig. 1).

Fatigue Crack Propagation Testing

FCP testing was performed using servo-hydraulic closed-loop testing systems. Environmental conditions were laboratary air at 23 ± 1 °C and 80 ± 1 °C, respectively. The applied waveform was sinusoidal with a constant load amplitude and a minimum/maximum load ratio, R, of 0.1. Values for the stress intensity factor range, ΔK, were obtained as

$$\Delta K = K_{max} - K_{min} = K_{max} \cdot (1 - R) \tag{2}$$

where K_{max} and K_{min} are the maximum and the minimum stress intensity factors, respectively, in a loading cycle. To avoid any effects of larger-scale hysteretic specimen heating, test frequencies were selected in the range from 1 to 10 Hz.

Two types of experiments were conducted. In the constant-ΔK experiments the loads were adjusted manually in crack increments of 0.1 to 0.2 mm, usually without interrupting the experiments. In the second set of experiments different specimen configurations were tested under constant load-range conditions. In this case the crack length was monitored after typical increments of 0.2 to 0.4 mm. Crack length measurements were performed with a travelling microscope.

Plastic zone size measurements

Craze dimensions in PMMA under fatigue loading conditions have been measured using miniature CT specimens and the optical interference method [2, 18]. For all other materials plastic zone dimensions were determined using optical microscopy on fracture surfaces (shear lip measurements in PC) [2] or on specimen side surfaces (craze or process zone measurements in PE-HD, PP, PB) [10]. In order to obtain information on cyclic effects on plastic deformation at crack tips, these experiments were performed at various R-ratios ($R = K_{min}/K_{max}$).

Further details as to the materials and the test procedures and the data reduction schemes are described elsewhere [2, 9, 10].

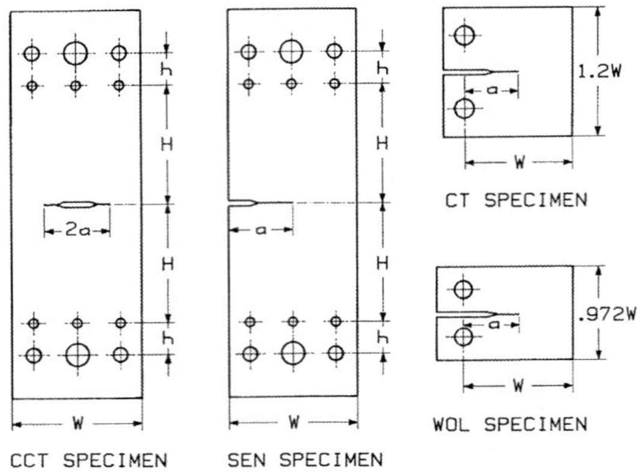

Fig. 1. Specimen types and configurations for FCP experiments (a - crack length, W - specimen width).

RESULTS AND DISCUSSION

Kinetics of Fatigue Crack Growth

Typical results of the constant-ΔK experiments with PC and PMMA are illustrated in Fig. 2 as log (da/dN) versus the normalized crack length a/W for various values of ΔK. The fact that macroscopic FCP rates under constant ΔK conditions are apparently unaffected by changes in the normalized crack length, gives a first indication of the validity of LEFM to these polymeric systems. Similar results in terms of the a/W-independence of da/dN values were obtained for PVC-U and PE-HD (PE 80 type) (Fig. 3). In the latter case, experiments were performed with CT specimens under constant load conditions. Equal ΔK levels for different a/W-ratios were obtained by using several specimens, each of which was tested at different load levels.

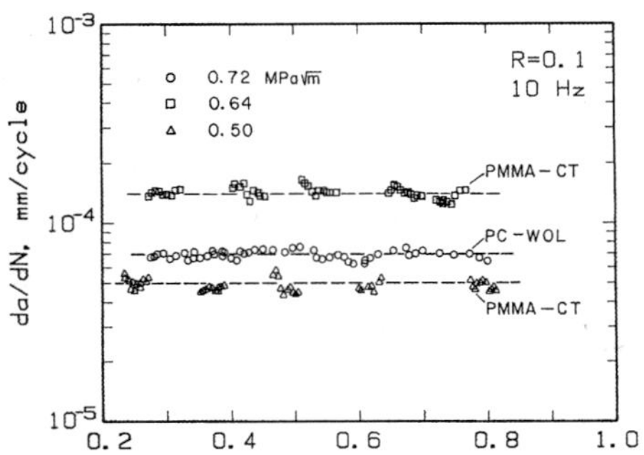

Fig. 2. Fatigue crack growth rates under constant ΔK conditions for PC and PMMA.

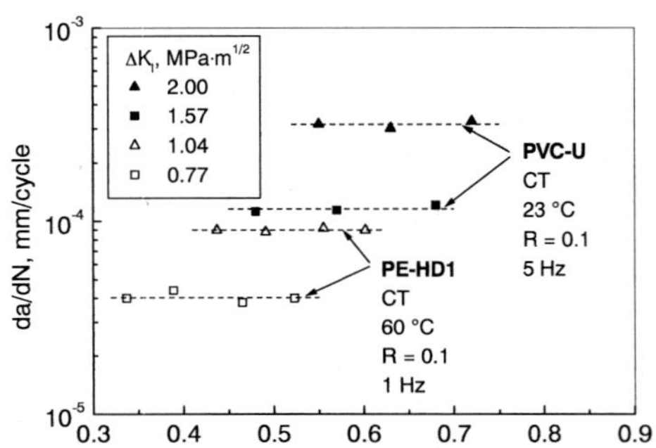

Fig. 3. Fatigue crack growth rates for equal ΔK conditions at various a/W-ratios for PVC-U and PE-HD.

Additional evidence for the applicability of LEFM to describe the kinetics of fatigue crack growth of the materials investigated is shown in Fig. 4 where the results of constant load-range experiments are plotted in the conventional way as log (da/dN) verus log (ΔK) for different specimen configurations. Again, crack growth rates are seen to be uniquely related to the stress intensity factor range independent of specimen type.

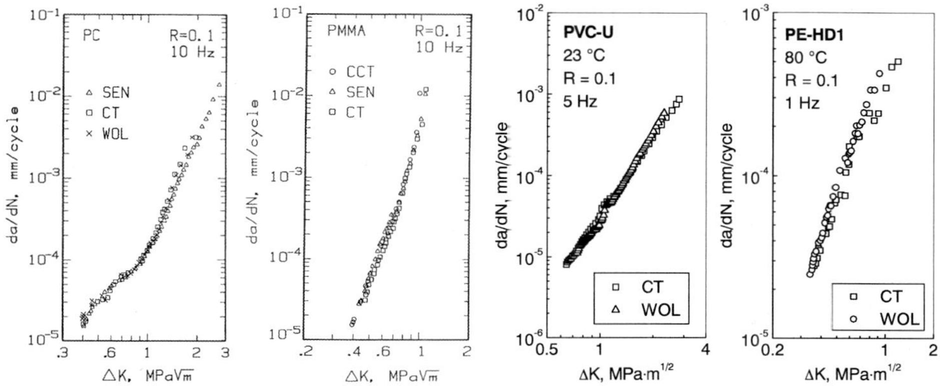

Fig. 4. FCP behavior of PC, PMMA, PVC-U and PE-HD for various specimen configurations.

Based on these results several important conclusions can be drawn. The fact that crack growth rates under constant ΔK conditions remain essentially unchanged for a/W values from around 0.2 up to about 0.8 clearly implies that even for very long cracks most of the then relatively short unbroken ligament still experiences stresses and strains in the linear viscoelastic range. In other words, viscoelastic nonlinearity which invariably exists in the high stress region outside the plastic zone, seems to be confined to a very small area around the crack tip. Moreover, even though the applied load levels for the various specimen types differed by as much as (and under some conditions even more than) an order of magnitude, the requirement for small-scale nonlinear viscoelasticity, which is crucial for the validity of LEFM to polymers, appears not to be violated even in the CCT and SEN specimens which experienced the highest overall stresses.

Of major importance, however, the results indicate that far-field memory effects in these materials have no influence on fatigue crack propagation rates. This conclusion is supported by both the FCP results obtained from different specimen types which obviously involved different far-field stress and deformation histories, and by the circumstance that growth rates under constant ΔK conditions remained unchanged despite the several hundred thousand loading cycles that were experienced by some specimens in the course of a single test.

And yet, possible changes in the thermodynamic and viscoelastic state of the material as it is being cycled in the far-field can still not be discounted. Rather these results suggest that far-field memory effects are erased entirely as a material element becomes part of the high stress region of the advancing crack, regardless of the number of cycles experienced before in the far-field. This hypothesis is consistent with the concept of fading memory that is frequently used in derivations of nonlinear theories of viscoelasticity [19, 20].

Finally it should be emphasized that these findings do not preclude any history related phenomena originating from a region in the immediate vicinity of the crack tip. Since the FCP results indicate that no additional requirements beyond the one that demands "linear viscoelastic - small-scale nonlinear viscoelastic behavior" are needed, it is clear, however, that any such memory effects have to originate from an extremely localized region within the crack tip near field over which K describes the stress distribution. Such localized memory effects and their influence on crack tip plasticity are discussed in more detail elsewhere [2, 17 ,18].

Crack tip plastic deformation
As pointed out above, two principal modes of plastic deformation - crazing and shear yielding - may occur in polymeric materials. A typical situation where craze formation and shear deformation at crack tips coexist, is during fatigue crack growth in PC (Fig. 5a). In the specimen interior associated with plane strain conditions, craze formation predominates. On the other hand, near the free surfaces where plane stress conditions prevail, crack propagation occurs via the formation of shear zones commonly referred to as shear lips. According to suggestions by Irwin [21] for metallic materials, several attempts have been made to relate the shear lip width, w_s, to half the plane stress plastic zone size, r_y, as follows:

$$w_s \approx r_y = \frac{1}{2} \cdot r_p = \frac{1}{2\pi} \cdot \left(\frac{K}{\sigma_{ys}} \right)^2 \tag{3}$$

The results of shear lip measurements on FCP fracture surfaces of PC tested at different R-ratios are plotted in Fig. 5b as a function of K_{max}. The data are seen to be in reasonable agreement with predictions based on equation 3 when using the maximum stress intensity factor as the relevant correlation parameter, which reflects the irreversible nature of plastic deformation in a loading cycle. The experimental values of the shear lip width apparently fall just between the plastic zone size estimates based on equation 3 when using the cyclic yield stress value of 37 MPa and the monotonic yield stress value of 65 MPa, respectively, of PC reported in [22]. Hence the consistent decrease in shear lip dimensions with increasing R-ratio is apparently related to the magnitude of unloading in a fatigue cycle which affects the cyclic softening behavior of the material deformed plastically by shear yielding at the crack tip [2].

In the case of PMMA craze formation occurs at crack tips during fatigue loading over the entire specimen thickness. As the shape of single crazes frequently assumes a wedge-like configuration similar to the line plastic zone considered by Dugdale [23], several authors have adopted the Dugdale plastic strip model to describe craze formation [17, 24, 25]. According to the Dugdale model we can expect the following form for the craze length, l_c, under cyclic loading conditions:

$$l_c = \frac{\pi}{8} \cdot \left(\frac{K_{max}}{\sigma_c} \right)^2 \tag{4}$$

where σ_c is the material´s craze stress (i.e., the stress level necessary to induce crazing). A schematic representation of a crack tip craze indicating the Dugdale stress distribution is illustrated in Fig. 6a.

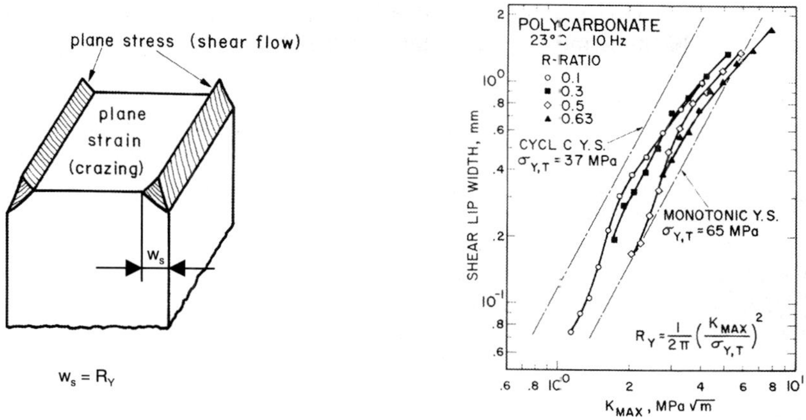

a) b)

Fig. 5. a) Schematic representation of fracture surface in PC; b) Shear lip width in PC for R-ratios as a function of K_{max} and comparison with plastic zone size predictions (Y.S. - yield stress).

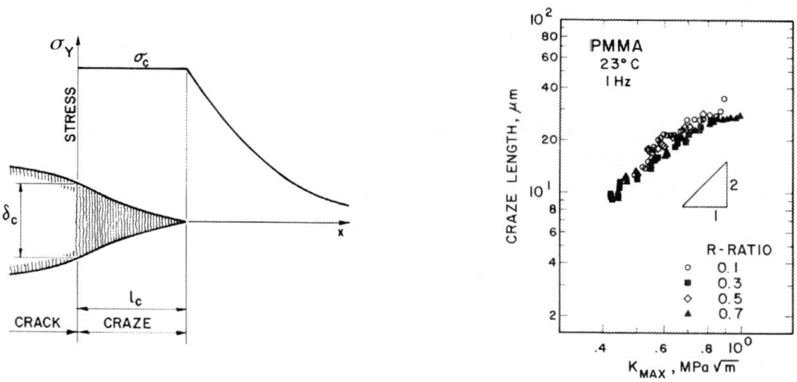

a) b)

Fig. 6. a) Schematic representation of a single craze at the tip of a crack showing characteristic geometrical parameters along with the crack tip stress profile according to the Dugdale model; b) Effect of R-ratio on the craze length at the crack tip in PMMA.

Experimental data for the craze length in PMMA as a function of K_{max} for various R-ratios are shown in Fig. 6b. Although the slope of the curves in the double logarithmic plot appears somewhat less than 2, no consistent or significant influence of R on the measured craze dimensions can be established. This finding is in sharp contrast to the above results for PC and apparently implies that crazing is less prone to cyclic softening than shear yielding. The inferred craze stress for PMMA from the results in Fig. 6b is about 90 MPa. For a more detailed discussion on local stress history effects in cyclically loaded PMMA see [2, 17, 18]

Plastic zone size results obtained from side surface observations (i.e., plane stress region) during FCP experiments with the semi-crystalline polymers PE-HD (PE 100) and PB are depicted in Figs. 7 and 8. While single crazes were observed in PE-HD (Fig. 7a), a crack tip process zone consisting of multiple crazes and/or shear bands was characteristic for PB (Fig. 8a). In terms of the plastic zone size dependence on K_{max}, for both materials slopes of larger than 2 were found on double logarithmic scales (Figs. 7b and 8b). The inferred craze or process zone stresses of about 49 MPa and 19 MPa, respectively, for PE-HD and PB are not entirely unreasonable considering the cyclic nature of the loading mode.

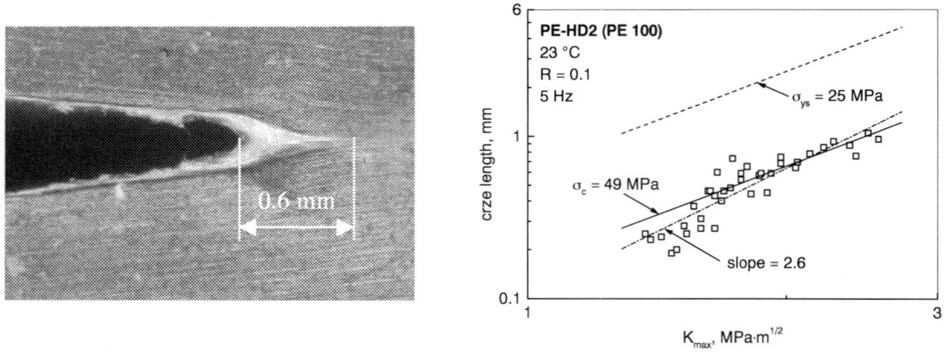

a) b)

Fig. 7. Crack tip craze formation at the specimen side surface during FCP in PE-HD; a) Macrophotographic example of the crack tip region; b) Craze length as a function of K_{max} and comparison with plastic zone size predictions.

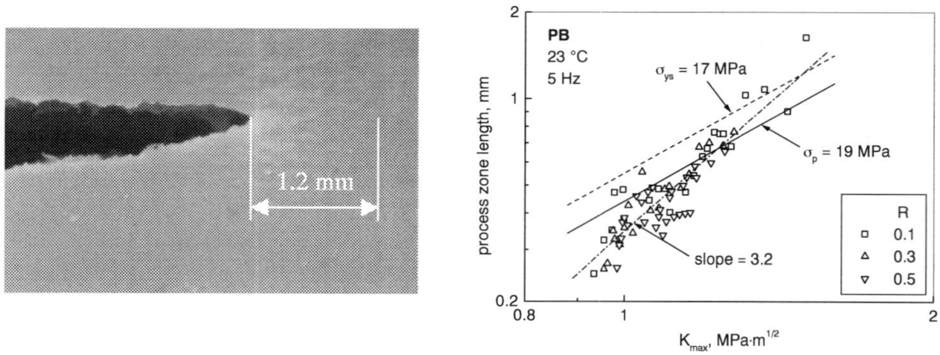

a) b)

Fig. 8. Crack tip process zone formation at the specimen side surface during FCP in PB; a) Macrophotographic example of the crack tip region; b) Process zone length as a function of K_{max} for various R-ratios and comparison with plastic zone size predictions.

CONCLUSIONS

Both macroscopic and microscopic aspects of the applicability of the linear elastic fracture mechanics (LEFM) approach to fatigue crack propagation (FCP) in polymeric solids have been investigated using several amorphous (PC, PMMA, PVC-U) and semi-crystalline polymers (PE-HD, PB) as model materials. In good agreement with expectations, the FCP response of these materials was found to be independent of specimen geometry and specimen configuration. Based on these observations it is concluded that viscoelastic nonlinearity is confined to a very small region around the crack tip, and that far-field memory effects have no influence on the FCP resistance of these materials.

On the microscopic scale, plastic zone size measurements have been performed, the results of which have been compared to predictions by appropriate crack tip plastic zone models. Based on experiments at various R-ratios it was found that the plastic zones in the cyclically loaded polymers were controlled rather by K_{max} than by ΔK, although some R-dependent tendency for cyclic softening was observed for shear yielding in PC. Moreover, plastic zone dimensions appeared to be in reasonable agreement with predictions of the relevant plastic zone size models (at least to a first approximation). However, the power exponent relating the plastic zone size to the ratio of the maximum stress intensity factor to the relevant plastic deformation stress level (K_{max}/σ_0) was found to be somewhat smaller than 2 for PMMA, at around 2 for PC and larger than 2 for PE-HD and PB. While these deviations for PMMA and the R-ratio effects on plastic zone sizes of PC are discussed and interpreted in more detailed manner elsewhere [2, 17, 18], a clarification of the findings for PE-HD and PB needs further investigations.

Finally, it should also be mentioned that the LEFM approach to fatigue in engineering plastics may be limited for certain polymer types and for specific test conditions. For example, the phenomenon of fatigue crack growth retardation in certain types of poly(propylene) (PP) was observed and reported by these and other authors [7, 8, 10, 26, 27]. While some data reduction and data presentation procedures have been proposed also for this case, no investigations providing clear evidence of specimen configuration independent FCP results have been reported so far. Furthermore, using different specimen configurations and test frequencies in experiments with poly(amide) 66 (moisture content of 4.7 m%), it was shown that nonlinear viscoelasticity and hysteretic heating on a larger scale result in a violation of basic assumptions of LEFM [2, 28]. Under these circumstances, which may in principle arise for any polymeric solid for certain combinations of crack size, and specimen or body configuration and loading conditions, macroscopic crack growth rates are no longer uniquely related to ΔK.

ACKNOWLEDGMENTS

A major part of this work was performed by one of the authors (R. W. Lang) at the Materials Research Center at Lehigh University, Bethlehem, Pennsylvania. Partial support by the Austrian American Educational Commission (Fulbright Commission) and by the Burgenländische Landesregierung (Austria) is gratefully acknowledged. This author is also grateful to Professors J. A. Manson and R. W. Hertzberg for many helpful discussions and suggestions. The remaining parts of the research were performed within the doctoral dissertation of W. Balika at the Institute of Materials Science and Testing of Plastics at the University of Leoben (Austria) and within research projects (W. Balika, G. Pinter) at the Polymer Competence Center Leoben GmbH (PCCL, Austria). The PCCL is funded by the Austrian Government and the State Goverments of Styria and Upper Austria within the K_{plus} program of the Austrian Ministry of Traffic, Innovation and Technology.

REFERENCES

1. Hertzberg, R. W.; Mason, J. A. (1980). "Fatigue of Engineering Plastics", Academic Press, New York.
2. Lang, R. W. (1984). Ph.D. Thesis, Lehigh University Bethlehem, USA.
3. Proc. Int. Conf. On Fatigue in Polymers (1983), The Plastics and Rubber Institute, London.
4. Proc. "Schwingfestigkeit von Kunststoffen" (1986), Deutscher Verband für Material-

prüfung e. V., Frankfurt/Main.
5. Williams, J. G. (1987). "Fracture Mechanics of Polymers", Ellis Horwood Limited, Chichester.
6. Döll, W.; Könzöl, L. (1990). In "Advances in Polymer Science - Crazing in Polymers" (H.-H. Kausch, ed.), Vol. 2, 137-214, Springer, Berlin Heidelberg.
7. Karger-Kocsis, J. (1995). "Polypropylene - Structure, blends and composites", Chapman & Hall, London.
8. Chudnovsky, A.; Moet, A.; Bankert, R. J.; Takemori (1983). J. Appl. Phys. 54 (10), 5562.
9. Pinter, G.; Balika, W.; Lang, R. W. (2002). In "Temperature-Fatigue Interaction " (L. Rémy and J. Petit, eds.), ESIS Publication 29, 267-275, Elsevier, Amsterdam.
10. Balika, W. (2003). Ph.D. Thesis, University of Leoben, Austria.
11. Castellani, L; Rink, M. (2003). In "Fracture Mechanics Testing Methods for Polymers, Adhesives and Composites" (D. R. Moore, A. Pavan, J. G. Williams, eds.), ISBN 008 0436897, Elsevier.
12. Rice, J. R. (1967). ASTM STP 415, 247.
13. Broek, D. (1989). "The Practical Use of Fracture Mechanics", Kluwer Academic Publichers, Dortrecht.
14. Anderson, T. L. (1991). "Fracture Mechanics - Fundamentals and Applications", CRC Press Inc. Boca Raton.
15. Saxena, A. (1998). "Nonlinear Fracture Mechanics for Engineers", CRC Press LLC, Boca Raton.
16. McClintock, F. A.; Irwin, G. R. (1965). ASTM STP 381, 84.
17. Lang, R. W.; Manson, J. A.; Hertzberg, R. W.; Schirrer, R. (1984). Polym. Eng. and Sci. 24 (10), 833.
18. Lang, R. W.; Schirrer, R. (1985). Proc. 6th Int. Conf. On Deformation, Yield and Fracture of Polymers, 1-4 April 1985, Chruchill College, Cambridge (The Plastics and Rubber Institute, London), p. 25.1.
19. Coleman, B. D. (1964). Arch. Ration. Mech. Anal. 17, 1.
20. Christensen, R. M. (1982). "Theory of Viscoelasticity - An Introduction", 2nd ed., Academic Press, New York.
21. Irwin, G. R. (1964). Appl. Mater. Res., 3, 65.
22. Rabinowitz, S.; Beardmore, P. (1974). J. Mat. Sci. 9, 81.
23. Dugdale, D. S. (1960). J. Mech. Phys. Solids, 8, 100.
24. Döll, W. (1983). Advances in Polymer Science, 52/53, 105.
25. Schirrer, R.; Schinker, M. G.; Könzöl, L.; Döll, W. (1981). Coll. & Polymer Sci., 259, 812.
26. Sandt, A.; Hornbogen, E. (1981). J. Mat. Sci. Lett. 16, 2915.
27. Karger-Kocsis, J.; Friedrich, K. (1989). Int. J. Fat. 11, 161.
28. Lang, R. W.; Hahn, M. T.; Hertzberg, R. W.; Manson, J. A. (1984). J. Mat. Sci. Lett. 3, 224.

The Application of Fracture Mechanics to Polymers, Adhesives and Composites
D.R. Moore (Ed)

COMPARISON BETWEEN MATERIAL TOUGHNESS AND COMPONENT TOUGHNESS

D.R. MOORE

INTRODUCTION

The application of fracture mechanics to the measurement of toughness for polymeric materials is now well advanced. Its use compared to more conventional measurements of toughness can be shown to provide advantages and additional insights. The European Structural Integrity Society (ESIS technical committee 4) has developed many protocols in the form of test methods [1]. Each of these protocols provides an experimental route to the measurement of material toughness.

Polymeric materials are used in the manufacture of components in the plastics, adhesives and composites industries. Therefore, it is worth asking the question as to whether there is a difference between polymer toughness and component toughness. Of course, the difficulty of this distinction is readily envisaged when a polymer is used as an adhesive or in a composite. In these circumstances, the component will be in the form of metal laminate, or structural joint or an engineering structure with fibre reinforcement and as such will be quite different from the properties of the polymer on its own. But what of those situations when the polymer is simply used in order to manufacture some particular geometric structure? For example, unplasticised poly vinyl chloride (PVC-U) can be used to prepare a plaque for a linear elastic fracture mechanics test and the same material can be used to extrude a pressure pipe that can be used in the water industry. Is there a correspondence between material and component toughness under these circumstances?

A singular or isolated view of this issue is seldom of interest. However, a comparative view is a significant issue and has relevance to material selection in that application. Consequently, a more pertinent question might be that of whether the comparative toughness of two materials is reflected in the comparative toughness of those materials in component form. This is the issue that will be discussed in this article. It is a further instance where application of fracture mechanics provides clarity and resolution.

The topic will be addressed first in a general sense and then by considerations of material selection for pressure pipe applications. It will also be noted that clarity and rigour are treated as conjugate variables in this article. As one is developed to assist the argument, the other tends to lose focus.

FRACTURE IN COMPONENTS

Components are designed not to fail in service! For a specific application the geometry and loading details of a structure are defined and then a stress analysis will establish a critical component dimension (e.g. thickness) in order to ensure that the article is strong enough for service. Usually, this prevents ductile failure in service. It is seldom the case that all contingencies are accommodated in such design since this would not be cost effective. Moreover, there are circumstances that can lead to crack-like fracture in plastics if a relatively simple approach is given to the engineering design of the component; circumstances such as dynamic loading, inadvertent scratching of a surface or perhaps an environmental attack following an unscheduled exposure to some chemical.

Consequently, it is of relevance to consider the general scenario that might lead to brittle fracture, even when an article has undergone a "strength of materials" approach to design. If such a component is subject to loading then linear elastic fracture mechanics (LEFM) suggests a simple criterion for failure by consideration of the size of the plastic zone associated with a deformation zone in the material:

Ductile fracture can occur when the plastic zone radius (r_p) is relatively large.
Brittle fracture can occur when the plastic zone radius is relatively small.

For these conditions to apply, it is implicit that the "flaw" size is small compared with the smallest specimen dimension or the size of the plastic zone. If this is not the case, then the article by Marshall on the application of fracture mechanics to plastic pipes describes the alternatives. For the purposes of this article, we will assume that the plastic zone and smallest specimen dimension are the critical parameters.
In the development of the LEFM test protocols, these relative sizes relate to the specimen dimensions that have been selected in order to provide plane strain fracture toughness conditions [e.g. 2]. Components are not designed to the same criteria and therefore the smallest dimension (for example, a thickness *t*) is selected instead. Therefore:
Brittle fracture occurs when $r_p << t$
Ductile fracture occurs when $r_p \approx t$
This leads to the conditions:

For brittle fracture $\frac{r_p}{t}$ is small e.g. < 0.1

For ductile fracture $\frac{r_p}{t}$ is large > 1

Consequently the term $\frac{r_p}{t}$ is a measure of component toughness. When its value is large (>1) the component is tough enough, when its value is small (< 0.1) brittle fracture may occur. In addition, if the value of $\frac{r_p}{t}$ is the same for two components but made from different materials, then their component toughness is similar.
Plastic zone radius is given by a number of different expressions of the form:

$$r_p = A\left(\frac{K_c}{\sigma_y}\right)^2 \quad (1)$$

The coefficient ***A*** can have a number of values depending on the stress field conditions. However, in this illustration we are primarily concerned with a comparison of component toughness. Therefore, the value of ***A*** need not matter so we will give it the value of unity.
The square of the ratio of fracture toughness to yield stress is called the ductility factor (***DF***) for a material. It is time and temperature dependent and can be used to describe material toughness; the larger the ductility factor, the tougher the material.

$$DF = \left(\frac{K_c}{\sigma_y}\right)^2 \quad (2)$$

Therefore, by invoking this further simplification it is possible to equate plastic zone size with ductility factor. Naturally, the qualifications associated with doing this need to be clear.

PRESSURE PIPE TOUGHNESS AND TOUGHNESS FOR PRESSURE PIPE MATERIALS.

Pressure pipes can be made from polymeric materials; PVC-U and polyethylene (PE) have been popular contenders. Suppose that it was necessary to comment on the relative toughness of pressure pipes manufactured from these two plastics. A specific size of pipe helps illustrate the discussion, so let us consider a 150 mm diameter class C pressure pipe. For the purposes of illustration only temperatures of 23 °C are accommodated and loading for short durations only will be considered. Needless to say, this makes this example somewhat academic, because long term loading is a key issue for pipes that are required to have lifetimes of up to 50 years. However, simplification of the service conditions will assist the principal focus of this article. (See *The Use Of Fracture Mechanics To Assess Plastics Pressure Pipe Materials* by G.P.Marshall and J.G.Williams for a more detailed description of the performance of pressure pipe in the context of the application of fracture mechanics.)

The first step is to establish the wall thickness required for each pipe. Assuming a thin-walled pipe, this is calculated as follows:

$$t = \frac{PD}{2\sigma_y} \qquad (3)$$

P is the pipe pressure, D its diameter and σ_y is the yield stress for the pipe material.

The relationship between P and D is given in standard documents (e.g. British standard BSI 3505). Therefore, if the polyethylene is a medium density material MDPE (density of 945 kg/m^3) and the PVC-U is an extrusion grade material, then the yield stresses at 23 °C and at Instron type test speeds (e.g 10 mm/min.) might be 28 MPa and 64 MPa, respectively.
This leads to two quite different pipe wall thickness values:

$$t_{PVC-u} = 6.6mm$$

$$t_{MDPE} = 14.2mm$$

The fracture toughness values (K_c) for these materials are taken as 5.2 MPam$^{1/2}$ for PVC-U and 3.4 MPam$^{1/2}$ for the MDPE. To some extent, these are artificial values. Although they are typical for those reported in the literature [e.g. 3] the failure mechanisms at Instron-type speeds of about 10 mm/min are not the same as those for long term loading. Of course, it is long term loading conditions that are important for pressure pipe applications. However, the purpose in this example is to highlight a concept, not to cite a specific practical case. Therefore, we will adopt the values above but with the warning that they should not be used for a real material selection case. In any case, the choice of actual plastic pipe grades is much wider than the specific instances discussed here.
By adopting the values of K_c and σ_y above the following values are determined for ductility factor from equation 2 and in turn the values to be used for plastic zone size.

$DF_{\text{PVC-u}} = 6.6 \text{ mm} = (r_p)_{PVC-U}$

$DF_{\text{MDPE}} = 14.2 \text{ mm} = (r_p)_{MDPE}$

The ductility factor is a measure of toughness; the higher its value the greater the toughness of the material. Therefore, the MDPE material is significantly tougher as a material than PVC-U. (This is in contrast to the simple measurement of fracture toughness!)

The component toughness for the two pipes can now be calculated:

$$\left(\frac{r_p}{t}\right)_{\text{PVC-u pipe}} = 1.00$$

$$\left(\frac{r_p}{t}\right)_{\text{MDPE pipe}} = 1.04$$

In other words the toughness of the pipes (within the conditions that have been defined) is the same, even though the material toughness of MDPE is significantly larger. Therefore, it should not be assumed that material toughness and component toughness are necessarily the same for plastics.

CONCLUDING COMMENTS

Linear elastic fracture mechanics has been used to provide a method for the definition of material toughness and component toughness for plastics. In both cases geometric complications in these assessments have been accommodated. Material toughness is linked to two material properties, namely fracture toughness (**K_c**) and yield strength. Component toughness also uses these two material properties together with an additional critical component dimension (usually the smallest dimension). It should not be assumed that comparative material toughness reflects the comparative component toughness.

REFERENCES

1 Moore D.R., Pavan A, Williams J.G. ed *Fracture Mechanics Testing Methods for Polymers Adhesives and Composites* ESIS Publication 28, Elsevier 2001
2 Chan, M.K.V., & Williams, J.G., Polym. Engng.Sci., 21 (1981) 1019-1026
3 Williams, J.G., *Fracture Mechanics of Polymers* Ellis Horwood, Chichester 1984

THE ENGINEERING DESIGN OF WATER STORAGE TANKS

D.R. MOORE

INTRODUCTION

Linear elastic fracture mechanics (LEFM) describes fracture in polymers and metals. This implies that with the knowledge of specimen geometry together with the loading conditions then the conditions of crack growth are fully described. Therefore, it should also be possible to apply LEFM to the avoidance of such fractures in components manufactured with polymers. Although this seems a reasonable notion it has proved a difficult task in practice.

Nevertheless, many artefacts are manufactured with polymeric materials and often an engineering requirement relating to significant load-bearing capability is required. It is important for safety and economic reasons that these artefacts should not fracture in service. In particular, the avoidance of a sudden fracture is especially important.

Water storage tanks, as used in domestic applications, that are cylindrically shaped is one such example. Polyethylene is used in the manufacture of these tanks by application of a rotational casting process. In this article we will examine how such tanks can be designed, in engineering terms, by the use of linear elastic fracture mechanics and also by standard strength of materials considerations.

ENGINEERING DESIGN OF CYLINDRICAL TANKS

The way in which a vertical cylindrical tank distorts when filled with a liquid can be investigated by a number of approaches. It would be possible to use a numerical approach, such as finite element analysis but it is also possible to use an analytical approach such as that determined by Timoshenko and Woinowsky-Krieger (1). This analysis was simplified by Forbes et al (2) when they applied it to the design of large liquid storage tanks fabricated from extruded sheet. With a modest degree of over design the engineering equation reduces even further to:

$$\frac{\sigma_D}{\rho} > \frac{g\,h\,r}{d} \qquad (1)$$

where σ_D is a design stress, ρ is the density of the liquid to be stored in the tank, d is the tank wall thickness, h the height of the liquid level and r the tank radius.

Moore and Gotham (3) used this equation in the design of rotationally cast polyethylene tanks for cold-water storage. The engineering problem was to define a magnitude for design stress for the polyethylene and hence determine the appropriate wall thickness for the tank of known height and radius in order to avoid the tank breaking during a required 5 years lifetime.

A "STRENGTH OF MATERIALS" APPROACH

The materials science aspects of the problem were tackled by a combination of creep, creep rupture and fatigue experiments on rotationally cast polyethylene samples.

Figure 1 shows tensile data for unnotched specimens (specimen geometry is also illustrated in the figure). Creep data are also included in the form of isometrics i.e. as stress versus log time at constant strain. The creep rupture experiments all indicate ductile fracture in the polyethylene samples, but cyclic loading was shown to introduce embrittlement after a few weeks testing. The cyclic loading involved 8 hours on load and 16 hours off load and the data are plotted as total time under load.

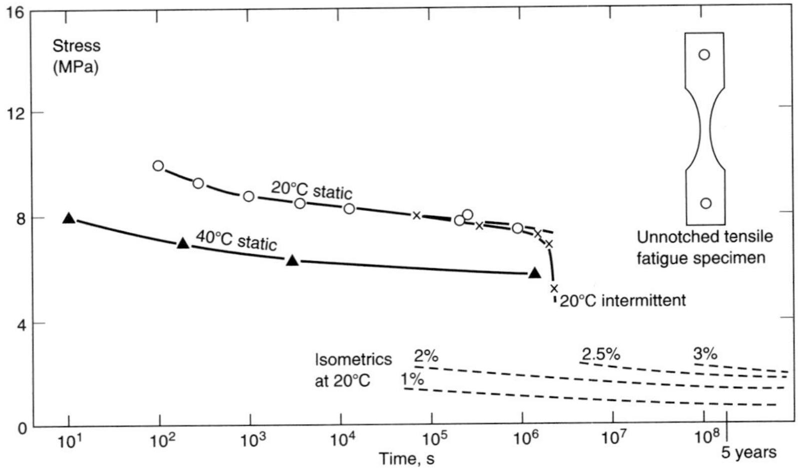

Figure 1 Fatigue (crosses), creep rupture (circles and triangles) and isometric creep data (dotted lines) for rotationally moulded polyethylene at 20^{0}C and 40^{0}C

The prospect of crack initiation in service could reduce the lifetime of the tank to less than 5 years and therefore additional fatigue experiments, involving notched specimens, gave an indication of the level of brittle strength. These data are shown in Figure 2. They show that a blunt notch (tip radius 250 μm) does not precipitate severe embrittlement but that as the notch severity is increased (to a sharp crack of tip radius 10 μm) then a ductile-brittle transition in fracture may occur. The brittle strength value was commensurate with a strain level of about 0.025, which had earlier been adopted as a limiting value for safe load-bearing design for many grades of polyethylene. Therefore that value was used in this particular case to obtain the 5 years design stress and a value at 20°C of 1.7 MPa is obtained.

It is possible that temperatures above 20°C will need to be accommodated in the service use of the tanks. Therefore, the 20°C design stress will have to be de-rated for temperature. Service use up to 40°C will be accommodated and reference 3 indicates how to adjust the design stress for temperature. A general design stress value of 1.13 MPa emerges.

This design stress (1.13 MPa) was applied to the engineering design formula of equation 1. Tank wall thickness was then established for a wide range of tank sizes. For liquids other than water the wall thickness would have been different according to the density of the stored liquid but additionally the possibility of environmental stress cracking would have been investigated.

This particular approach to solving the design problem, an approach that we have termed "strength of materials", provides an acceptable solution. However, there is a need for a significant experimental work programme together with a broad and detailed knowledge of polymer engineering.

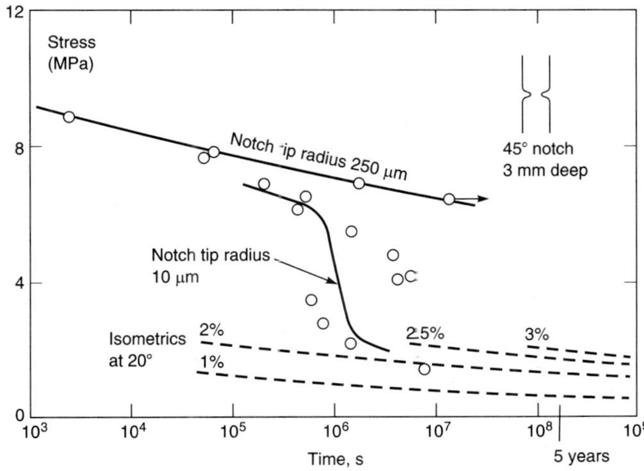

Figure 2 Fatigue (squares and circles) on notched rotationally moulded polyethylene at 20 °C (8 hours on load: 16 hours off load), together with isometric data (dotted lines).

DESIGN WITH FRACTURE MECHANICS

Griffith (4) introduced fracture mechanics concepts to describe the brittle strength of glass. He postulated that materials contain intrinsic flaws that act as crack-like defects in an appropriate stress field. For quasi-brittle materials, Orowan (5) and Irwin (6) modified the theory in order to accommodate the plastic deformation that could occur around the intrinsic flaw. They expressed the now well known relationship between fracture toughness (K_c), flaw size (a) and fracture stress (σ_F):

$$K_c = \sigma_F Y(a)^{1/2} \tag{2}$$

This can be applied to rotationally cast polyethylene tanks where the fracture stress is a hoop stress and is designated as the design stress (σ_D), Paris and Sih (7) suggest that Y=2 and a becomes the intrinsic flaw size. Design of the tank should ensure no crack-like fractures and field experience with tanks in successful service for a number of years with a wall thickness of 2.1 mm suggests that an intrinsic flaw size of 2.1 mm should be helpful (8). Therefore, adapting equation 2 in order to calculate the design stress gives:

$$\sigma_D < \frac{K_c}{2a^{1/2}} \qquad [3]$$

K_c should be obtained on rotationally cast polyethylene samples under long term loading where brittle fractures are observed. This can be achieved from the results in Figure 2 for the notched specimens (tip radius 10 μm) where a value of brittle strength of 2.5 MPa is observed. The results in Figure 2 show some scatter therefore some confirmatory experiments were conducted on some compression moulded samples of a lower molecular weight polyetheylene grade in an air and silicone oil environment (3). In these experiments a ductile brittle transition was observed for both the air and silicone oil environments and a common value of brittle strength at long times under load was observed. The confirmatory experiments gave confidence in using the data of Figure 2 for the calculations of K_c but also suggested that accelerated tests at 20 °C may be possible for other routes in the determination of K_c.
Returning to the brittle fracture strength value of 2.5 MPa from Figure 2 with the aim of calculating Kc. Reference 7 suggests the following equation for the determination of the critical value for the stress field intensity factor (fracture toughness):

$$K_I = \sigma_R(\pi a)^{1/2}\{\frac{2b}{\pi a}\tan(\frac{\pi a}{2b})\}^{1/2} h(\frac{a}{b}) \qquad [4]$$

where σ_R is a stress at fracture remote from the double notch, a is the notch depth, b is half the specimen width and $h(\frac{a}{b})$ is a function of specimen dimensions (7).

We used σ_R = 2.5 MPa, a = 2 mm and b = 12.7 mm. This generated a value of K_c at 5 years under load (which is commensurate with K_I for plane strain fracture) of 155 MPam$^{1/2}$.
The design stress at 20 °C is then calculated from equation 3 using a K_c value of 155 MPam$^{1/2}$ and a flaw size (a) of 2.1 mm giving a value of 1.7 MPa. This agrees with the value at 20 °C from the “strength of materials” approach. If a temperature de-

rating factor needs to be applied then a similar approach to that discussed earlier should be used.
An alternative approach for the determination of K_c might reduce the effort in application of fracture mechanics in this design problem. For example, a relatively large thickness beam specimen should enable K_c to be determined at short loading times and still provide plane strain conditions. Compression moulding technology will provide large thickness specimens of polyethylene. A time de-rating factor will then need to be applied to the short-term value of K_c and ample time dependent strength data are available for polyethylene.

CONCLUDING COMMENTS

It has been possible to demonstrate the successful application of fracture mechanics concepts to the design of cylindrical water storage tanks. In the calculation of design stress for 5 years loading times, LEFM provides a similar value to that obtained from consideration of the creep, creep rupture and fatigue behaviour of the polyethylene used for the application. In this particular example, a comprehensive approach has been used in generating long-term properties for both approaches. However, it is possible to envisage that the fracture mechanics approach could be conducted in a much more time efficient manner for other similar applications.

REFERENCES

1 Timoshenko, S., and Woinowsky-Krieger, S., "Theory of Plates and Shells" McGraw Hill 1959
2 Forbes, K ., McGregor, A and Turner, S., *Brit.Chem. Eng.* Oct 1970
3 Moore, D. R, and Gotham, K.V., *Plast. & Poly* Feb 1975 p30.
4 Griffith, A.A., *Phil. Trans.R.Soc.*A221,1920,163
5 Orowan, E., *Trans.Inst. Engrs. Shipbuilders, Scotland*, 1945, 89, 165
6 Irwin, G. R., Inter.Congr. *Appl. Mech. VIII. Paper 101, (11), University of Brussels* 1957, 245
7 Paris, P.C., and Sih, G.C., *Fracture toughness testing and its applications* ASTM 1964, 30-8
8 British Standards Institute, BS 4213: 1967

The Application of Fracture Mechanics to Polymers, Adhesives and Composites
D.R. Moore (Ed)

CRITICAL PRESSURE for RAPID CRACK PROPAGATION in THERMOPLASTIC WATER PIPE

P.S. LEEVERS and C.J. GREENSHIELDS

INTRODUCTION

Tough polymers such as those used for water pipe pose a real challenge to Linear Elastic Fracture Mechanics. Under constant crack driving force, glassy polymers, e.g. PMMA, sustain stable crack propagation which is slow enough to be observed directly using simple methods, but fast enough to provide data within the duration of a research project. For this reason they served a vital role in the development of LEFM for plastics, but are seldom used for load-bearing components. The ductile crystalline polymers used for many engineering components and structures do sometimes suffer LEFM-compatible brittle crack propagation at stresses well below yield, but they usually do so only at very short or very long loading times. In both cases the mechanism is probably a combination of disentanglement and of crystalline disintegration around tie chain anchor points. At low temperatures and short times this seems difficult to believe, since any Eyring-type slip process would be frozen. However it has been argued [1, 2] that the separation process is mobilised by adiabatic heating so localised that the surrounding bulk polymer remains cold and the crack remains sharp.

For crystalline polymers the adiabatic decohesion model yields a prediction of the *dynamic fracture resistance*, G_D, under which a brittle crack will propagate, as a function of crack speed. At less than 10 m/s or more than a few hundred m/s this resistance becomes very high. At some intermediate speed G_D passes through a minimum value given simply [2] by

$$G_{D,min} = 3.06 s_w \rho H_f, \qquad (1)$$

where s_w is the size of a structural unit within the polymer melt (interpreted as the extended length of a weight-average chain), ρ is the mass density of the polymer and H_f is the total heat to melt unit mass of polymer from the test temperature. Note that earlier developments [1] used an expression in which specific enthalpy and latent heat contributions were treated separately, but less rigorously.

The model is restricted to conditions sufficiently deep inside the material for *plane strain* conditions to generate a flat, brittle fracture. Nearer to the free surfaces, plastic flow is less constrained and the polymer is effectively tougher. To devise a conservative test method it may be necessary to score the surfaces along the crack path so that this free-surface region is removed.

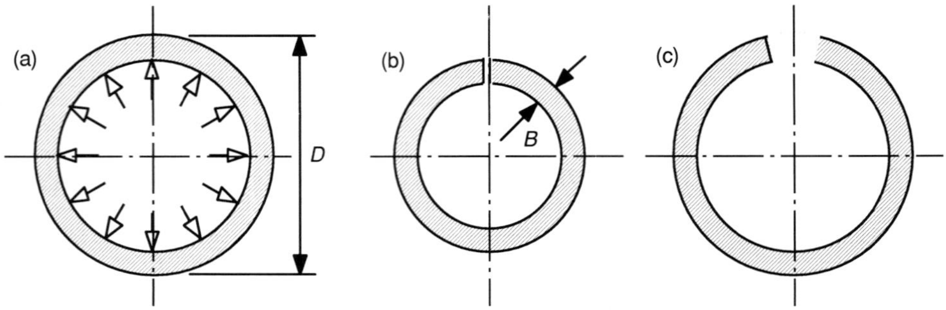

Fig. 1. If a liquid-pressurised pipe (a) fractures rapidly, the wall hoop stress can relax only either (b) with axial flow or (c) with circumferential contraction.

Under these conditions we can also write, as usual,

$$K_{\mathrm{D,min}} = \left(EG_{\mathrm{D,min}}\right)^{1/2}, \tag{2}$$

so that the model can be used to provide equally conservative predictions of failure loads, stresses or pressures. In practice, there are few situations to which G_{D} is directly useful for predicting failure, because most loaded plastic components are small. It is not worth worrying about stopping a running crack rather than avoiding its initiation. A pressurised pipeline is an exception, as first shown by Irwin [3] in an unpublished report dealing with the problem of rapid crack propagation (RCP) in *steel* pipelines carrying pressurised *gas*. The analysis was a classical application of LEFM and we will review it in full.

THE CRACK DRIVING FORCE IN WATER-PRESSURISED PIPE

The pipe section shown in Fig. 1 has an outside diameter of D and a wall thickness B, the ratio of these being D^* (whose nominal value is the *standard dimension ratio* or SDR of the pipe). The wall material has a dynamic modulus E and, if D^* is large enough to define the pipe as thin-walled, an internal pressure p will give rise to an average hoop stress

$$\sigma_{\mathrm{h}} = \tfrac{1}{2}p\left(D^* - 2\right)$$

The strain energy stored by this hoop stress per unit volume of pipe wall is given by

$$\frac{\sigma_{\mathrm{h}}^2}{2E} = \frac{p^2\left(D^* - 2\right)^2}{8E}$$

and since the volume of the pipe wall per unit length is

$$\pi\left(\frac{D}{D^*}\right)^2\left(D^* - 1\right)$$

we find that the strain energy stored per unit length of pressurised pipe is given by

$$\frac{\pi}{8}\frac{p^2}{E}\left(\frac{D}{D^*}\right)^2\left(D^* - 2\right)^2\left(D^* - 1\right)$$

Now if a crack runs axially along unit length of pipe wall as in Fig. 1(b) and releases all of this energy, it will have cut through an area $B_{\mathrm{c}}D/BD^*$ of material, where $B_{\mathrm{c}} \leq B$ may be reduced by surface notching as described above. The crack driving force G generated by release of all (and only) the stored elastic energy is therefore given by

$$G = \frac{\pi}{8}\frac{p^2}{E}\frac{D}{D^*}\frac{B}{B_c}\left(D^* - 2\right)^2\left(D^* - 1\right). \tag{3}$$

It is true that some of the released strain energy is converted to kinetic energy. However, a unusual and convenient feature of this system is that for a long pipe section a *steady state* will develop in which the kinetic energy of the region around the crack remains constant.

For a pipe wall material of known crack resistance G_D, the critical pressure needed to sustain steady axial crack propagation can now be predicted by setting $G = G_D$:

$$p_c = \left[\frac{8}{\pi}\frac{E_D G_D}{D}\frac{B_c}{B}\frac{D^*}{\left(D^* - 2\right)^2\left(D^* - 1\right)}\right]^{1/2}. \tag{4}$$

A pipe pressurised by gas contains an enormous reservoir of energy, some proportion of which can contribute to sustaining the crack driving force by driving the cracked pipe wall outwards. Determining what this proportion is — and hence what is the resulting reduction in critical pressure — is a daunting problem. However, if the pipe is pressurised by water the contained energy is much smaller and its contribution can, if necessary, be accounted for. On the other hand, if the fractured pipe wall simply contracts in the hoop direction without contracting radially (Fig. 1(c)), it contains some bending strain energy which is also unaccounted for. On balance, Eqn. (4) is probably more than sufficiently accurate.

THE PLUGGED S4 TEST

In practice, the critical pressure calculated using Eqn. (4) is doubly conservative for a pipe filled only with water. When a crack has extended far enough along such a pipe, the wall can move *outwards* slightly to relax the internal pressure, even before there is any leakage. A decompression wave will now propagate axially through the water, overtake the crack, rob it of its driving force and arrest it. It has been shown by Greig [4] using full-scale testing and Greenshields [5] using small-scale testing that the case of most concern for water pipe is an air-water mixture, in which there is enough air in the pipe to slow down the decompression wave and hold up the pressure, but not enough to contribute substantially to driving the crack.

Greenshields devised a method of RCP testing for water pipe by simulating this situation in a well controlled and conservative test, without introducing pressurised air. The method was based on the ISO 13477 'S4' (small scale steady state) RCP test method for pipe. A loose-fitting internal solid (steel) mandrel is inserted into the pipe, and the annular space between pipe and mandrel is pressurised with water (Fig. 2). This reduces the fluid decompression wave speed to about 150 m/s, at which the crack easily outstrips it.

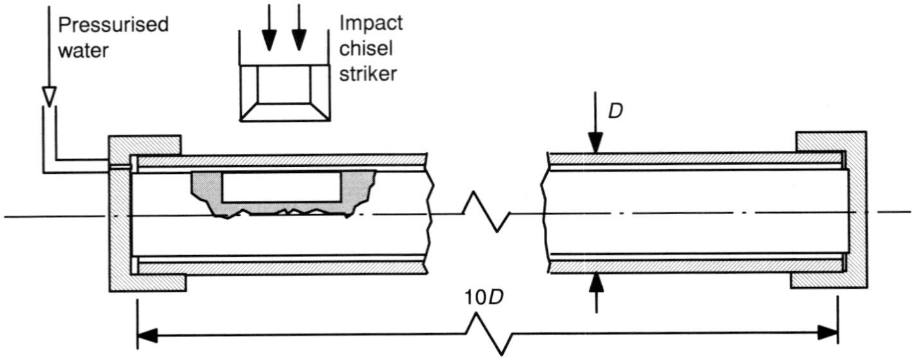

Fig. 2: The 'Hydrostatic S4' pipe test: the pipe specimen contains a close fitting solid mandrel and is pressurised by only a thin skin of water.

In a pipe the plane-stress surface layer referred to above forms primarily at the bore, sometimes because the material there has cooled at a different rate and partly because it is differently stressed during fracture. The ductile bore layer thickness tends to remain independent of pipe wall thickness, so that thinner pipe walls benefit more. This layer is removed, when necessary, by notching the bore to a depth of 1–2 mm, and accounting for the effect through the ratio B_c/B in Eqns. (3-4).

The pipe is pressurised and fracture is initiated from one end, as in a conventional ISO 13477 S4 test, by impact of a sharp, chisel-ended striker. To establish conservative test conditions the crack must run for a one-diameter length even if the pipe is unpressurised, but at all pressures below a well-defined critical value p_c the crack hardly propagates any further. Only at pressures above p_c is the wall stress able to drive the crack on to the end of the specimen. A series of tests at various pressures and constant temperature — here 3°C, regarded as a standard worst-case condition for buried water pipe — can determine p_c to any required precision.

RESULTS

The results of this test method are interesting in several respects. Firstly, once the effect of any inner notch has been accounted for they provide what is arguably an absolute minimum critical pressure for a water-pressurised pipe. Secondly, the Irwin-Corten analysis suggests that this critical pressure can be scaled in a conservative way to other sizes through the assumption that G_D is a material constant. Finally, they allow the calculated G_D to be compared with the G_{Dmin} value predicted using the thermal decohesion model. Materials not used for pressurised pipe but available as 50 mm diameter tube (the size of the main test rig) were therefore tested, while for the principal pipe materials (PE and PVCu) other pipe sizes were available and these were tested as well.

The critical pressure results have been discussed in detail elsewhere [5]. Amongst the pipe materials PVCu and PVDF do not come out of this test well: both sustain RCP below their rated pressures. PE of both strength classes, on the other hand, must be notched in order to sustain RCP at all in these sizes, and even then they resist RCP above their rated pressures. It is important to re-emphasise that this test represents what is absolutely a worst case scenario in every respect: the pipe has been impacted at high speed, its decompression has been artificially prevented, the wall material is at its minimum service temperature and in some cases any residual plane-stress ductility has been suppressed by notching. It certainly seems unnecessary to apply any further safety factor to the results.

Table 1 shows the results of Greenshields' [5] tests on pipe extruded from a number of polymers: polyethylene (PE) of two strength classes, unplasticised PVC (PVCu), an undefined polypropylene copolymer (PPco), polyvinylidene fluoride (PVDF), polybutene (PB), nylon 6 (PA6) and amorphous polycarbonate (PC). The structural dimension shown in Table 1 is derived from the weight-average molecular weight, as determined by gel permeation chromatography, using the molecular weight and fully-extended length of a repeat unit. The molecular weight of PVDF could not be determined and a manufacturer's estimate was used. The decohesion temperature is taken to be that of the glass transition for PC and that of full crystalline phase fusion for the other polymers, including PVCu. The enthalpy H_f from 3°C to the decohesion temperature can usually be inferred from readily available DSC data (here, each material was independently tested) and from PVT (pressure, specific volume, temperature) plots.

Material	Pipe dimensions diameter, D (mm)	SDR, D^*	Extended chain length, s_w (μm)	Decohesion temperature, T_{dc} (°C)	Mass density ρ (kg/m^3)	Enthalpy to T_{dc} H_f (kJ/kg)	RCP resistance, $G_{D,min}$ predicted [Eqn. (1)] (kJ/m^2)	measured [Eqn. (4)] (kJ/m^2)
PE-80	50	11	1.6	130	944	480	2.2	2.6
PE-80	125	11	1.6	130	942	480	2.2	2.3
PE-100	50	11	1.8	138	960	490	2.6	3.4
PVCu	50	13.5	0.4	231	1350	270	0.45	0.64
PVCu	114	19	0.4	231	1420	290	0.50	0.64
PPco	50	17	3.6	165	905	420	4.6	3.1
PVDF	50	17	0.3	174	1785	274	0.45	0.40
PB	50	11	3.1	127	939	323	2.9	4.1
PA6	50	10	0.4	220	1135	436	0.61	1.9
PC	50	25	0.1	138	1253	180	0.69	0.64

Table 1: Pipe specimen dimensions, structural and physical parameters and predicted and measured fracture resistance for several thermoplastics.

DISCUSSION and CONCLUSIONS

At first sight, the Irwin-Corten analysis for axial fracture in pressurised pipe provides an unusually clear-cut demonstration of the power and value of fracture mechanics. The outcome of the analysis is a simple closed-form equation for critical pressure: Eqn. (4). These long-running fractures can only occur at high speeds at which most of the polymer grades used are indeed linearly elastic.

For the two materials available at different pipe sizes, the assumed material property G_D does indeed emerge as a geometry-independent material property. However, if Eqn. (4) is used only to scale for pipe diameter and thickness then neither E nor G_D need even be evaluated. If not, things get rather more complicated: it would be tempting to use G_c data from an impact fracture test method such as ISO 17281, but experience and the arguments presented in a companion paper [2] suggest that this would be non-conservative.

For materials available in tube form, the Hydrostatic S4 serves as a very convenient test method for rapid crack propagation resistance G_D. This opens up a wider field of applications. Here we have gone on to demonstrate correlation between the measured G_D and that predicted from very basic properties using a thermal decohesion model. The correlation is not perfect and (for example) the use of extended weight-average chain length as a structural dimension is contentious. However, there are few problems in which such a leap in scale from microstructural parameters to infrastructural integrity can be made with such an encouraging outcome.

REFERENCES

1. Leevers, P.S. (1995) *Int. J. Fracture* **73**, 109.
2. Leevers, P.S. (2003), 'Why ductile thermoplastics sometimes fracture under impact', this volume.
3. Irwin, G.R. and Corten, H.T. (1968), 'Evaluating the feasibility of basing pipeline operating pressure on in-place hydrostatic test pressure'. Report to Northern Natural Gas Comp[any and El Paso Natural Gas Company, USA.
4. Greig, J.M. (1988), in Proc. 7th Int. Conf. 'Plastics Pipes', Bath, UK, Institute of Materials, p12/1-7.
5. Greenshields, C.J. (1997), Plastics Rubber Comps. Processing Appls. **26**, 387

The Application of Fracture Mechanics to Polymers, Adhesives and Composites
D.R. Moore (Ed)

THE USE OF FRACTURE MECHANICS TO ASSESS PLASTICS PRESSURE PIPE MATERIALS

G.P. MARSHALL and J.G. WILLIAMS

INTRODUCTION

Thermoplastics pressure pipes have been in common use for nearly 50 years in a wide variety of applications. Materials such as PVC-U and Polyethylene (PE) have long been used for water and sewage transmission and distribution all over the world. Both materials are relatively 'cheap' in terms of raw materials costs and can be readily processed from powder and granular feedstock using straightforward extrusion equipment. In both cases pipes up to 1m diameter may be made and in the case of PE, pipes up to 1.6m diameter have been installed. They are almost 'commodity' materials for pressure pipes and are certainly the most widely used polymers in the pipe industry.

Traditionally, water utility applications have been dominated by the use of cast iron (CI) and asbestos cement (AC) pipes and in many areas of the world, ductile iron (the modern successor to CI) is still used as the main choice of pipe material. In most pipe 'systems', the availability and ease of assembly of pipe fittings is of prime importance. Plastics materials may readily be moulded into a wide range of complex shapes with ease and fittings are generally more widely available. This has made plastics a very attractive alternative to CI or DI and Polyethylene in particular has found great favour since the flexible nature of PE allows pipes to be bent around corners etc without the use of transition fittings. Whilst the pipe material itself may be more expensive, the lower installation costs often compensate. With PVC-U, the main selling point is the inherent low cost of the raw material combined with a relatively high strength which allows the use of thinner pipe walls.

Where there is danger of pipe failure causing major safety problems – as with compressed air and gas pipelines – the materials most commonly used are ABS (for air lines) and PE (for gas). In both cases, a prime reason for use has been the ability to weld the pipe lengths together and so ensure a leak-free joint since gas leakage from socket a spigot cast iron systems has historically led to a number of major gas leaks which have subsequently led to explosions. The downside of having a continuously welded pipeline is that if a pipe failure were to occur by unstable rapid crack propagation (RCP) then the crack may run at high speed for many hundreds of metres. Thus high toughness and excellent resistance to both slow crack growth and RCP are prime requisites for pipe materials for such critical applications. Most gas utilities in the world use PE pipes as their preferred material for distribution systems.

Whilst failures of water and sewage pressure pipes are generally held to be less critical than failures of gas pipes where there is always a risk that lives may be lost if there is an explosion, there can be serious consequences. Pipes are frequently laid under main roads and motor accidents can occur. There is also major inconvenience associated with loss of service of trunk mains – a factor which caused many utilities to shun the use of PVC-U in the U.K. when pipes laid in the 1960-70 period failed with great regularity. The lack of appreciation of the causes of failure by the pipe suppliers and in particular the reasons for transitions in the failure mode from the expected ductile yielding to highly brittle crack growth followed by RCP created a situation where there was confusion. The standard tests employed by suppliers were made on perfect pipes which contained few defects and used loading conditions which did not simulate what actually happened in service. Pipes are seldom subject to only pressure loading. Variations in installation can create many situations where bending moments are applied and under such loading, even small defects can generate critical stress intensity factors which subsequently may cause slow crack growth.

The use of Fracture Mechanics testing methods and analysis provides the only rational basis for assessment of pressure pipes which must be sufficiently robust to withstand the rigours of installation and operation in the real world.

MECHANICAL PROPERTIES RELEVANT TO PRESSURE PIPE DUTIES

The three main mechanical properties that are relevant for pipe applications are:

Strength: This is clearly of major consequence because low strength materials such as plastics generally need larger wall thicknesses than their metallic counterparts. This has a major cost implication since plastics pipes costs are directly proportional to the weight of material used.

The long-term strength that a given plastic pipe may sustain without rupture is determined using standard long-term pressure tests where pipes are held under different pressures/stresses until rupture occurs. A power law relationship is then used to predict the value of stress to cause failure after 50 years pressurisation. This value of stress is then adjusted using a safety factor to derive a design stress which is used to calculate the wall thicknesses required to sustain different values of constant pressure. For PVC-U materials, standard bodies throughout the world have generally used a safety factor of 2. With PE pipes, safety factors of as low as 1.25 are used. The reasons for favouring PE are not transparent. There has been a general feeling that PE is a more predictable, ductile material.

Impact Resistance: This is particularly relevant for PVC-U pipes where there have been frequent reports of pipes being damaged by impacts caused during initial installation and subsequent excavations onto mains by 3rd party contractors. However, most PVC-U standards use the drop weight test not to simulate practical circumstances but as a process control test. If a pipe is made containing small defects in the bore region they will propagate in an unstable manner in the impact test. It is almost impossible to cause PE pipes to fail in standard drop weight impact tests and no standards specify this type of test. However, in the U.K., the Water Industry has specified a notched Charpy impact test for many years (ref. 1) – in the belief that a high toughness (Gc) indicates good resistance to RCP.

Crack Growth Resistance: In early standards for pipes for water and sewerage duties, there was perceived to be no need to show that products had any resistance to crack growth. It was simply assumed that there was no need and even in 2003, CEN committees composed largely of pipe suppler representatives tend to resist the imposition of Fracture Mechanics testing for either PE or PVC-U pipes. In the USA, no such tests have ever been proposed or accepted for inclusion in standards for water duties.

With the gas industry the attitude has always been different. The consequences of RCP in an all-welded PE system are dramatic. There could be loss of life in heavy urban areas if gas mains explode. With the PE systems, British Gas and their Engineering Research Station (ERS) carried out fundamental RCP assessment before launching PE pipes as the prime distribution system. They also persuaded their pipe materials suppliers to carry out their own R&D into the slow crack growth resistance of different PE materials – although the Fracture Mechanics tests and requirements for high fracture toughness were never codified in standards. Instead, the standards required pressure testing at 80°C, which had historically been shown to discriminate materials with different levels of ESC resistance.

FRACTURE MECHANICS APPROACH

The long-term ductile-brittle transition in failure mode which undoubtedly can occur with a number of thermoplastics, can be explained more rationally by invoking a Fracture Mechanics approach (ref. 2). The analysis considers the conditions which control the initiation and subsequent growth of a crack and has two major benefits:

(a) If the material fails in a brittle manner at stresses well below that which is needed to cause ductile yielding, then a ‘Toughness’ (Kc) can be determined which can be used to determine the critical conditions required to produce cracking. The toughness can also be used as a useful parameter to compare different materials.

(b) If the result of testing notched samples is that the material fails in a ductile manner, then the exercise has generated confidence that the pipe is capable of withstanding failure from the worst-case scenario.

The way in which cracks affect the strength when acting as stress concentrations (generating brittle failure) and alternatively as a means by which the net section is reduced (generating ductile failure) is illustrated in Figure 1.

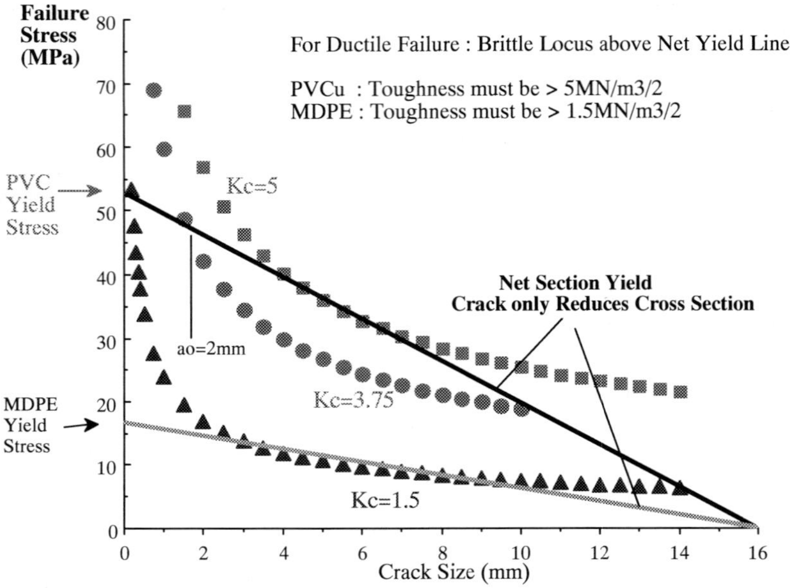

Figure 1: Effect of Crack Size on the Stress needed for Brittle and Ductile Failures

There are two major effects of cracks on the pressure bearing properties of pipes:

1) The presence of a crack clearly reduces the cross section of the pipe wall; since less material is present to sustain the pressure the stress needed to cause failure will decrease. For tensile loading (pressure), the wall decreases linearly with crack size and so does the gross failure stress, as shown in Figure 1. This is the net section yielding locus.

2) A stress concentration is produced at the crack tip which causes the crack to grow when the stress in the general pipe is at a modest level. For a constant stress intensity factor at the onset of failure, the stress to cause brittle crack growth decreases with $1/\sqrt{(\text{crack size})}$, shown as the curves in Figure 1. By altering the processing conditions for PVC or the basic raw material for PE, it is possible to alter the resistance to cracking. This gives different levels of stress intensity needed to initiate cracking which are defined by the levels of the curves and in turn characterised by single values of the 'Fracture Toughness' Kc.

The mode of cracking will generally be brittle for failures controlled by the stress concentration. Where it is the reduction in net section which causes low stress failure, the material will yield across the ligament ahead of the crack and the mode will be ductile.

There is clearly a critical crack size (a_o), where for cracks smaller than a_o there will be failure by ductile net section yielding and for larger cracks there will be brittle failure. For PVC-U, a_o will be 2mm if the Fracture Toughness is $3.75MN/m^{3/2}$ If the

Toughness is raised to 5 MN/m $^{3/2}$ (e.g via better control of processing) there will be no value of a_o which will cause brittle failure.

If the strength of a polymer is increased without a commensurate increase in toughness, the net section yield locus alters. There is then a risk that there will be a brittle regime for a wider range of initial crack sizes-since the brittle locus will be below the yield condition. Hence, 'strong' materials are generally found to be more prone to brittle failure.

It should be noted that total ductility of a PE pipe grade may be achieved with a modest value of Toughness (approx 1.6 MN/m $^{3/2}$ in the example shown in Figure 1) - because of the low yield stress of PE. Such a low level would give serious problems for a PVC-U material and indeed such low levels have been measured on PVC-U pipes which have failed in service.

ACCELERATION OF BRITTLE FAILURE OF PE AT 80°C

World-wide standards committees consistently require the specification of accelerated tests at 80°C - even though there are few theories which can rationally explain the transition from ductility to crack growth in terms which relate to service conditions. It is difficult to translate rate process theories and activation energy explanations into terms which are explicable to the average water engineer. What in effect happens is that all brittle failures are initiated from tiny 'inherent' defects which may be present with the feedstock granules. These stay in the molten phase and eventually are located close to the pipe bore, since this is the last place to cool during production (e.g. see lower RHS photo in Figure 2). At ambient temperatures, these defects are so small that they cannot generate a critical stress intensity factor unless the gross stress is raised above the yield stress. Thus, brittle failure is unlikely. When the pressure test temperature is increased however, both the yield stress and the fracture toughness of the PE decrease. With PE, the toughness decreases at a greater rate than the yield stress and with some higher density materials, small flaws are above the critical size and are able to grow at a critical stress intensity factor which is generated by gross stresses below the yield stress.

It is to be noted that it is possible to accelerate the onset of a ductile-brittle transition in PE by introducing larger sizes of cracks into pipe samples – this gives a better reflection of the material performance since there is less sensitivity to variation in defect size and location. Also, creating 'artificial 'damage' is far more likely to represent the real situation where pipe is scored during handling and installation. This procedure was adopted as standard procedure in the mid 1980s by the UK water and gas industries in specifications for PE pipes (e.g. ref. 1). This was one of the first moves towards using concepts of Fracture Mechanics for materials evaluation by a specifying authority. Whilst Fracture Mechanics analytical procedures have not been applied to try to predict critical pressures of pipe containing defects, the testing methods using notched samples has allowed suppliers and clients to be confident that damage commensurate with that likely to be found in service will not compromise long term performance. There is much greater confidence that long term ductile-brittle failure mode transitions will not occur.

USING FRACTURE MECHANICS TESTING FOR PIPE APPRAISAL

The first changes in standards towards the use of a F.M. approach were made following a high profile series of failures of large diameter PVC-U pressure pipes in the UK in the 1970s. The common characteristics of the pipe failures was that the mode was always a period of slow crack growth initiated from an 'inherent' defect in the pipe bore – followed by RCP cracking through the complete length of the pipe (usually 6m or 12m). Typical failures are shown in Figure 2.

Figure 2: Origins of Failures and Subsequent Consequences of Brittle Failure in PVC-U (top) and PE Pipes (bottom)

Top: LH: PVC-U origin and slow crack growth: RH: RCP Failure - PVC-U Pipe
Bottom: LH: Slow crack growth in HDPE. RH: RCP of large diameter HD PE Pipe

Also shown are examples of relatively rare brittle failures of PE pipes which have occurred in service. There is clearly concern that slow crack growth from small inherent flaws in pipe walls or in butt fusion welds may subsequently cause failure by RCP. The risk that an RCP failure may propagate along complete welded sections which may be some hundreds of metres in length is not to be ignored – especially for gas distribution systems (see Figure 2, lower RHS photo).

With PVC-U, conventional pressure tests did not show any significant differences between pipes that subsequently failed and those which have remained in service without problems. The initial thoughts were that 'surge and fatigue' problems had led to failures (ref. 3), but a programme of work undertaken by Joseph (ref. 4) showed that although the long-term properties of PVC-U were seriously reduced by cyclic loading, there was no evidence to show that any one pipe would be more sensitive to failure than another (see also ref. 5). This was a major drawback for the argument that cyclic loading was the source of problems especially since PVC-U pipes performed very well elsewhere in the world where pumping of water was far more commonplace than in the UK. Also, it was noted that the UK failures occurred just as frequently on gravity mains which are seldom subject to major pressure cycles. There was clearly a case that processing of the pipes had a major influence on critical toughness/crack resistance properties and in 'conventional' standards there were no tests or analyses in common use to discriminate a poor pipe from a good one.

From on-site forensic investigations, it was determined that the prime cause of failure was overloading caused by laying of pipes on stones and boulders which generated localised high stress concentrations via loading in bending – a situation not modelled in any of the standard tests. Following a long work programme involving testing of many new and exhumed PVC-U pipes made by different suppliers (ref. 6), it was found that:

1) The only tests that discriminated between pipes that had known poor service history and those that performed satisfactorily were those where the long-term toughness properties were determined using pre-notched samples.

2) The derivation of a Fracture Toughness Kc based on LEFM was the best way to assess performance characteristics in short term quality control testing.

3) The most suitable test was to pre-notch a ring sample, cut a section from the ring opposite to the notch and then load the system in bending by applying a dead weight load at the cut section (C Ring test).

It was concluded that for successful service experience pipes should be required to have sufficient toughness to withstand at least three factors which may go awry.

1) It must be assumed that pipes will be damaged and defects will be introduced.

2) The service loading will seldom be static pressure alone, since additional bending moments are frequently applied.

3) Backfill, bedding and surround materials may differ from those specified and will often be compacted inefficiently. There may also be adverse ground conditions with waterlogged trenches.

The UK Water Research Centre (WRC), acting on behalf of the UK water industry, introduced the C ring test into UK water industry Specifications (WIS) in 1982 and this test was subsequently adopted in BSEN 3505 (ref. 7)

C RING ASSESSMENT OF LONG-TERM CRACK GROWTH RESISTANCE

Although it is realistic to model pressure failures using notched pipes which are pressurised, it is also possible to closely simulate the actions of point loads etc by testing the C-shaped ring specimen (ref. 7). Such samples can be used to measure the toughness Kc and typical results on pipes made from various plastics are shown in Figure 3. All the samples were subjected to dead weight loading at different loading levels until failure occurred. It can be seen that the modified MPVC material has higher resistances to crack growth than PE. Indeed, poor PVC-U has almost identical fracture toughness characteristics as good HDPE pipe. It was also found that when PE pipes are made from some HDPE materials which have low intrinsic toughness, the data from this simple test show that problems of brittle crack growth may be expected to be worse in such grades of PE than poor PVC-U materials.

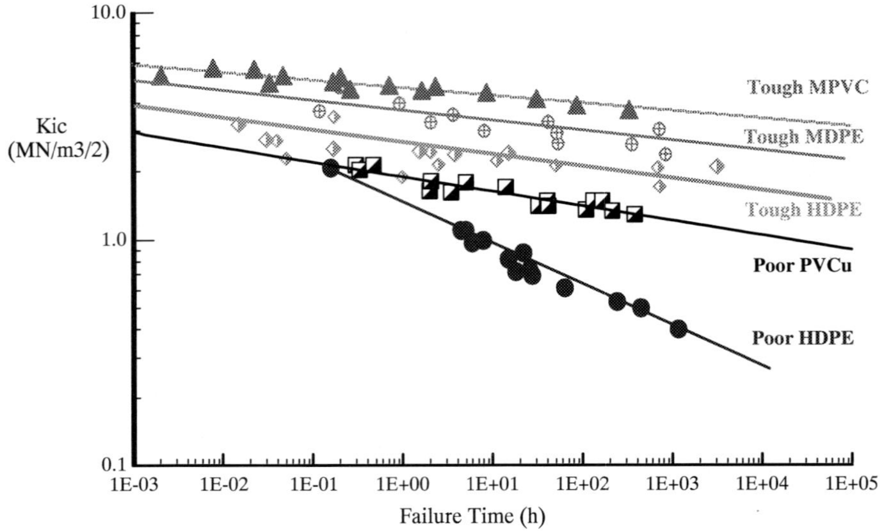

Figure 3: Toughness data from Notched C ring tests for PVC and PE materials

That PE appears to be always less tough than PVC-U is, in fact an illusion. PE is always less 'strong' than PVC and will thus tend to yield and deform in a ductile manner at lower stress levels. When the toughness has been increased to a level where the stress on the net section ahead of the notch exceeds yield before the brittle condition can be initiated, then the sample will eventually fail by ductile collapse. Because PE materials have lower yield stresses than PVC, the net section yield condition occurs at lower gross stress levels. It is thus not possible to measure a very high true toughness since any result is invalidated by yielding. This is the situation for all the data shown in Figure 3 except for the cases labelled "Poor PVC/Poor HDPE" where as noted above, both these pipes were indeed of inferior quality.

All pipe which has been identified as having poor crack growth resistance by this type of test has been found to have poor service performance. Conversely, very few PVC –U pipes discerned to be of high toughness have failed in the UK as a consequence of long-term slow crack growth. This is shown in burst data statistics gleaned from a U.K. Water Company where details of the burst behaviour of pipes were collected as a function of the year in which pipes were laid. The burst rates shown in Figure 4, clearly illustrate that there was a very significant decrease in burst rates for PVC-U pipes laid after 1982 – which coincides with the date at which all suppliers were required to test pipes to ensure that they had a high level of toughness via the introduction of the C ring F.M. test in B.S. 3505.

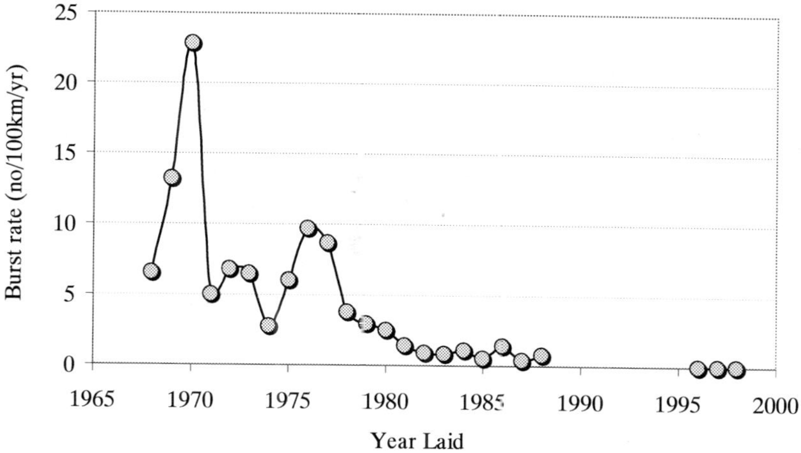

Figure 4: Burst Rates of PVC-U Pipes in a UK Water Company (1998 data)

CRACK RESISTANCE, J-R CURVES

When either the strength has been reduced or toughness increased such that the crack grows in a ductile manner, conventional analysis to determine a Fracture Toughness (K_{1c}) by measuring stresses to cause failure, is no longer a valid approach. Rather, it is necessary to measure the energy needed to promote a given increase in crack growth. The energy/unit area parameter is known as the Crack Resistance (J) and the resistance curve is J versus the crack growth Δa.

Often increasing toughness by altering processing is hard to achieve, and it may be more rewarding to reduce the yield stress so as to lower the net section yield line below the brittle locus (Figure 1). This could be considered the case with both modified PVC and MDPE and for both materials it is not possible to verify whether there has also been a commensurate toughness increase using Linear Elastic Fracture Mechanics methods, since net section yielding intervenes. The failure mode in notched sample tests is via ductile tearing and thus a J analysis is necessary to characterise crack growth.

Much effort has been expended during recent years in developing appropriate test methods by working groups in universities and industry throughout Europe and the USA via ESIS and ASTM joint programmes. Results from notched 3 point bending tests to the new procedure (ref. 8) are shown for a CPE modified PVC in Figure 5. Considering the wide range of test centres, the consistency of data is excellent and it is to be noted that no group managed to induce embrittlement.

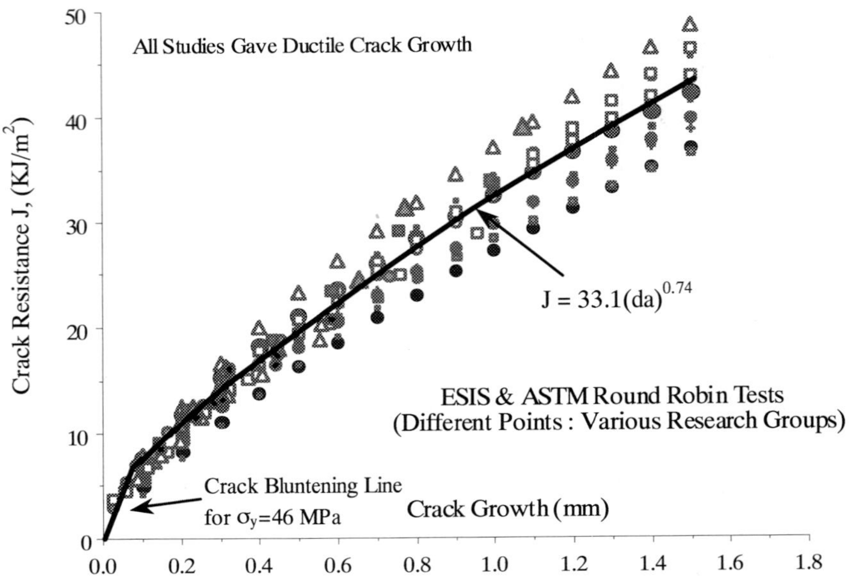

Figure 5: Crack Resistance Data for a CPE modified PVC Pipe Material

Similar results for three types of PE pipe grades are shown in Figure 6. In this case, the test was successful in distinguishing differences between the different grades of material. The MDPE pipe has markedly better crack growth resistance than either of the two HDPE materials. For short lengths of crack extension, the initial sharp crack tip in the MDPE becomes blunt as the material at the crack tip deforms and yields – forming multiple crazes in the process.

Where an initial sharp crack becomes blunt because of crack tip yielding, analysis (ref. 2) predicts that the slope of the R curve should be proportional to the material's yield stress. That MDPE has a higher slope is indicative of a higher yield stress. However, in simple tension, MDPE has a lower yield stress than HDPE. This contradiction can be rationalised by proposing that there is an increase in constraint on yielding caused by the triaxial stress field at the crack tip. With the HDPE materials, there is only marginal constraint because the formation of a single craze at the crack tip relieves the triaxial stress field and the slope is close to that expected from the yield stress. This is a potentially dangerous condition, since slow crack growth can readily occur via void coalescence in the craze. The HDPE 2 material for which results are shown in Figure 6 had been known to suffer a number of brittle service failures at stress concentrations generated at changes of section in stub flanges and so the test results are not inconsistent with field behaviour.

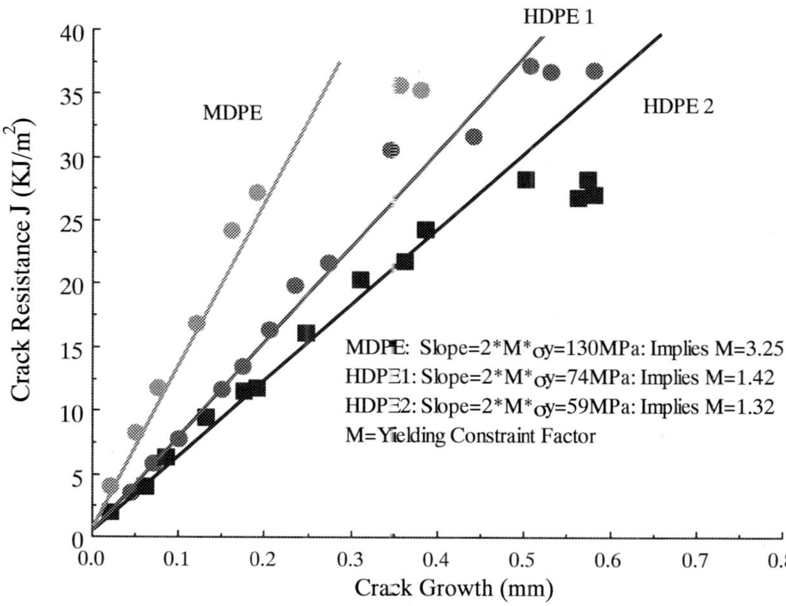

Figure 6: Crack Blunting Data for PE Pipe Grades

A MOVE TO PERFORMANCE SPECIFICATIONS

The research that has been conducted to determine the mechanisms that produce failures in plastics pipe systems over the last 30 years, has shown that pressure alone is seldom the cause of problems. Additional bending loads are frequently applied and these must be considered and catered for in establishing performance specifications. It has also been established that 'perfect' products seldom fail. All pipes and fittings will be subjected to some degree of 'damage' during handling and installation. Stress concentrations arising from defects or sharp changes in section represent the most common cause of problems with pipes and moulded fittings. It is believed that the client needs to be able to specify the provision of products that are fit for purpose and for plastic pipe systems.

A Fracture Mechanics approach using pre-notched samples to assess whether ductile-brittle failure modes can be induced and to measure acceptable levels of K_{1c} is believed to provide the most rational approach for identifying plastic materials which are likely to have superior service performance.

When new materials are being considered for pipe applications, Fracture Mechanics assessment is the most useful method of predicting whether long term properties will be compromised by ductile-brittle failure mode transitions. Test samples are simple to make and the loading conditions can be varied with ease to simulate real-life situations. The limits of satisfactory performance for a pipe can be assessed quickly and both the client and supplier can have confidence that the worst case conditions have been simulated. It is to be hoped that the current drive by standards committees to simplify all testing in order to achieve consensus does not mean that pipe assessment will return to the unsatisfactory state which existed 30 years ago.

REFERENCES

1. U.K. Water Industry Specification for MDPE Pipes ; WIS 4-32-03.
2. Williams, J.G., 'Fracture Mechanics of Polymers', Ellis Horwood, 1980
3. Kirby P.C., Plastic Pipes 5, PRI, York, (1981)
4. Joseph S. H., Plast.& Rubb. Proc. & Appl., 4, pp.325 (1984).
5. Mai, Y.W., Cotterell, B; Int. J. Fract., 24, 229, (1984)
6. Birch, M.W., Marshall, G.P. Plastics Rubber Processing 2 (4) p369 (1982).
7. British Standard for PVCu Pressure Pipes : B.S.3505
8. Hale J., Testing Protocol for Conducting J- Crack Growth Resistance Tests on Plastics, The Welding Institute, (May 1995).

MATERIALS SELECTION USING FRACTURE MECHANICS

DONALD D. HUANG, STEPHEN LI and J. GORDON WILLIAMS

INTRODUCTION

Materials selection in the design process is determined by many criteria but that of toughness is important in many applications. It is in such cases that Fracture Mechanics (FM) has a role since it provides a sound basis for choosing the material that is least likely to fracture. The methods embodied within FM are capable of providing detailed design schemes to define safe loads and limiting flaw sizes [1] but, to date, most experience has been with ranking materials by toughness. In order to properly rank materials, the end use conditions must be considered. This may require testing, for example, at high rates, at non-ambient temperatures or under fatigue loading [2].

This article describes the use of FM in the materials selection process in finished parts. The first case describes the evaluation of several toughened nylons as possible replacements for a filled phenolic resin used in making commercial casters. In the second, three short glass fibre filled nylon resins are compared for a chair base application. In the third case Ultra High Molecular Weight PE (UHMWPE) is examined for use in artificial hip joints. These examples illustrate some of the considerations that must be taken into account during laboratory testing as well as the limitations in data interpretation.

CASTERS [3]

Commercial casters can be made in very thick moldings using burlap filled phenolic resin. A typical size can be 20 cm in diameter and up to 5 cm thick. Four nylon materials of different toughness were explored as possible replacement materials in this design. They were nylon 6 (N6), nylon 6,6, (N66), a medium toughened grade of nylon 6,6, (MTN66) and a rubber toughened nylon 6,6 (RTN66). Casters were injection molded from these materials and then tested along with the phenolic caster, by dropping a 445N weight onto the rim from a height of 70 cm. The phenolic specimen passed the test but, as shown in Table 1, two of the nylon samples passed and two failed. Also shown in Table 1 are the conventional mechanical property test data for the materials.

Table 1. Mechanical Properties and Impact Test Results

Material	Impact Test	Yield Strength (MPa)	Strain to break %	Izod J/m
N6	Passed	85	230	69
N66	Failed	83	60	53
MTN66	Failed	65	78	186
RTN66	Passed	51	49	>800

There is no apparent correlation of any of these properties with the caster Impact Test results since the two materials with the lowest and highest properties (N6 and RTN66) both passed (as did the phenolic) while the middle two did not. In fact the N6 caster was partially damaged and in reality should be counted as a failure. Thus only the strongly toughened material passed the test along with the phenolic which would usually be regarded as a relatively brittle material. Some light was thrown on this when the moldings were sectioned. The phenolic was sound throughout but all the nylon parts contained fabrication voids approximately 2 cm in length. These voided moldings would require very high fracture toughness to prevent failure, as was observed.

These phenomena are embodied in the Linear Elastic Fracture Mechanics (LEFM) relationship,

$$K_c^2 = EG_c = \sigma_c^2 Y^2 a \tag{1}$$

where K_c is the critical stress intensity factor
E is Young's Modulus
G_c is the fracture toughness
σ_c is the critical stress
Y is a geometrical factor (≈ 4)
and a is the flaw size.

The toughness appears in the left hand side of the equation while the failure stress, and hence energy, appear with the inherent flaw size on the right hand side. The voided samples have a high a and hence a high G_c (or K_c) is required to give a sufficiently high energy, proportional to $\sigma_c^2 / 2E$, to prevent failure.

LEFM may be applied to N6, N66 and the phenolic composite and tests were performed according to appropriate standards [4,5] on all three. The values are given in Table 2 and show a clear ordering of N6 > N66 > phenolic. Since the fracture toughnesses of the nylons were higher than the phenolic, one route to fracture prevention would be through optimized molding conditions to decrease the void size. Using (1) and the finding that the N6 caster was marginal in performance, FM predicts that a reduction of void size by 1/3 in the N66 caster should lead to successful impact performance.

Table 2. LEFM Results

Material	$K_c, MPa\sqrt{m}$	$G_c, kJ/m^2$
N6	5.0	6.9
N66	4.0	4.5
Phenolic	2.7	1.1

The toughened materials undergo plastic deformation prior to fracture which results in stable crack growth and non-linearity which violates the LEFM criteria. The toughness in these circumstances can be evaluated using Elastic-Plastic Fracture Mechanics (EPFM) through the use of the J integral [6,7]. In this case the toughness is characterised by J, as a function of crack growth which is usually expressed in a power law form,

$$J = J_o(\Delta a / \Delta a_o)^N, \quad \Delta a_o = 1\ \text{mm}, \quad J \text{ in } kJ/m^2, \quad \Delta a \text{ in mm.} \tag{2}$$

Generally $N \approx 0.6$ for these tough materials so the curve is initially steep and tends to level off at higher Δa values. For less tough materials N is smaller leading to an earlier plateau and for $N = 0$ there is a single value, $J_o = G_c$, the LEFM parameter. Because of the rising curve it is difficult to define an appropriate initiation value of J to compare to G_c. The current ISO standard arbitrarily selects the J value at a growth of 0.2 mm, $J_{0.2}$, for comparative purposes. These values along with the values of J_c and N are given in Table 3 to illustrate these concepts. Note that the $J_{0.2}$ values correctly rank the toughness of the two toughened nylons. However, both $J_{0.2}$ values are much higher than G_c for N6 yet the MTN66 failed the impact test suggesting that the direct comparison between $J_{0.2}$ and G_c does not appear to be useful.

An alternative is to define initial blunting via a blunting line given by [8],

$$J = 2\sigma_Y . \Delta a_b \tag{3}$$

Table 3. EPFM Results

Material	J_o	N	$J_{0.2}$, kJ/m^2	J_{init}, kJ/m^2	Δa_b mm
MTN66	25	0.48	11.5	5.5	0.042
RTN66	48	0.71	15.3	7.6	0.075

and to find where this intersects the $J - \Delta a$ curve and define this as the initiation toughness. This is given by,

$$J_{init} = \left[\frac{J_o}{(2\sigma_Y)^N} \right]^{\frac{1}{1-N}} \tag{4}$$

and the appropriate values with the values of Δa_b at initiation are also given in Table 3. The initiation using this definition is at a much lower apparent crack growth, i.e. about 0.05 mm compared to the 0.2 mm used in the standard. Although the ranking of the J_{init} and G_c values correspond with the impact test results of the nylon materials, it is an open question whether the EPFM values can be used quantitatively for fracture prevention.

CHAIR BASES [9]

A set of chair bases was made from three types of glass filled resins (GFR). These were a 33% GFR nylon copolymer (GNCOP), a 33% GFR nylon 6,6 (GN66) and a 30% GFR PET (GPET). The bases were mounted on a screw driven Instron testing machine and loaded at 2.5 mm/min until they failed. The failure was typically in a leg near the central boss in a section about 4 mm thick and in a direction transverse to the flow direction. Table 4 shows the average failure loads for each material. Also shown are those loads normalised to the value of the GNCOP material.

Table 4. Average failure loads

Material	Load kN	Normalised	Tensile Strength	Izod (J/m)
GNCOP	15.3	100	183	120
GN66	13.2	86	186	117
GPET	10.4	68	158	101

Also shown in Table 4 are tensile strength values and Izod numbers measured on standard 3 mm thick specimens. The results are similar with no strong correlation to the chair base failure loads.

It is well known that there is a strong thickness effect in these materials in terms of structure. Because of the flow during moulding, there are highly oriented surface skins and an isotropic core. The ratio of surface to core thicknesses can be measured by sectioning and Table 5 gives values for the three materials for 3, 6 and 12 mm thicknesses in addition to the chair bases.

Table 5. Surface/Core ratio %

Material	3mm	6mm	12mm	Chair Base
GNCOP	85/15	50/50	50/50	50/50
GN66	90/10	40/60	45/55	50/50
GPET	80/20	50/50	25/75	25/75

Although the chair bases were around 4 mm thick the structures are more closely approximated by the 12 mm thick plaques.

LEFM K_c tests were then performed on 12 mm thick samples in the transverse direction and the values are given in Table 6 together with their normalised

Table 6. Average K_c values (MPa $\sqrt{m}$)

Material	12 mm	Transverse direction Normalised value %
NCOP	7.2	100
N66	6.2	86
PET	4.4	61

form. Comparison with the normalised loads in Table 4 shows that K_c is a good predictor of failure providing the appropriate crack orientation and specimen thickness are used.

HIP JOINTS [10]

The socket, or cup part of artificial hip joints are now mostly made of UHMWPE. They generally perform well in practice and failures are uncommon. When failures occur, they are often attributed to excessive wear and sometimes cracking. Although the primary loading mode is compression, local tensile stresses can lead to fatigue cracking. Improved materials based on UHMWPE have been proposed to overcome these problems. The prediction of in-vitro performance is very difficult involving long time scales and environmental effects, so various mechanical property tests have been evaluated as possible predictors. Screening parameters such as Young's Modulus, Yield Stress, Failure Strain from tensile tests, and density have not been particularly successful. The most promising test appears to be J-Δa tests and fatigue tests performed on the cups.

The J_c tests are performed on compression molded samples and used to explore such variables as radiation dosage. An example is given in Fig. 1 in which J-Δa curves are given for a range of dosage levels A pronounced trend is apparent with drastic reductions in toughness with increased dosage.

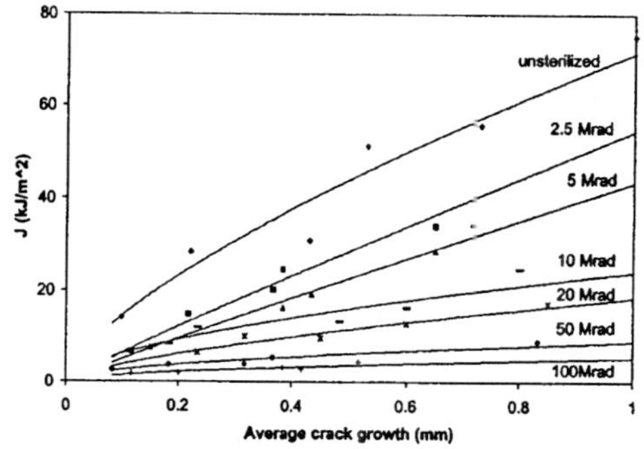

Fig. 1 Resistance curves for UHMWPE with difference radiation dosages [10]

Fatigue crack growth tests were performed on experimental hip cups by inserting a sharp crack in a high tensile region and then subjecting it to cyclic loading. The amount of crack growth per million cycles was then measured. This is a semi-empirical test in that neither K nor G is computed. Since the original crack lengths, test loads, and specimen geometry are nominally constant, the initial applied K is also constant. Under these conditions, this test provides a useful comparison of crack growth performance. Table 7 gives a set of data exploring the effect of radiation dosage and a subsequent heat treatment (HT) process of 5 hrs at 150°C. The basic material showed a growth of 0.32 mm/cycles × 10^6 for a load of 4.2 kN. A dosage of 2.5 MRad increased this slightly, as expected from the J data, but the heat treatment caused a huge increase of a factor of 10^4 suggesting that such a process is best avoided. Similar effects are seen at higher dosage rates.

Table 7

Treatment	Load (kN)	da/dN mm/cycle × 10^6
None	4.2	0.32
2.5 MRad	4.2	0.5
	2.2	0
2.5 MRad HT	2.2	5,533
5 MRad	2.2	n/a
5 MRad HT	2.2	150
10 MRad	2.2	0.27
10 MRad HT	2.2	25,900

CONCLUSIONS

The three examples given demonstrate the utility of the FM parameters K_c, G_c and J_c in materials selection. The need for J_c is determined by the level of ductility in the material and it is possible to gain some insight from the J-Δa, or resistance R, curves as they are sometimes known. Such R curves can be measured for LEFM conditions as, for example, in composite laminates, but usually an initiation value is determined. When there is a range of materials being assessed which span the LEFM-EPFM range then it may be useful to have initiation values for J_c , as illustrated for the nylons. For the glass filled systems only, K_c was needed and for the UHMWPE, only a comparison of the resistance curves was necessary.

It is also apparent from these examples that great care must be taken to include all the relevant factors. For the nylons the presence of voids was crucial to the performance and for the glass filled systems the structure is a vital ingredient in the performance. When such factors are included, however, the FM parameters work well.

In these examples the fracture data was reviewed after the event. In practice, of course, such test data would be of value in deciding on materials at the design stage.

REFERENCES

1. D. Broek, *The Practical Use of Fracture Mechanics*, Kluwer Academic Publishers, Dordrecht, The Netherlands, **1989**.

2. J. G. Williams, *Fracture Mechanics of Polymers*, Ellis Horwood, Ltd, West Sussex, UK, **1984**.
3. D. D. Huang, *Polymer Engineering and Science*. **1996**, *36*, no 18, 2270.
4. ASTM D5045, Standard Test Methods for Plane Strain Fracture Toughness and Strain Energy Release Rate of Plastic Materials
5. ISO 13586-1: Standard Test Methods for "Determination of Fracture Toughness (G_c and K_c) – Linear Elastic Fracture Mechanics (LEFM) Approach", **(2000)**
6. ASTM D6068 Standard Test Method for Determining J-R Curves for Plastic Materials
7. G.E. Hale, F. Ramsteiner, "J-Fracture Toughness of Polymers at Slow Speed" in Fracture Mechanics Testing Methods for Polymers and Adhesives – Ed D.R. Moore, A. Pavan and J.G. Williams, pp123-157. ESIS Publication 28, Elsevier **2001**, ISBN 0080436897.
8. D. D. Huang, "Fracture Toughness Testing of Toughened Polymers", in *Toughened Plastics I*, C.K. Riew and A. J. Kinloch eds, Advances in Chemistry Series, **1993**, vol 233.
9. D. D. Huang, *Polymer Composites*, **1995**, *16*, 10.
10. A. M. Gillis, J. J. Schmieg, S. Bhattacharyya, S. Li, Trans Soc for Biomaterials, **1999**, 216.

Application of Fracture Mechanics to Adhesives

PEEL STRENGTH AND ADHESIVE FRACTURE TOUGHNESS

D.R. MOORE

INTRODUCTION

Peel strength is commonly used in order to monitor adhesive strength. This relates to adhesives in many applications and a number of test methods have been used, including fixed arm peel [1], T-peel [2], climbing drum peel [3] and floating roller peel [4]. As industrial tests, the measured peel strength is used synonymously with adhesive strength. However, a global energy analysis [1] using fracture mechanics concepts indicates that a number of energy contributions make up the input energy used in pulling apart the laminate in a peel test:

$$G_c = \frac{dU_{ext}}{Bda} - \frac{dU_s}{Bda} - \frac{dU_{dt}}{Bda} - \frac{dU_{db}}{Bda} \quad (1)$$

where G_c is adhesive fracture toughness (adhesive strength), U is an energy term with the suffixes *ext*, *s*, *dt* and *db* relating to external energy, strain energy, dissipated energy in tension and dissipated energy in bending, respectively. B is the specimen width and da is an increment of crack extension (by peeling in this case).
In defining peel strength as the adhesive strength, several of the terms in equation (1) are removed leaving:

$$G_c = \frac{dU_{ext}}{Bda} \quad (2)$$

Therefore an important issue to resolve is whether equation (1) or (2) is needed in order to define adhesive strength. Alternatively, the issue is whether the application of fracture mechanics concepts contributes value to analysis of the measurement of adhesive strength.
It would be a vast study to examine this issue comprehensively and to include all of the geometry configurations referred to in the references above [1-4]. Therefore this article will limit consideration to a fixed arm peel test conducted on a polymeric laminate system.

FIXED ARM PEEL OF A PETP/PVC LAMINATE.

The polymeric laminate.

A polyethylene teraphthalate/ polyvinyl chloride (PETP/PVC) laminate system based on 125 μm coated PETP film, an ink and a PVC plastisol (500 μm) is used for this study. (The ink is used for pattern printing). A fixed arm ("L-type") peel test is used as a quality control in assessing the peel strength of this system where it is usual for peel fracture to occur between the PETP and the ink. In this industrial test, the PVC arm is fixed and the

PETP arm is the peel arm. Overall, there is a desire to achieve adequate adhesion between the PETP and the ink/PVC system.

Fixed arm peel test.

The configuration of the fixed arm peel test is shown in Figure 1. An analysis of the energy associated with peel is given elsewhere [1].

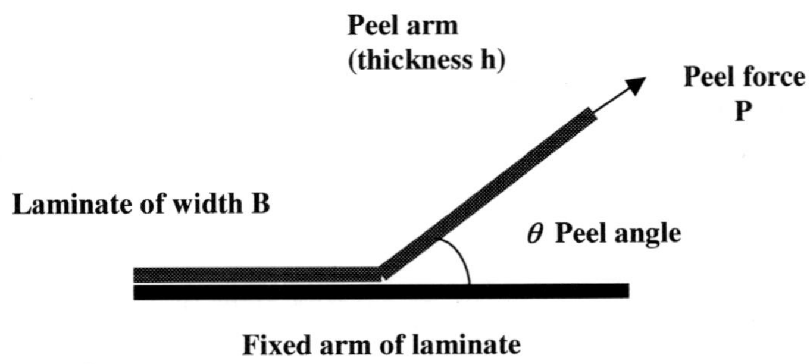

Figure 1 Configuration in a fixed arm peel test.

Peel strength (*P/B*) is measured in the test for a specific peel angle θ. It is possible to obtain equipment that enables the peel angle to be varied in the range 30° to 180° [5].

The fixed arm peel test can be supplemented by a tensile stress-strain measurement on the peel arm in order to calculate the elastic and plastic deformations that occur in the peel arm [1, 5].

If there is no strain in the peel arm (known as a peel arm of infinite modulus) then Kinloch et al [1] show that the interfacial work of fracture is given by:-

$$G_c = G_A^{\infty E} = \frac{P}{B}(1 - \cos\theta) \quad (3)$$

In practice however, this condition is not realised. If there is elastic deformation in the peel arm and dissipated tensile deformation, then the interfacial work of fracture is given by:-

$$G_c = G_A^{eb} = \frac{P}{B}(1 + \varepsilon - \cos\theta) - h\int_0^{\varepsilon} \sigma d\varepsilon \quad (4)$$

Where h is the thickness of the peel arm and σ and ε are the stress and strain in the peel arm.

When there is dissipated energy in bending as well then the interfacial work of fracture is given by:-

$$G_A = G_A^{eb} - G_A^{db} \tag{5}$$

Where the dissipated energy term is given by:-

$$G_A^{db} = \frac{dU_{db}}{Bda} \tag{6}$$

The dissipated energy term is a complex function but has been derived analytically by using elastic-plastic large displacement theory [1].

The details of these calculations are given elsewhere [1,5] but essentially conform to the global energy analysis concepts in equation (1).

Peel strength of the PETP/PVC laminate.

Using a fixed arm peel apparatus the peel strength was measured on laminates of width 20 mm at a test speed of 5 mm/min. First, the PVC material was fixed to the peel table and the PETP film was the peel arm (designated PETP/PVC) and second the PETP film was fixed to the peel table and the PVC was the peel arm (designated PVC/PETP). In these preliminary experiments, two peel angles were used, namely, 45° and 90°. Peel strength (i.e. *P/B*) measurements are shown in Table 1, where in all cases the peel fracture occurred between the PETP and the ink.

	Peel Strength (N/mm) 45 ° Peel angle	Peel Strength (N/mm) 90 ° Peel angle
PETP/PVC	3.95	1.84
PVC/PETP	1.74	0.74

Table 1 Peel strength measurements at 23° C & 5 mm/min on PETP/PVC laminates

The results in Table 1 demonstrate the problem in attempting to use peel strength to monitor adhesive strength. At a specific peel angle the peel strength of the laminates is not the same; it depends on which arm of the laminates is the peel arm. In other words, it depends on the deformational properties of the peel arm. In addition, for a specific laminate (either will do), it is apparent that the peel strength is dependent on the peel angle.

Therefore, peel strength is not an adequate measurement of the adhesive strength and consequently equation (2) is an inadequate description of adhesive fracture toughness. Expressed in different terms, equations 2 relates to the energy required to pull the laminate apart, whilst what is required is a measure of how well they are bonded together.

Adhesive fracture toughness from the fixed arm peel test.

The fixed arm peel test, with PETP as the peel arm, has been conducted for a wide range of peel angles. Peel strength has been measured as well as the tensile stress versus strain behaviour of the PETP peel arm. Consequently equations 3-6 can be used in order to calculate the input energy with and without elastic corrections, the dissipated energy and the adhesive fracture toughness. These four terms are plotted against peel angle as shown in Figure 2.

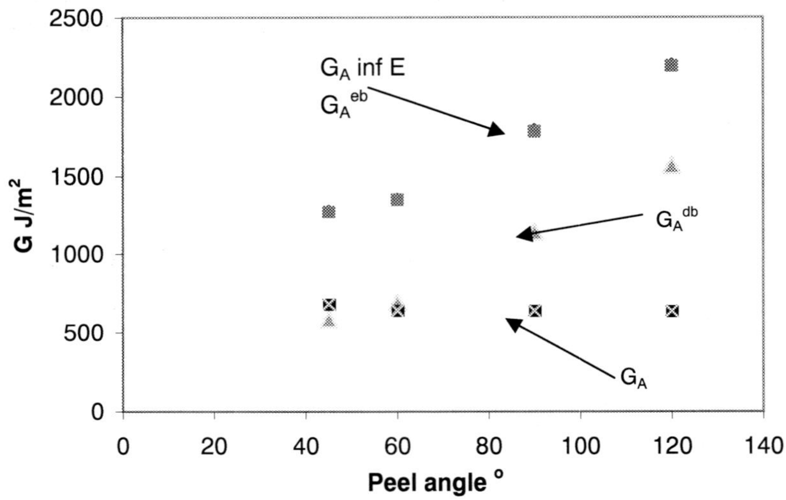

Figure 2. Fracture toughness (G) versus peel angle for PETP/PVC laminates with PETP as the peel arm. Total input energy, input energy adjusted for elastic deformation, plastic energy and adhesive fracture toughness values are included.

Several important observations emerge from these results.

(i) The elastic corrections are very small since the terms $G_A^{\infty E}$ and G_A^{eb} are virtually the same.

(ii) The terms $G_A^{\infty E}$ and G_A^{eb} are both strongly dependent on peel angle. It would be unreasonable for adhesive fracture toughness to be peel angle dependent and therefore neither of these terms can be adequate measurements of adhesive fracture toughness.

(iii) $G_A{}^{db}$ is also peel angle dependent.

(iv) The use of equation 5 where the plastic energy is subtracted from the input energy (with corrections for elastic deformation) provides an adhesive fracture toughness that is independent of peel angle. This is an objective measurement of adhesive fracture toughness.

(v) The plastic contribution to the input energy is up to 70% (from the data in Figure 1). Therefore, to peel apart the laminate, the level of correction necessary to convert input energy to true adhesive fracture toughness can be large.

Adhesive fracture toughness for different peel arms

The PETP/PVC laminate can be tested in a manner so that either the PETP can be the peel arm or the PVC can be the peel arm. Peel strength results for these two conditions were shown in Table 1 and peel strength was not the same despite the plane of fracture being the same.

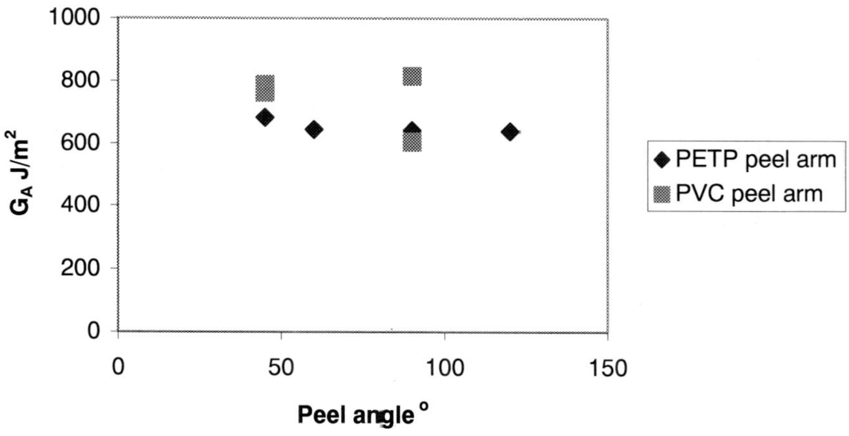

Figure 3 Adhesive fracture toughness versus peel angle for PETP/PVC laminate, with a PETP peel arm and then a PVC peel arm.

However, the tensile stress versus strain behaviour for both peel arms can be measured. Then equation 5 can be used to subtract the plastic energy from the input energy (with elastic corrections) in order to determine the adhesive fracture toughness. Figure 3 shows this adhesive fracture toughness (from equation 5) plotted against peel angle.

It can be seen that the adhesive fracture toughness is now the same and independent of which arm of the laminate is used as the peel arm. The adhesive fracture toughness is also independent of peel angle.

This is further and convincing evidence for why it is important to account for plastic deformation in the input energy required to peel apart a laminate

CONCLUDING COMMENTS

Although peel strength has been used widely for many years as a measurement of adhesive strength, the application of fracture mechanics concepts has shown that this historic approach is potentially in error. The fracture mechanics approach has shown for fixed arm peel measurements on a flexible polymeric laminate system, that correction must be made for the plastic deformations in the peel arm. When this is done an objective determination of adhesive fracture toughness can be made.

It is probable that this principle will apply to many other forms of peel geometry although not all of the evidence has yet been compiled.

REFERENCES

1 Kinloch, A.J., Lau, C.C. and J G Williams (1994), *Int. J. Fract.* 66, 45-70
2 ASTM D 1876-95 9 (1995) *Standard Test Method for Peel Resistance of Adhesives (T-Peel)*
3 ASTM D 1781-93 (1993) *Standard Test Method for Climbing Drum peel for Adhesives*
4 ASTM D 3167-97 (1997) *Standard Test Method for Floating Roller Peel Resistance of Adhesives*
5 Moore, D.R., Pavan, A. and Williams, J.G. (2001) *Fracture Mechanics Testing Methods for Polymers, Adhesives and Composites* ESIS pub 28, Elsevier, ISBN 80436897

The Application of Fracture Mechanics to Polymers, Adhesives and Composites
D.R. Moore (Ed)

PREDICTION OF DUCTILE-BRITTLE TRANSITIONS IN COATING MATERIALS

D.R. MOORE

INTRODUCTION

Polymeric resins are widely used to coat various substrates. The paint industry is the biggest single example for which there are decorative and functional applications. The use of polymers to provide protection is a particularly important functional application and this can only be effectively achieved if the coating does not crack.

A description of brittle fracture (cracking) is the domain of linear elastic fracture mechanics (LEFM) for polymeric materials [1]. Therefore the application of LEFM to situations where brittle fracture is to be avoided would seem to be a likely topic for discussion. Surprisingly, reality does not bear out this logic mainly because a wide range of practical assessments ensures that pragmatism wins over persuasion.

Nevertheless, there are cases where the application of LEFM can be particularly helpful. Consider the requirements of predicting the thickness of a protective polymeric coating over a wide temperature range where brittle fracture had to be avoided in impact loading situations. The large number of variables implies many practical assessments and the use of LEFM can considerably shorten a long programme of practical tests.

A common scenario to address for fracture of a coating material relates to the possibility that it can behave in either a ductile or a brittle manner, dependent on the service conditions. When the coating is required to provide protection, then it will be important to design the laminate in such a way that brittle fracture is avoided.

Impact loading of a laminate on the non-coated surface is such a case. The requirement is for the coating to be applied at such a thickness that brittle cracks do not occur over a range of laminate temperatures. There are four specific issues to solve:-

(a) Will the coating fracture in a ductile or brittle manner?
(b) How will the thickness of the coating influence this?
(c) How will temperature influence this fracture scenario?
(d) Can one predict when a transition will occur from ductile to brittle fracture in the coating material?

In order to illustrate how application of LEFM can be used to solve this problem, we have conducted work on a polycarbonate substrate with an epoxy coating. These materials have been used for illustration purposes only. The polycarbonate provides a ductile substrate under impact loading whilst the epoxy coating can render both ductile and brittle fracture dependent on the test conditions (its thickness and the test temperature)

THE CONCEPTS OF DUCTILE AND BRITTLE FRACTURE

The general case for ductile- brittle transitions in the coating material is given in Figure 1. Absorbed impact energy versus temperature is shown in the top diagram, where it can be seen that low energy absorption accompanied by brittle fracture will occur at low temperatures. On the other hand, high energy absorption with ductile fractures will occur at high temperatures and a ductile-brittle transition will occur at some intermediate temperature. We now require to convert this to an energy versus coating thickness plot at some fixed temperature. This is achieved in the lower diagram in Figure 1.

Brittle fractures can be expected in the coating material when the coating thickness is large. This is because the thickness will be large compared with the plastic zone around any inherent flaw in the coating material. This maps the low energy and high thickness portion of the diagram. As the coating becomes thinner a stage will be reached when the plastic zone size around an inherent flaw will be comparable with the coating thickness. For such conditions, ductile fracture can be expected. This maps the high energy and low coating thickness portion of the diagram. Therefore, a ductile- brittle transition in the fracture of the coating material can be expected between these two regions at an intermediate thickness. The lower diagram in Figure 1 shows the full shape of energy versus coating thickness plot, at a specific temperature.

In the energy versus thickness diagram, it is known that the thickness of the coating at the start of brittle fractures will be at least an order of magnitude larger than a plastic zone size. Moreover, under impact conditions it can be expected that the laminate will experience relatively large out-of-plane deflections. Therefore, we can associate the fracture with plane stress conditions for which the plastic zone radius (r_P) is given by:-

$$r_P = \frac{1}{2\pi}\left(\frac{K_C}{\sigma_Y}\right)^2 \tag{1}$$

where K_C is a plane strain value for fracture toughness and σ_Y is a tensile yield stress.

With a knowledge of those two material properties for the coating material it would be possible to calculate the plane stress plastic zone radius and from that determine the thickness condition for a ductile-brittle transition in the coating material, when the laminate is impacted on the substrate side. If that is known at one temperature it can also be made available at a range of temperatures and hence the temperature dependence of the laminate in impact can be predicted.

PREDICTION OF COATING THICKNESS AND COMPARISON WITH EXPERIMENT

Table 1 summarises the fracture properties for an epoxy coating material for a range of temperatures. These fracture properties can be obtained using existing test methods [2]. Yield strength is measured in compression [3] and then converted to tension by dividing by an empirical factor of 1.3.

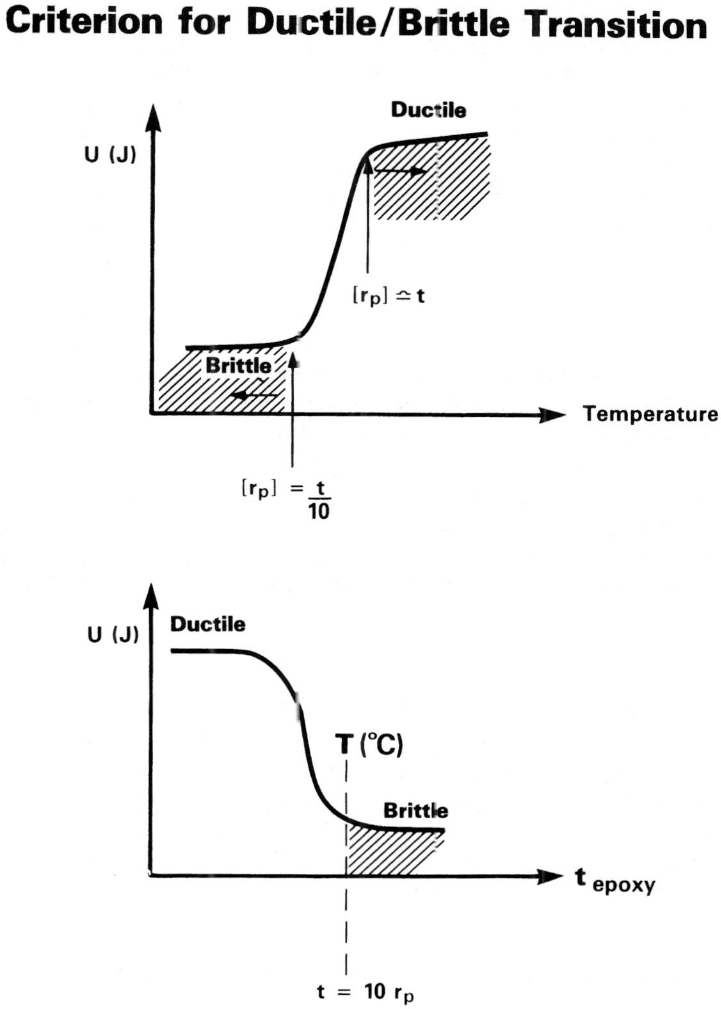

Figure 1 Criteria for ductile-brittle transitions in coating material.
Top curve on the basis of energy versus temperature.
Bottom curve on the basis of thickness versus temperature.

Temperature (^oC)	K_C ($MPam^{1/2}$)	Yield strength (MPa)
0	0.9	82
23	0.82	71
33	1.64	61
40	3.11	51

Table 1. Fracture properties at 1 m/s for the epoxy coating material.

Laminates with different thickness of coating have been impacted under drop weight conditions at a range of temperatures. The polycarbonate side was impacted and various impact energies were used in order to achieve coating fractures that ranged from ductile to brittle. Impact energy versus coating thickness is plotted in Figure 2. The plane stress plastic zone size is determined from the fracture data in Table 1. A vertical line at (10 x r_p) is drawn onto the graphs in Figure 2 at the various temperatures. These lines mark the transition to brittle fracture for the coating material. A comparison between measurement and calculation for the transition to brittle fracture (and the coating thickness at which this occurs) is shown in Figure 2 and the agreement between measured and predicted coating thickness for the ductile - brittle transition is encouragingly good.

CONCLUDING COMMENTS

LEFM has been used with considerable success in order to calculate the coating thickness (as a function of temperature) for avoiding brittle fracture in the coating material under impact conditions. The alternative approach to LEFM involves a large and specific experimental programme on coated laminates. Such an experimental study would be considerably more costly compared with the LEFM predictions and entirely specific to a set of stress and environmental conditions. On the other hand, the LEFM material properties and approach are comprehensive and adaptable.

REFERENCES

1 Williams J.G., *Fracture Mechanics of Polymers* 1984 Ellis Horwood

2 Moore D.R., Pavan A, Williams J.G. ed *Fracture Mechanics Testing Methods for Polymers Adhesives and Composites* ESIS Publication 28, Elsevier 2001, Ch 1, p27

3 Davies M., Moore D. R Comp. Sci. & Tech., 40, 1991, p131.

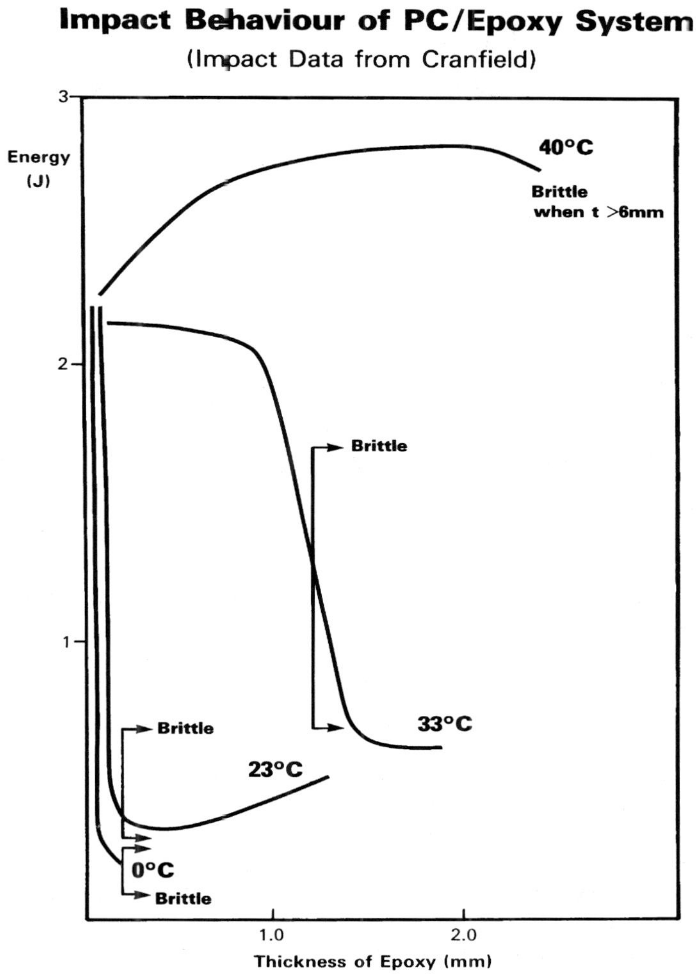

Figure 2 **Energy versus coating thickness at various temperatures where a comparison is made between predicted and experimental ductile-brittle transitions**

The Application of Fracture Mechanics to Polymers, Adhesives and Composites
D.R. Moore (Ed)

THE DURABILITY OF ADHESIVE JOINTS IN HOSTILE ENVIRONMENTS

B.R.K. BLACKMAN & A.J. KINLOCH

INTRODUCTION

The strength and toughness of adhesively bonded joints is known to be very sensitive to environmental exposure, especially the presence of moisture [1, 2]. Indeed, joint strength and toughness values can decrease significantly upon exposure to water and, as this represents the most common service environment, techniques to assess the long term performance of joints in this environment, i.e. to assess the durability of the joint, are clearly needed. Such techniques should allow engineers to assess the level of deterioration of adhesive joint performance with respect to the service environment and to enable the optimisation of an adhesive joint system, i.e. the combination of the substrate, adhesive and surface pre-treatment, to provide the best durability in a given environment.

Fracture mechanics has proved to be a very valuable tool in the study of adhesive joint durability. In such an approach, well defined toughness parameters may be measured before, during and after exposure to the hostile environment. Any deterioration in joint toughness can thus be quantified and this information can be coupled with a detailed examination of the fracture surfaces after the test to identify the failure path and the likely mechanisms of environmental attack. Two approaches followed to assess the durability of adhesive joints using fracture mechanics methods are briefly described here. The first is a residual toughness approach, where an adhesive joint is manufactured and then immersed in distilled water in an unstressed state. Joints are then periodically removed from the immersion tank and tested using standard fracture mechanics tests, e.g. DCB or TDCB, and the residual toughness determined together with the locus of joint failure. Such tests are frequently accelerated by increasing the temperature of the immersion tank to a value greater than would be expected in service, to increase the rate of water diffusion into the joint and thus accelerate the degradation. Some dangers are introduced by this temperature acceleration, and it very important that different failure mechanisms are not induced by the higher temperatures that would not exist at the service temperature. The second approach described here is the use of cyclic-fatigue tests using a standard fracture mechanics TDCB test specimen with the test performed in the service environment. Such an approach has the advantage that it is not necessary to increase the temperature greatly above that expected in service, as the fatigue loading ensures that long term data can be obtained in a relatively short time period e.g. weeks rather than the months or years required for 'accelerated unstressed ageing.' Also, by the use of a Paris-Law approach in which the rate of crack growth per cycle is plotted against the maximum value of G, i.e. G_{max}, usually on a logarithmic scale, then a threshold value of G, termed G_{th}, can be identified, below which no crack growth occurs. Such a parameter has obvious potential for predicting the service life of joints [3, 4].

ACCELERATED FRACTURE MECHANICS DURABILITY TESTS

Objective

In this example [5], an accelerated durability test was performed using bonded unidirectional carbon-fibre reinforced composite joints in order to compare the performance of four different adhesive joint systems. Both an epoxy based composite (U/C epoxy) and a thermoplastic polyetheretherketone based composite (U/C PEEK) were employed.

Experimental

The U/C epoxy substrates were lightly abraded and wiped with an acetone soaked cloth prior to bonding to manufacture double cantilever beam test specimens. Bonding was effected using either a rubber toughened epoxy-paste adhesive (EA9309) from Hysol Dexter or by using a rubber toughened epoxy-film adhesive (FM73M) from American Cyanamid. The U/C PEEK substrates were lightly abraded and wiped with an acetone soaked cloth and additionally were surface treated using a corona discharge apparatus [6] to a level of 20J/mm^2 prior to bonding with the epoxy-paste adhesive and to a level of 10J/mm^2 prior to bonding with the epoxy-film adhesive. These levels have been established to be sufficient to yield a tough, cohesive in-the-adhesive failure path in the unexposed joints. Joints were submerged in distilled water at 50°C for periods of up to one year. Joints were periodically removed from the environment and tested at a constant displacement rate of 2mm/min to measure the value of G_c for the joint, and to observe the locus of joint failure. The testing procedure and data analysis was as outlined in the recently published British standard [7] The values of G_c measured for these joints were then expressed as a % retention of the 50°C "dry" G_c value. A number of unexposed "dry" joints had been post-cured at 50°C for periods of up to 18 days to determine the fully post-cured values of G_c. These joints were then tested at 23$\pm$1°C and 55% relative humidity.

Results

The results for the joints bonded with the epoxy-paste adhesive and exposed for periods of up to one year are shown in Figure 1(a). All failures remained cohesive in-the-adhesive layer and the retained toughness values after one year of exposure were 70% for the U/C PEEK substrates and 50% for the U/C epoxy substrates. The equivalent results for the epoxy-film adhesive are shown in Figure 1(b). All failures in the joints manufactured with the U/C epoxy substrates were cohesive in-the-adhesive layer, and the retained toughness values after one year of exposure were 50% for these joints. However, the joints employing the U/C PEEK substrates showed a much steeper decline in toughness values and this decline was accompanied by a change in the locus of failure from cohesive in-the-adhesive to interfacial along the adhesive-substrate interface. The approximate % cohesive failure observed on the fracture surfaces are indicated in Figure 1(b). After just five months of exposure, the retained toughness was 28% for these joints, after which no further deterioration was seen and fully interfacial failure (0% cohesive) was then always subsequently observed.

These results clearly demonstrate the very deleterious effect of water immersion on the toughness of all joints studies here. What is also evident from this limited study is that the relative dependence of the durability upon substrate material was clearly different (and reversed) for the two adhesives employed here. The results allow the identification of combinations of adhesive-substrate-surface pre-treatment which yield the poorest durability in this most demanding exposure condition. Effects such as adhesive plasticization, swelling, irreversible chemical ageing and the breaking down of the intrinsic adhesion forces across the adhesive/substrate interface may all contribute to the durability of the joints in a complex manner. However, what is clear is that there is a strong need to reduce the time required for durability testing, without introducing failure mechanisms that are not observed in-service. The cyclic fatigue test discussed below represents one such attempt to overcome this problem.

CYCLIC-FATIGUE FRACTURE MECHANICS DURABILITY TESTS

Objective

In this example [8], cyclic-fatigue tests were performed using adhesively-bonded aluminium alloy TDCB specimens to order to determine the optimum surface pre-treatment for the aluminium alloy prior to bonding with a hot-curing, rubber toughened adhesive and prior to exposure in a wet environment.

Experimental

An aerospace grade of aluminium alloy was employed (BS EN 2014A) and this was bonded with a hot-curing toughened epoxy adhesive (EA9628 UNS) from Hysol Dexter Inc. Four different surface pre-treatments were employed, as typically used in the aerospace industry. These were: (i) grit blasting followed by degreasing (GBD), (ii) chromic-acid etching (CAE), (iii) phosphoric-acid anodising (PAA) and (iv) phosphoric-acid anodising followed by the application of a primer (PAAP). Tapered double cantilever beam specimens were then prepared with substrates pre-treated as described above. Initially, specimens were tested monotonically, with a constant rate of displacement of 1.0mm/min being employed. The test environment was 23$\pm$1°C and approximately 55% relative humidity. The value of G_c for the joint was determined, as now outlined in the British standard [7]. Joints were then loaded in cyclic fatigue on a servo-hydraulic testing machine. Tests were conducted using both a "dry" environment, 23$\pm$1°C and approximately 55% relative humidity and a "wet" environment, by immersing in distilled water at 28$\pm$1°C . Tests were performed using a sinusoidal waveform at a frequency of 5Hz and a displacement ratio $\delta_{min}/\delta_{max}=1/2$. A range of maximum displacement values were employed to cover the complete range of applied G_{max} values. The growth of the crack was monitored by visual examination through an optical microscope at various intervals during the test and thus the crack growth per cycle, da/dN, was determined. The corresponding values of G_{max} were determined via:

$$G_{max} = \frac{P_{max}^2}{2b} \cdot \frac{dC}{da} \tag{1}$$

where P_{max} was the maximum applied load in the fatigue cycle, b the width of the joint, C the compliance (displacement/load) and a the crack length.

Results

The values of the adhesive fracture energy measured from the monatomic tests are shown in Table 1 for joints prepared with substrates subjected to the four different surface pre-treatment techniques. These results indicate that the failure path in the joints consisting of the GBD substrates was a mix of interfacial and cohesive and the value of G_c measured was 600J/m^2. However, all the acid based pre-treatments resulted in cohesive failure through the adhesive layer, with a correspondingly higher value of G_c approximately equal to 1650J/m^2.

Table 1. Values of the adhesive fracture energy, G_c

Surface Pre-treatment	G_c (J/m^2)	Range (J/m^2)	Locus of failure ([a])
GBD	600	$\pm$100	M
CAE	1550	$\pm$50	C
PAA	1650	$\pm$100	C
PAAP	1700	$\pm$50	C

([a]) Visually assessed locus of joint failure: M mixed- via cohesive fracture through the adhesive layer and via fracture along the adhesive/substrate interface; C: cohesive fracture through the adhesive layer.

The results obtained from the cyclic-fatigue tests in either a "dry" or "wet" environment are shown in Figures 2-3. In these figures, the experimental relationship between *da/dN* and G_{max} are shown. It is clear that for each condition, a threshold strain energy release rate, G_{th}, exists, below which no cyclic fatigue crack growth occurs. The value of G_{th} was shown to be dependent upon the choice of surface pre-treatment used for the aluminium alloy prior to bonding, and the test environment. The "dry" cyclic fatigue results indicate that the values of G_{th} are low when compared to the values of G_c. However, the locus of failure in these dry fatigue tests was always the same as observed in the monotonic tests. Values of G_{th} measured were 125J/m^2 for the GBD joints and approximately 200-240 J/m^2 for the acid-based pre-treatments. The effects of testing the joints in an aqueous environment was very pronounced, as clearly seen in Figures 2(a) and 3(a) for the GBD and the PAA surface pre-treatments. For these joints, the linear region of the *da/dN* versus G_{max} plot is shifted towards the left, so that for a given value of G_{max}, the rate of crack growth is far more rapid in the presence of water, compared to the "dry" environment. In addition, the threshold values for the GBD and PAA joints are greatly reduced. This reduction in the value of G_{th} was accompanied by the locus of failure becoming completely interfacial (as assessed via visual inspection). For the CAE joints (Figure 2(b)), the wet environment had negligible effect on the value of G_{th} and the locus of failure, which remained cohesive through the adhesive layer. Also, for the PAAP joints (Figure 3(b)), no reduction in the fatigue threshold value was measured, although the locus of joint failure was found to have changed from cohesive in-the-adhesive to along the primer/adhesive interface.

CONCLUSIONS

In the two examples described in the present work, fracture mechanics tests have been used to assess the performance of adhesively bonded joints in a wet environment. Fracture mechanics parameters, i.e. G_c in the monotonic accelerated durability tests and G_{th} in the cyclic-fatigue durability tests, have been determined. These parameters were shown to depend strongly on the test environment and also upon the adhesive joint system employed, i.e. the combination of adhesive, substrate and surface pre-treatment used. The measurement of these parameters, when combined with detailed surface analytical techniques, enables considerable progress to be made towards identifying the mechanisms of environmental attack.

ACKNOWLEDGEMENTS
We wish to thank the EPSRC for an Advanced Research Fellowship (AF/992781) and a Platform Grant.

REFERENCES

1. Kinloch, A.J., *Adhesion & Adhesives: Science & Technology*. 1987, London & New York: Chapman Hall. 441.
2. Comyn, J., *Ch. 3 "Kinetics and Mechanisms of Environmental Attack*, in *Durability of Structural Adhesives*, A.J. Kinloch, Editor. 1983, Applied Science Publishers: London and New York. p. 85-129.
3. Curley, A.J., J.E. Jethwa, A.J. Kinloch & A.C. Taylor, *The fatigue and durability behaviour of automotive adhesives. Part III: Predicting the service life.* Journal of Adhesion, 1998. **66**: p. 39-59.
4. Curley, A.J., H. Hadavinia, A.J. Kinloch and A.C. Taylor, *Predicting the service-life of adhesively-bonded joints.* International Journal of Fracture, 2000. **103**: p. 41-69.
5. Blackman, B.R.K., *The Fracture Behaviour of Bonded Polymeric Fibre-composites.* 1993, Imperial College, University of London.

6. Kodokian, G.K.A. and A.J. Kinloch, *The adhesive fracture energy of bonded thermoplastic fibre-composites.* Journal of Adhesion, 1989. **29**: p. 193-218.
7. BSI, *Determination of the mode I adhesive fracture energy, GIC, of structural adhesives using the double cantilever beam (DCB) and tapered double cantilever beam (TDCB) specimens.* 2001. **BS 7991**.
8. Kinloch, A.J., M.S.G. Little, and J.F. Watts, *The role of the interphase in the environmental failure of adhesive joints.* Acta Materialia, 2000. **48**: p. 4543-4553.

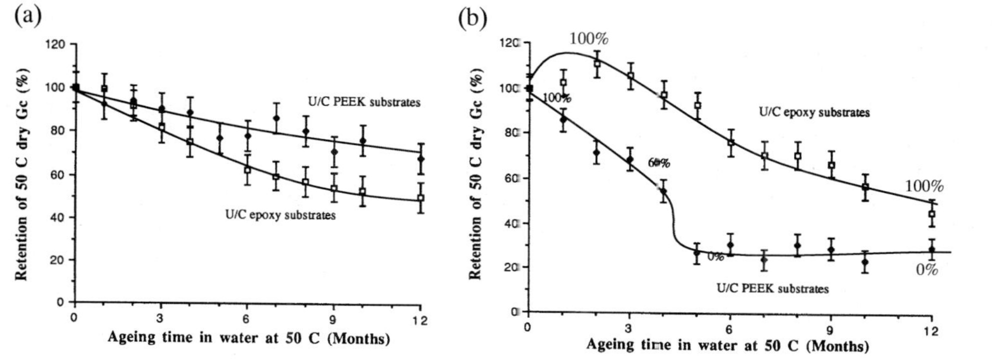

Figure 1. Retention of the 50°C dry G_c values versus ageing time for: (a) the joints bonded with the epoxy-paste adhesive and (b) joints bonded with the epoxy-film adhesive. Square points indicate U/C epoxy substrates; Diamond points indicate U/C PEEK substrates. (All failures for (a) were 100% cohesive failure in the adhesive. The % values indicated on (b) are the % cohesive failure observed on the fracture surfaces).

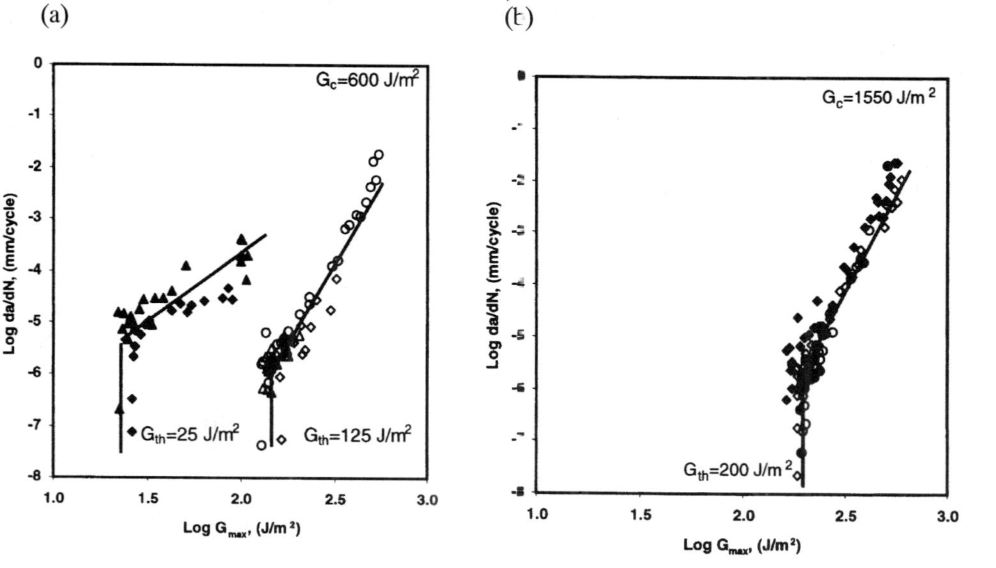

Figure 2. Logarithmic crack growth rate per cycle, da/dN, vs logarithmic G_{max} for the: (a) GBD-pre-treated joints and (b) CAE-pre-treated joints. ("Empty" symbols are for tests conducted in the dry environment at 23°C and 55% RH and "filled" symbols for tests conducted in a "wet" environment of distilled water at 28°C [8].

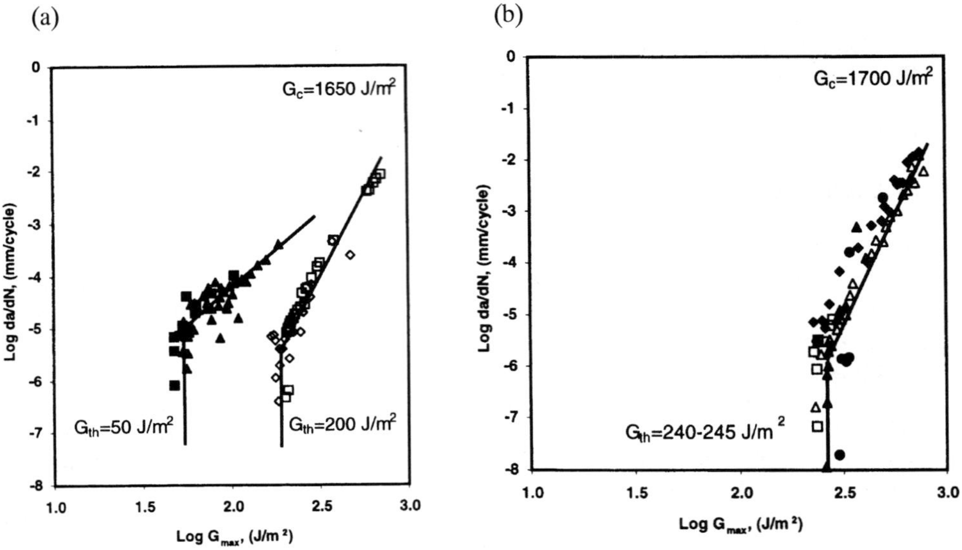

Figure 3. Logarithmic crack growth rate per cycle, *da/dN*, vs logarithmic G_{max} for the: (a) PAA-pre-treated joints and (b) PAAP-pre-treated joints. ("Empty" symbols are for tests conducted in the dry environment at 23°C and 55% RH and "filled" symbols for tests conducted in a "wet" environment of distilled water at 28°C [8].

The Application of Fracture Mechanics to Polymers, Adhesives and Composites
D.R. Moore (Ed)

THE INFLUENCE OF ADHESIVE BOND LINE THICKNESS ON THE TOUGHNESS OF ADHESIVE JOINTS

A.J. KINLOCH and D.R. MOORE

INTRODUCTION

The use of substances to bond materials together is well established. Egyptian carvings dating back more than 3,000 years show the gluing of thin veneers to planks of sycamore. Bitumen, tree pitches and beeswax were used as sealants and adhesives in ancient and medieval times. The gold leaf of illuminated manuscripts was bonded to paper by egg white and wooden objects were bonded with glues from fish, horn and cheese. The technology of animal and fish glues advanced during the 18th century. In the 19th century rubber and nitrocellulose-based adhesives were introduced. However, decisive advances in adhesives technology awaited the 20th century, during which time natural adhesives were improved and many synthetic polymers were developed. The demand for adhesives that had a high degree of structural strength and were resistant to both fatigue and environmental conditions led to the development of many high-performance materials.
Throughout these times there was always a requirement for adequate adhesive strength albeit that these requirements varied widely. Application of the adhesive substance always provided choice in terms of the amount required for satisfactory performance. This was a pragmatic and variable process driven by a compromise between economics and "good" design. There would certainly be a knowledge in experienced hands that too much adhesive did not necessarily lead to good performance.
The growing demands of high performance for structural applications and the use of relatively expensive speciality products in high technology components (aerospace, automotive and electronic) suggested a need for better understanding on the amount of adhesive required. In particular, the influence (performance and cost) of bond-line thickness on adhesive toughness. In this context, the application of fracture mechanics concepts in demonstrating fracture types (e.g. ductile to brittle) and showing the effectiveness of the adhesive bond became an important technological tool. This is investigated in this article.

THE USE OF FRACTURE MECHANICS TO MEASURE ADHESIVE FRACTURE TOUGHNESS.

Tapered Double Cantilever Beam Joints.
Kinloch and Shaw [1] applied fracture mechanics principles to measure the adhesive fracture toughness of joints and the resin systems used to prepare the joints. They were following the work of Bascom et al [2] who noted the influence of bond-line thickness on adhesive strength. Kinloch and Shaw [1] were investigating some toughened epoxy adhesives with the aim of aiding the design of joints. The fracture toughness of a tapered double cantilever beam (TDCB) adhesive joint with mild steel substrates [1,3] was investigated as a function of bond line thickness and test speed. (This particular test geometry has recently been studied in detail and adopted as the British Standard for conducting such tests [3,4]). Some of their results are shown in Figure 1 where the

toughness versus bond-line plots are observed to rise to a maximum before falling to some plateau value of toughness that is independent of bond line thickness. This is observed at both test speeds and fracture was always cohesive in the adhesive system. These measurements were supplemented with fracture toughness (G_c) measurements of the resin systems in compact-test geometry configurations, together with modulus (E) and yield stress measurements (σ_y), and again as a function of test speed.

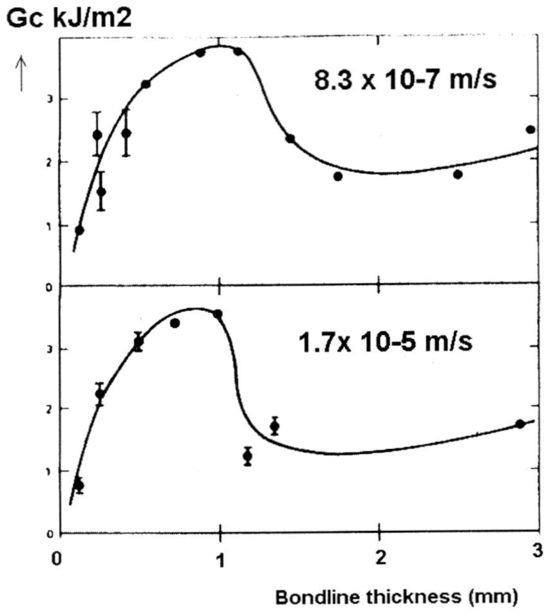

Figure 1 Fracture Toughness versus bond line thickness at 23 °C at two test speeds for TDCB joints [1].

The shape of the adhesive fracture toughness versus bond line thickness plots can be explained in terms of the plastic zone size at the crack tip relative to the bond-line thickness (t) of the adhesive. The plastic zone size (r_p) is given by:

$$r_P = \frac{1}{2\pi}\left(\frac{EG_C}{\sigma_y^2}\right) \qquad \text{plane-stress conditions} \qquad (1)$$

$$r_p = \frac{1}{6\pi}\left(\frac{EG_c}{\sigma_y^2}\right) \qquad \text{plane-strain conditions} \qquad (2)$$

and Kinloch and Shaw argued that, since the value of r_p is greater at the edge of the joint (where plane stress conditions act), then equation 1 was the more applicable for quantitative studies and direct comparison to the values of bond-line thickness (t).

The three regions of the toughness versus bond-line thickness plots can be accounted as follows:

(i) At small values of bond-line thickness, the plastic zone is relatively large. Therefore when it cannot develop fully within the adhesive resin the toughness of the resin also cannot be fully developed. The resin is of insufficient thickness to accommodate the full development of a plastic zone at the crack tip. In consequence the adhesive fracture toughness is small but increases with increases in bond line thickness. This is shown in Figure 2.

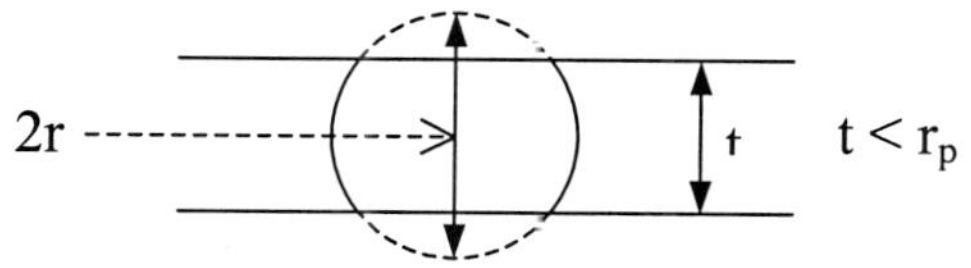

Figure 2 Small bond line thickness

(ii) When the plastic zone just fills the resin then twice its radius is equal to the bond line thickness. However, there can be distortion of the stress field ahead of the crack due to the presence of the substrates. This can alter the shape of the plastic zone from a circular section to that shown in Figure 3 and as a consequence the maximum toughness of the joint might exceed the plane-strain value of toughness (G_c) of the bulk resin [1,5]. Nevertheless, the joint will then have developed its maximum toughness.

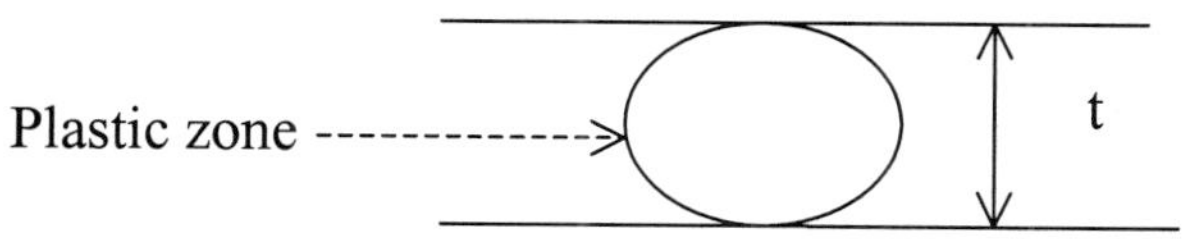

Figure 3 The plastic zone just fills the resin region.

(iii) When the bond line thickness increases beyond the situation described in Figure 3, then the plastic zone fits inside the bond-line thickness of the adhesive. Consequently, as the bond-line thickness then increases further in size, so the plastic zone eventually becomes small in comparison. At this moment, the adhesive fracture toughness becomes independent of bond-line thickness, and approximately equivalent in value to the plane-strain toughness of the bulk resin. This is shown in Figure 4.

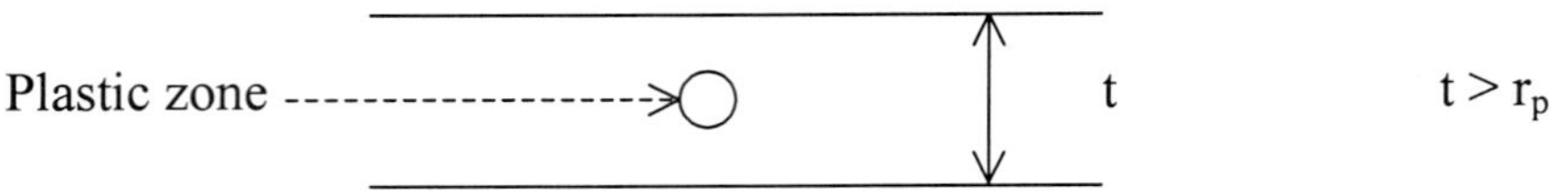

Figure 4 Bond line is larger than twice the plastic zone

The results presented by Kinloch and Shaw [1] fit this overall explanation for their tapered double cantilever beam geometry.

Compact tension adhesive joints

A compact tension joint (aluminium substrates) was used by Daghyani et al [6] in order to investigate the influence of bond line thickness on a rubber-modified epoxy. Tests were conducted at 0.3 mm/min. Fracture was cohesive in the adhesive resin region. Fracture toughness was proportional to the crack initiation load for this geometry and results are shown in Figure 5, where the value for the neat resin toughness is also included.

The pattern of behaviour for the compact tension geometry is consistent to that reported for the TDCB geometry and a similar explanation will account for the results. Daghyani et al [6] continued the work in order to use finite element analysis of the behaviour and emerged with a plausible account based on fracture resistance (J_c) versus bond line thickness.

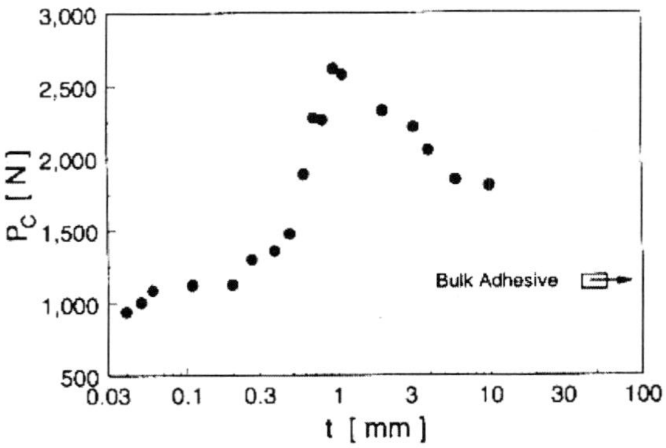

Figure 5 Fracture initiation load (proportional to fracture toughness) versus bond line thickness for rubber toughened epoxy [6] for compact tension geometry.

SINGLE EDGE NOTCHED BEAM (SENB) GEOMETRY

Perhaps the simplest geometry from a practical view-point is the SENB specimen geometry. In some unreported experiments, Moore and Hardy [7] used the SENB specimen by loading in three-point bending and from its dimensions (width = 10 mm and depth 25 mm) and associated force-displacement curve the LEFM parameters of K_C and G_C were determined. The adhesive used for this work was a toughened acrylic formulation, with a cure cycle of 12 hours at 23 °C. All specimens were stored dry prior to testing.

A number of adhesive bond specimens with adhesive layers of different thickness were made. The thickness of the adhesive layer (t) was varied from 0.3 mm to over 3 mm. Each specimen was manufactured from two aluminium blocks bonded together with the adhesive, using a jig that held the blocks rigid during the cure cycle and controlled the adhesive thickness. Prior to application of the adhesive, the faces of the blocks were grit blasted and degreased. After curing, any excess adhesive was removed from the specimen and the notch sharpened by machining, then tapping a razor blade into the adhesive. The adhesive layer and notch were examined and measured using a video microscope. SENB joint tests were conducted at 1 mm/min at 23 °C.

The fracture toughness was determined using the standard protocol [8] and results are shown in Figure 7 as a plot of fracture toughness (K_c) versus bond-line thickness. The force-displacements plots for these bonded specimens were not linear and the extent of the non-linearity varied with bond-line thickness. This results in some scatter in the data.

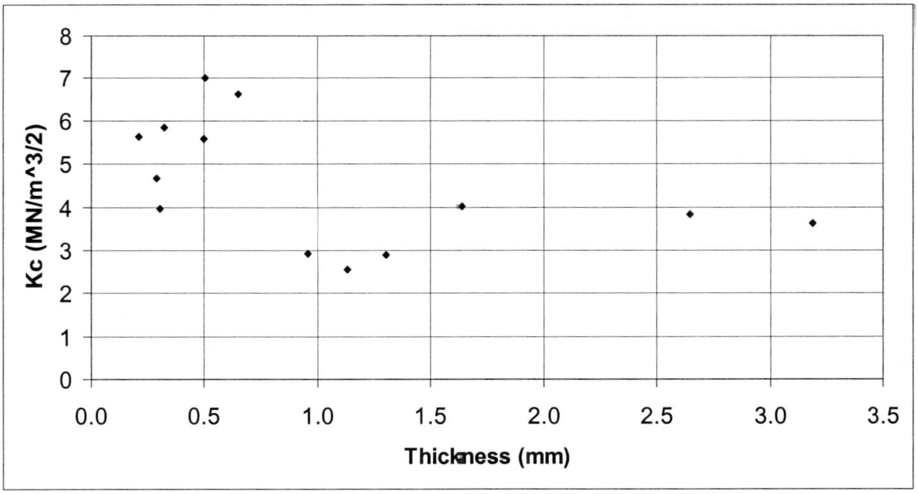

Figure 6 Fracture toughness K_c versus bond line thickness for SENB geometry at 23 °C for rubber toughened acrylic resin with aluminium substrates.

Plaques (5 mm thick) of the rubber toughened acrylic resin were used in order to measure the fracture toughness and yield stress. Property values of σ_y =34 MPa and K_C =2.1 $MPam^{1/2}$ were obtained. Resin fracture tests conducted at 23 °C did not generate valid plane strain data, but lower temperature tests enable an approximate plane strain value at 23 °C to be generated. A value of the plane-stress plastic zone size was determined from equation 1 (using K_c^2 =EG_c). resulting in a value of 0.6 mm. Therefore twice this value (i.e. 1.2 mm) should define the region in Figure 6 where the maximum value of adhesive fracture toughness occurs. This agrees only approximately with observation. The difference between prediction and observation is likely due to imperfect conditions for generating the plane-strain value of fracture toughness for the neat resin and the non-linearity for the force-displacement curves for the joint specimens.

CONCLUDING COMMENTS

The application of fracture mechanics principles have been helpful in monitoring the manner in which adhesive fracture toughness can be influenced by bond-line thickness. This is achieved through consideration of the size of the plastic zone and provides an account both for the qualitative and quantitative effects of the bond-line thickness. A reasonable predictive capability has therefore been established which would not exist without fracture mechanics. Other mechanisms have not been included in our considerations. For example, other workers have examined the interaction between substrate and adhesive. They refer to an interphase effect where a crack might grow as opposed to fracture in the bulk of the adhesive [9]. Nevertheless, this predictive capability from fracture mechanics, although not perfect, is geometry independent and this has been demonstrated by illustration of adhesive fracture toughness measurements for three quite different geometric arrangements for the joints.

REFERENCES

1 Kinloch, A.J., and Shaw, S. J (1981), *J. Adhesion*. Vol 12, p59-77

2 Bascom, W.D., Cottington, R.L., Jones, R.L. and Peyser, P., *J. Appl. Polym.Sci.* (1975) **19**, 2545

3 Moore, D.R., Pavan, A. and Williams , J.G., (Ed*) Experimental Methods in the Application of Fracture Mechanics Principles to the Testing of Polymers, Adhesives and Composites* ISBN 008 043689 7 2001 Blackman, B., and Kinloch , A., *Fracture Tests on Adhesive Joints* p203.

4 BSI 7991:2001 *Determination of the mode I adhesive fracture energy G_{IC} of structural adhesives using DCB and TDCB specimens* (2001)

5 Hunston, D.L., Kinloch, A.J. and Wang, S.S (1989) *J. Adhesion* Vol 28, p103-114

6 Daghyani, H.R., Ye, L. and Mai, Y-W (1995) *J. Adhesion* Vol 53, p149-162

7 Moore, D.R., and Hardy, R.S., (2000) Private communication

8 Moore, D.R., Pavan, A. and Williams , J.G., (Ed*) Experimental Methods in the Application of Fracture Mechanics Principles to the Testing of Polymers, Adhesives and*

Composites ISBN 008 043689 7 2001 Williams, J.G., *Kc and Gc at Slow speeds for Polymers* p11
9 Bentadjine, S., Petiand, R., Roche, A.A. and Massardier, V., *Polymer* 42, 14, 2001. 6271-6282.

The Application of Fracture Mechanics to Polymers, Adhesives and Composites
D.R. Moore (Ed)

ADHESIVE JOINT STRENGTH PREDICTION FOR COMPOSITE DESIGN

J.P.SARGENT AND P. DAVIES

INTRODUCTION

Assemblies are frequently the weak link in composite structures. The stress singularities associated with free edges make their analysis complex by traditional finite element analyses and for the prediction of joint behaviour a fracture mechanics approach appears attractive. There is a range of common engineering joints, but probably the most widely used is the lap-shear joint. Many authors have derived analysis for both the single/lap-shear, the double lap-shear and the cracked lap-shear; this includes both finite element (for example, [1-5]) and closed form analytical solutions [5-11]. These solutions may include geometrical and material non-linearity, and also the mode-mixity resulting from the combination of induced peel and shear forces. A useful introduction to the principles of a fracture mechanics approach to strength prediction for adhesive joints, including crack initiation and propagation for tough and ductile adhesives, may be found in the paper given by Fernlund et al. [12].

This chapter will describe an example of the use of data from fracture mechanics tests to predict the strength of bonded composite joints. First, the tests used to obtain the mixed mode fracture envelopes for two adhesives are presented and test results are given. Then the analysis of the single-lap shear specimen, based on the work of Papini, Fernlund and Spelt [11-13], will be described. Finally examples of correlations between predictions and lap shear test results are presented.

FRACTURE MECHANICS TESTS FOR ADHESIVELY BONDED JOINTS.

It should first be emphasised that there is currently no standard test method to determine the mixed mode fracture envelope for adhesively bonded composite joints. Many studies have examined mixed mode fracture, since the early work of Mostovoy, Ripling and colleagues [14] over 30 years ago, but the standardization activities in this area have so far concentrated on composites. A mode I DCB (double cantilever beam) test method has been developed recently for adhesives [15], and for mixed mode (I/II) loading the MMB (mixed mode bending) specimen is the most widely used [16]. Mode II shear testing remains controversial and few results are available for bonded composite assemblies.

In the present work fracture mechanics tests were performed on glass/epoxy composites assembled by two adhesives. The reinforcement is quasi-unidirectional E-glass, the resin is an amine hardened DGEBA-based epoxy (SR1500/2505 from Sicomin). Samples were produced by hand lay-up. The first adhesive is the same resin as the matrix, the second is a tough commercial structural adhesive (Redux 420). DCB and MMB specimens were employed. Full details of the tests have been presented elsewhere [17,18]. Figure 1 shows the mixed mode fracture envelopes obtained.

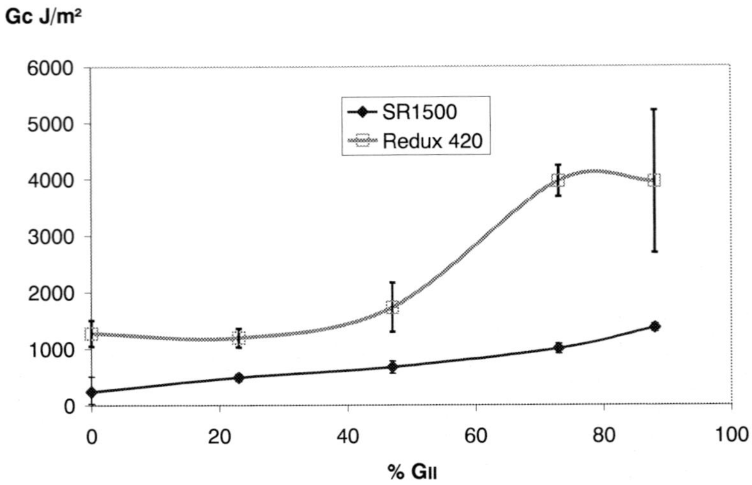

Figure 1. Mixed mode fracture envelopes, glass/epoxy composite and bonded composite specimens, NL (non-linearity) values. Error bars show ± one standard deviation.

ANALYSIS OF LAP SHEAR SPECIMENS

Papini, Fernlund and Spelt [11-13] used large deformation beam theory (permitting modelling of non-linear geometric behaviour) together with the J formulation of the critical strain energy release rate to derive fracture load predictions for the lap-shear joint. This had the merit that they were accessible, and could be applied relatively easily to a range of different geometries based on different adherend lengths and thickness. For the single lap shear specimen strength predictions were made by repeatedly calculating trial strain energy release rates and mode ratios using a system of 12 linear equations in 12 unknowns until these matched the critical strain energy release rates and mode ratios determined from the measured mixed mode fracture envelope. For the 3mm thick adherends the mode ratio Ψ (=atan($\sqrt{G_{II}/G_I}$)) was 56°. For the double lap shear specimen a simpler closed form solution was proposed:

$$G = \frac{P^2}{8E}\left(\frac{h_1}{h_2(h_2+h_1)}\right) \tag{1}$$

where P is the load, E the modulus, h_1 the outer adherend thickness and h_2 half the central adherend thickness. In this case previous mode partitioning studies indicated that the phase angle was 52°.

The solutions assumed that the adhesive layer was sufficiently thin such that the global deformation of the joint was determined entirely by the adherend material. Assuming no influence of adhesive is probably reasonable given long free adherend lengths. However, if the adherends were short, giving rise to a greater adherend deformation dependence on the crack tip stiffness contribution, or if the adhesive layer was thick, then the adhesive would make a significant contribution to the global deformation, and the analysis would be approximate. In addition, it should also be noted that the analysis did not deal explicitly with the consequences of any non-linear plastic behaviour of the adherends (e.g. as a result of damage or yielding) or adhesive, nor of the presence of an adhesive fillet. In spite of these limitations, however, the authors did note good agreement between theory and experiment for a rubber toughened adhesive system with very thick (12.7 mm) aluminium adherends with long free adherend lengths (~140 - 260 mm).

LAP SHEAR TESTS

Tests have been performed on single (SLS) and double (DLS) lap shear specimens, Figure 2. Overlap length and adherend thickness have been varied.

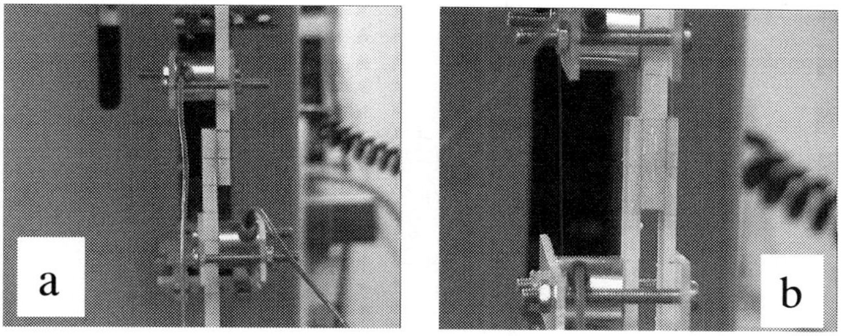

Figure 2. Lap shear specimens , a) SLS, b) DLS

The failure loads can be recorded, but also other parameters such as the visual appearance of first damage and the first acoustic emission. Damage development during tests performed on SLS specimens has been followed using a microscope, Figure 3. The presence or absence of a spew fillet makes a significant difference to results for the tough adhesive, as has been observed elsewhere [20,21].

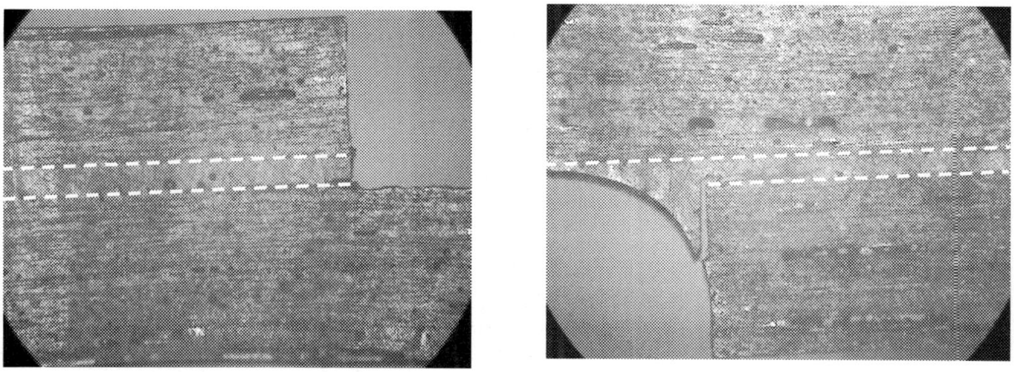

Figure 3. End of bonded overlap just before final failure, a) no fillet, b) with fillet
Dashed white lines indicate approximate location of adhesive.

TEST-PREDICTION CORRELATION

Figures 4 and 5 show examples of the comparison between test results and predictions, for SLS and DLS specimens respectively. Error bars indicate ± one standard deviation.

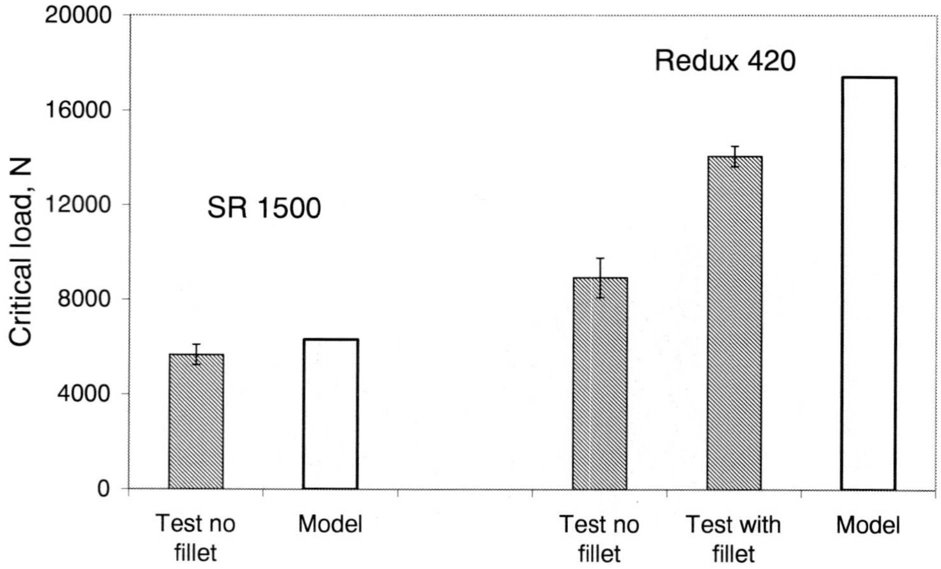

Figure 4. Comparison test-prediction, SLS, 20 mm width and overlap.

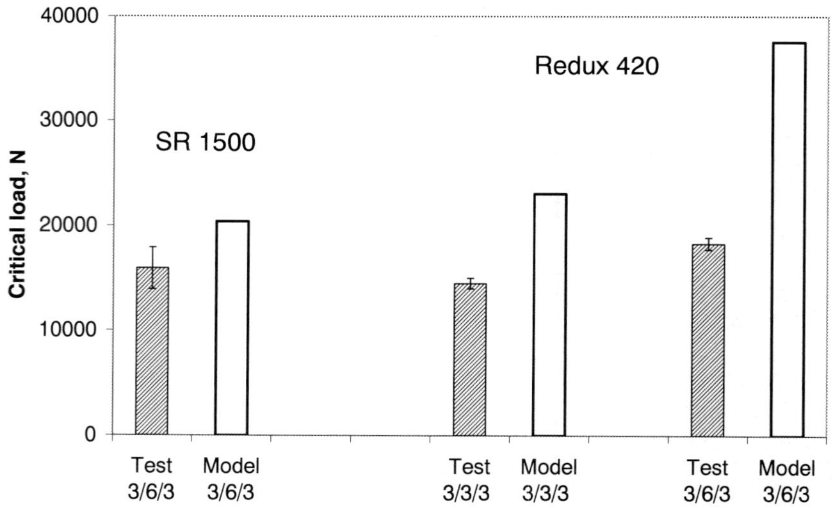

Figure 5. Comparison test-prediction DLS, different adherend thicknesses, all with 20mm width and overlap.

More details of these tests are given elsewhere [19]. The correlation is quite good for the more brittle (SR1500) adhesive. For the tougher adhesive, however, the model overestimates the critical load. One reason for this may be that first damage occurs in the composite, not in the adhesive layer. It may also be noted that an extensive damage zone develops in the lap shear specimens, which involves the initiation of shear microcracks ahead of the crack tip. Many of these are located at small voids in the adhesive layer. This is shown below, Figure 6, and is not accounted for in the measurement of the fracture envelope nor in the analysis of the lap shear specimens. A more complex model, based on damage mechanics and considering adherend damage, may prove more suitable to model assemblies involving tougher adhesives.

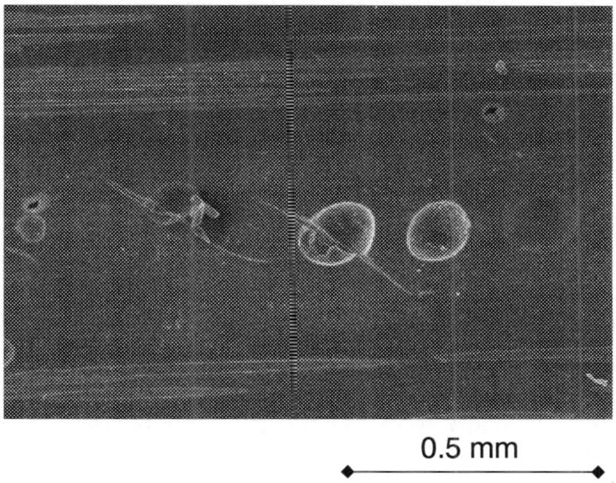

Figure 6. Shear microcracks in tough adhesive layer, DLS specimen unloaded before failure.

CONCLUSIONS

Fracture mechanics characterisation tests have been performed to determine the mixed mode fracture envelope of glass/epoxy composite bonded with two adhesives. Analysis of lap shear geometries has shown that for the relatively brittle adhesive reasonable first estimations of failure loads can be obtained. Damage development and accumulation have been observed before final fracture. This is more pronounced in the tougher adhesive and must be taken into account if better predictions are required.

REFERENCES

1) Brussat TR, Chiu ST and Mostovoy S, “Fracture Mechanics for Structural Adhesive Bonds, - Final Report”, AFML-TR-77-163, 1977.
2) Dattaguru B, Everett RA, Whitcomb J D, Johnson WS, “Geometrically Nonlinear Analysis of Adhesively Bonded Joints”, Journal of Engineering Materials and Technology, 106, 1984, 59.
3) Johnson WS and Mall S, “A Fracture Mechanics Approach for Designing Adhesively Bonded Joints”, ASTM-STP-876, 1985, 189.
4) Jangblad D, “Prediction of the failure loads of adhesive single lap joints using a fracture mechanical evaluation technique”, In Mechanical Behaviour of Adhesive Joints, Ed. Verchery G and Cardon EM, Pluralis (Paris) 1987, 293-306.
5) Curley AJ, Hadavinia H, Kinloch AJ, Taylor AC, “Predicting the service life of adhesively bonded joints”, Int. J. Fracture, 103, 2000, 41-69
6) Kendall K, “Crack propagation in lap-shear joints”, J. Phys. D:Appl. Phys., 8, 1975, 512.

7) Anderson GP, Bennett SJ, DeVries KL, "Analysis and Testing of Adhesive Bonds", Academic Press, NY, 1977.
8) Edde F and Verreman Y, "On the fracture parameters in a clamped cracked lap shear adhesive joint", Int. J. Adhesion and Adhesives, 12, No.1, 1992, 43
9) Hu GK and Francois D, "Non Linear Fracture Mechanics for Adhesive Lap Joints", J. Adhesion, 37, 1992, 261.
10) Lai Y-H, Rakestraw MD and Dillard DA, "The Cracked Lap Shear Specimen Revisted – A Closed Form Solution", Int. J. Solids Structures, 33, No. 12, 1996, 1725.
11) Papini M, Fernlund G & Spelt JK, "Effect of crack growth mechanism on the prediction of fracture load of adhesive joints", Comp. Sci. & Tech., 52, 1994, 561.
12) Fernlund G, Papini, M, McCammond D and Spelt J K, Fracture load predictions for adhesive joints, Composites Science and Technology, 51, 1994, 587.
13) Papini M, Fernlund G & Spelt JK, "The effect of geometry on the fracture of adhesive joints", Int. J. Adhesion & Adhesives, 14, 1, 1994, 5.
14) Ripling EJ, Mostovoy S, Corten HT, Fracture mechanics: a tool for evaluating structural adhesives, J. Adhesion, 3, 1971, 107-123.
15) BS7991 (2001) Determination of the mode I adhesive fracture energy G_{Ic} of structural adhesives using the double cantilever beam (DCB) and the tapered double cantilever beam (TDCB) specimens.
16) ASTM D6671-01 Standard Test Method for Mixed Mode I-Mode II Interlaminar Fracture Toughness of Unidirectional Fiber Reinforced Polymer Matrix Composites.
17) Ducept F, Gamby D, Davies P, A mixed mode failure criterion from tests on symmetric and asymmetric specimens Comp. Sci & Technology, 59, 1999, p609-619.
18) Ducept F, Davies P, Gamby D Mixed mode failure criteria for a glass/epoxy composite and an adhesively bonded composite/composite joint, Int. Journal of Adhesion & Adhesives, 20, 3, 2000 p233-244.
19) Davies P, Sargent JP, Fracture mechanics tests to characterise bonded glass/epoxy composites, Proc. 3rd ESIS TC4 conference, September 2002.
20) Adams RD, Comyn J and Wake WC, "Structural Adhesive Joints in Engineering", Chapman and Hall, 1997.
21) Hildebrand M. "Non-linear analysis and optimisation of adhesively bonded single lap joints between fibre-reinforced plastics and metals", Int. J. Adhesion and Adhesives, 14, No. 4, 1994, 261.

The Application of Fracture Mechanics to Polymers, Adhesives and Composites
D.R. Moore (Ed)

THE ROLE OF THE INTERPHASE AND ITS RESIDUAL STRESSES IN THE DETERMINATION OF FRACTURE TOUGHNESS

J. BOUCHET and A.A. ROCHE

INTRODUCTION

The application of organic layers has been well developed for many engineering structural components in aircraft, automotive and modern space industries in the last few decades. Usually, irrespective of the organic layer thickness, organic materials applied to various substrates were considered as a bi-layer system. It was well known that when epoxy prepolymers were applied onto metallic substrates and cured, internal stresses were developed within both the organic layers and the substrate. These stresses may reduce the work of adhesion and induce buckling [1], cracks in the coated materials or coating debonding [2-3]. In fact for coated systems, a third inner-layer called the interphase or interfacial region, having different chemical, physical and mechanical properties from those of the bulk coating, was created between the substrate and the remaining part of the organic material having bulk properties leading to a tri-layer system [4]. The polymer/substrate interphase was a region where several reactions take place that induce gradients of residual stresses and rearrangement of the structure. In the following, it was called "bulk coating" that part of the organic material which has bulk properties. Thus, the entire organic layer can be divided in two parts: the interphase and the bulk coating. Usually, when interfacial failure of the organic layer was observed, the work of adhesion was characterized by an appropriate yield criterion such as ultimate load (lap-shear, pull-off, and flexure tests [4-5]). Unfortunately, this kind of parameter was strongly dependent on the specimen geometry. Therefore, a criterion based on the fracture energy rather than strength was more suitable for describing the work of adhesion [6-7]. To study the crack (or failure) propagation, the linear elastic fracture mechanics theory provides an energy criterion: the critical strain energy release rate (G_c) which represents the sum of all energy losses incurred around the crack tip and which was, therefore, the energy required to increase the crack by unit length in a specimen of unit width [8-9]. Thus an interfacial failure was generally associated with mode-I (tensile) and/or mode-II (shear) fractures along the debonded interface. Hence several test specimen geometries such as the tapered double cantilever beam (TDCB) [10-11] have been designed to determine the fracture energy of adhesive joints [12]. The TDCB test, proposed by Mostovoy and Ripling [13] can be easily adapted to induce "adhesional" fractures in a pure cleavage or tensile-opening mode (mode I) which was the most critical mode for adhesively bonded structures [10-11]. The G_{Ic} from a TDCB test, using specimens with symmetric arms, can be calculated from the following equation:

$$(G_{Ic})_{\exp} = \frac{F_{\max}^2}{2b_m}\left(\frac{\partial C}{\partial a}\right) = \frac{F_{\max}^2}{2b_m}\frac{8M_{\exp}}{E_m b_m} \tag{1}$$

Where C was the compliance of the TDCB specimen, M_{exp} was the experimental geometrical parameter (the procedure for the determination of M_{exp} has been described previously [10]), b_m was the width of the metallic beam, E_m was the substrate Young's modulus, and F_{max} was the ultimate load.

The aim of the present work was to determine the critical strain energy release rate (G_c) from a three-point flexure test, taking into account both the residual stress profile within the tri-layer system (bulk coating/interphase/substrate) and the gradient of mechanical properties (Young's modulus) within the interphase region. Then the critical role of interphase properties can be emphasized. Similarly, other authors [14] have considered residual stresses with other geometries. The geometry of the specimens used with the three-point flexure test allows the residual stress state to be easily highlighted from the curvature of the specimens. To validate our model a comparison between the critical strain energy release rate obtained from a TDCB and three-point flexure tests before residual stresses were included will be made.

THEORETICAL DEVELOPMENT

Here, the interphase was considered as an inner layer of the coating/substrate system in which an initial field of residual stress, denoted (σ), was developed. Moreover, it was considered a gradient of the Young's modulus within the interphase as observed experimentally [15-17]. The theoretical development was carried out by assuming that all materials of tri-layer coated systems were elastic. Under the curing cycle, tri-layer systems were subjected to strains, which can be of either chemical or thermal origins. These were referred to as adhesional strains $(\varepsilon^{ad}(y))$ because they can depend on the various phenomena resulting from both the adhesion and the curing cycle of the coating. Intrinsic stresses were produced as a result of the mismatch between active sites of the metallic substrate and the organic network and/or during the formation of polymer network. It was quite evident that this kind of stress depends on the work of adhesion between the polymer and the substrate. Thermal stresses were developed during the cooling process and were the result of the thermal expansion mismatch between the metallic substrate, the interphase and the bulk polymer. In this work, total stresses, which were equal to the sum of intrinsic stresses and thermal stresses were considered. Our main objective was to identify these adhesional strains by using the radius of curvature (R_1) of the tri-layer system, and to find the resulting mechanical strains $(\varepsilon^{mech}(y))$ in order to determine the residual stress profile in the tri-layer system. Then, it was assumed that the total strain $(\varepsilon^{tot}(y))$ in the tri-layer system as represented in fig. 1, was:

$$\varepsilon^{tot}(y) = \varepsilon^{mech}(y) + \varepsilon^{ad}(y) = \frac{y - y_0}{R_1} \tag{2}$$

where: y_0 was the position where the total strain was equal to zero ($\varepsilon^{tot} = 0$), y was the coordinate distance and R_1 was the radius of curvature of the coated sample.
Considering the geometry and the size of the tri-layer systems studied (fig. 2), the beam theory can be used. Based on these assumptions, final uni-axial residual stresses (σ) in the x-direction of the tri-layer system (bulk coating/interphase/substrate) were given by:

$$\varepsilon^{mech}(y) = \frac{\sigma}{E(y)} = \frac{y - y_0}{R_1} - \varepsilon^{ad}(y) \tag{3}$$

where: $E(y)$ was Young's modulus at the y coordinate.

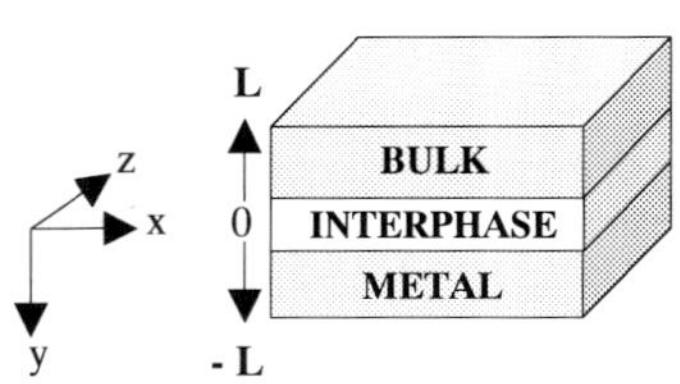

Fig. 1: The tri-layer system.

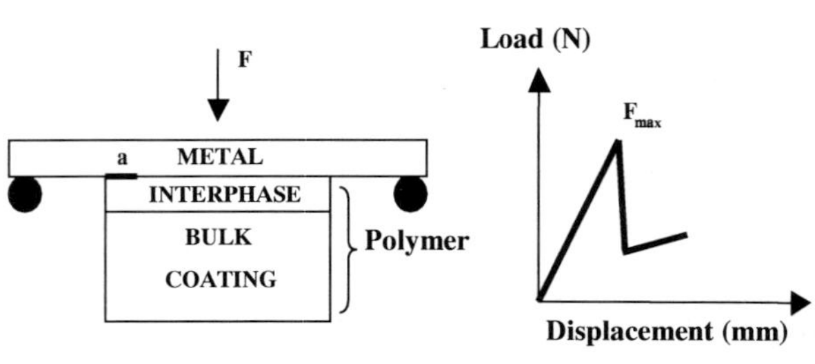

Fig. 2: The three-point flexure test (ISO 14679-1997).

To determine the distribution of residual stresses in the tri-layer system from equation (3) requires a knowledge of the radius of curvature (R_1) and the position of the zero deformation (y_0). Therefore, it was considered the equilibrium conditions for the force (N) and the moment (M) for any cross section area, S, of the coating/interphase/substrate system:

$$N = \int_S \sigma dS = 0 \quad and \quad M = \int_S y\sigma dS = 0 \tag{4}$$

For each layer, l, comprising the tri-layer system as represented in fig. 1, geometrical parameters and mechanical properties were known. To simplify writing it was adopted the following notation: the n-order moment of a function f(y) was denoted as:

$$\mu_n^f = \int_S f(y) y^n dS \quad and \quad y_{0f} = \frac{\mu_1^f}{\mu_0^f} \tag{5}$$

For the force (N), Eqs. (2) and (3), yield:

$$N = \int_S \sigma dS = \sum_l \int_{Sl} \sigma_l dS = \sum_l \int_{Sl} E_l (\frac{y - y_0}{R_1} - \varepsilon_l^{ad}) dS = \frac{\mu_1^E - y_0 \mu_0^E}{R_1} - \mu_0^{\varepsilon^{ad} E} = 0 \tag{6}$$

For the bending moment (M), the rearrangement of Eqs. (3) and (4), yields:

$$M = \int_S \sigma y dS = \sum_l \int_{Sl} y \sigma_l dS = \sum_l \int_{Sl} E_l y (\frac{y - y_0}{R_1} - \varepsilon^{ad}) dS = \frac{\mu_2^E - y_0 \mu_1^E}{R_1} - \mu_1^{\varepsilon^{ad} E} = 0 \tag{7}$$

Now, y_0 and R_1 were obtained as:

$$R_1 = -\frac{\mu_2^E \mu_0^E - (\mu_1^E)^2}{\mu_0^{\varepsilon^{ad} E} \mu_1^E - \mu_0^E \mu_1^{\varepsilon^{ad} E}} \quad and \quad y_0 = \frac{\mu_2^E - y_{0\varepsilon^{ad}E} \mu_1^E}{\mu_1^E - y_{0\varepsilon^{ad}E} \mu_0^E} \tag{8}$$

It was now possible to determine the adhesional strain evolution law by using the radius of curvature. To measure the work of adhesion of coatings to metals, Roche *et al.* [18] introduced a three-point flexure sample geometry (Fig. 2). This three point flexure test has already been extensively described elsewhere [19] and standardized (AFNOR T 30010 and T 76 143 and ISO 14679-1997). The formed block polymer acting as a mechanical stiffener can be considered as an interphase and a part of polymeric material having bulk properties. As a first approximation, the three-point flexure specimen can be analyzed by using the classical beam theory. It was assumed a crack length (a) at the interface between the stiffener (i.e. interphase + bulk coating) and the substrate and also it was considered that the total strain within the tri-layer system was expressed as:

$$\varepsilon_x^{tot}(x, y) = \varepsilon_x^{mech}(x, y) + \varepsilon^{adh}(y) = \frac{y - y_0(x)}{R_I(x)} \tag{9}$$

The radius of curvature and the position where the total strains were equal to zero were determined by writing the two equilibrium conditions for the force (N) and the moment (M) for any cross-section (area S) of the bulk coating/interphase/substrate system:

$$N(x) = \int_S \sigma(x, y) dS = 0 \text{ and } M(x) = \int_S y\sigma(x, y) dS \tag{10}$$

The rearrangement of equations (9) and (10) allows the radius of curvature and the position, where the total strains were equal to zero, to be expressed as a function of the bending moment ($M(x)$).

$$R_1(x) = \frac{\mu_2^E \mu_0^E - (\mu_1^E)^2}{-\mu_0^{\varepsilon^{ad} E} \mu_1^E + \mu_0^E (\mu_1^{\varepsilon^{ad} E} + M(x))} \quad and \quad y_0(x) = \frac{\mu_1^E (\mu_1^{\varepsilon^{ad} E} + M(x)) - \mu_2^E \mu_0^{\varepsilon^{ad} E}}{\mu_0^E (\mu_1^{\varepsilon^{ad} E} + M(x)) - \mu_1^E \mu_0^{\varepsilon^{ad} E}} \tag{11}$$

Moreover, experimental ultimate parameters F_{max} and d_{max} (i.e. the maximal load and maximal displacement respectively at the initiation of the fracture) were associated with a critical crack size inducing a significant variation of the sample stiffness. This point was generally associated with the initiation of the fracture because it corresponds to the initial size of the experimental crack, which can be observed before the propagation of the fracture. In fact, at this ultimate point, the fracture has already been initiated [20]. It was assumed that, at this point, the theory of linear fracture mechanics was applicable since this point corresponds to a crack length (a), tending to zero (a→0). When $a \to 0$, it was assumed that the beam theory assumptions were still valid. The failure in our system was always nearly instantaneous, so, Eq. (12) will be written at constant displacement (δ). Then, the critical energy release rate, in the case of the three-point flexure test, can be defined as:

$$G_c^{flexure} = -\frac{1}{b}\left[\frac{\partial}{\partial a}\left[\int_0^L \left(\int_{S(x)} \frac{1}{2}E(x,y)\left[\frac{y-y_0(x)}{R_1(x)} - \varepsilon^{ad}(y)\right]^2 dS\right)dx\right]\right]_\delta \quad (a \to 0) \qquad (12)$$

In previous work [21], using a finite element model for both aluminium and titanium substrates, the stress pattern for such tri-layer systems was calculated. They shown that mode I dominates mode II. If mode I was dominating the failure initiation it can be stated: $G_c^{flexure} = G_{Ic}^{flexure}$.

RESULTS AND DISCUSSION

In recent works [21-24], the formation of 600 μm thick interphase was reported when an epoxy (DGEBA, DER 332 from DOW) monomer was mixed with an isophorone diamine monomer (IPDA from FLUKA) as hardener and applied onto a 600 μm thick chemically etched titanium alloy (Ti6Al4V) and cured. When the same epoxy prepolymer was applied onto a 500 μm thick degreased or chemically etched aluminum alloy (5754) panel and cured, the interphase thicknesses were found to be 200 and 250 μm respectively. Reminding that to identify the adhesional strains, a knowledge of Young's modulus as a function of the coating thickness within the interphase was necessary, figure 3 represents the Young's modulus as a function of the coating thickness for both chemically etched titanium and degreased aluminium substrates as reported in [21].

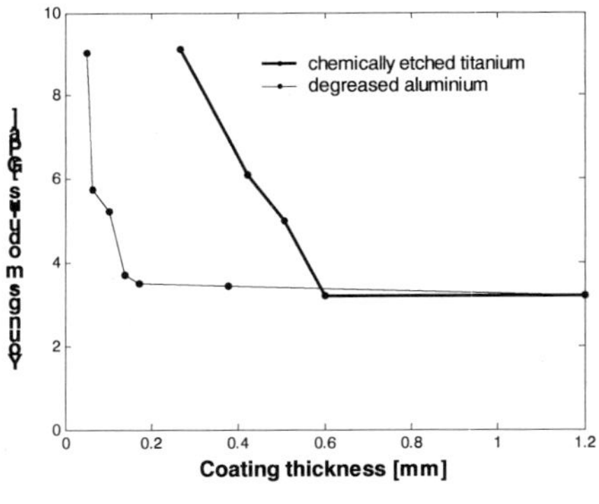

Fig. 3: Variation of the Young's modulus as a function of the coating thickness for both chemically etched titanium and degresed aluminium.

The data obtained from figure 3 suggests that Young's modulus was a piecewise linear function. Thus the interphase was divided into 3 linear regions *i* of *dhi* thickness irrespective of metallic substrates. At the interphase/metal interface, Young's modulus of the metallic substrates (i.e. 120 GPa for titanium and 70 GPa for aluminum) was taken and at the bulk coating/interphase interface, Young's modulus of the bulk epoxy (i.e. 3.2 GPa) was used. The adhesional strains within the interphase were considered linear and continuous. Adhesional strains were considered as constant within both metal and coating having bulk properties. Results obtained in terms of residual stresses were represented in figure 4. The maximum stresses within the tri-layer system were at the interphase/metal interface irrespective of the metal. In mechanical terms this means that the failure of such systems should take place at the interphase/metal interface where residual stresses were at their maximum. To illustrate this point, figure 5 represents an SEM observation of the failed surface for a chemically etched aluminum substrate [21]. The failure initiation and propagation area can be easily distinguished. No trace of polymer was observed at the initiation area in contrast to the failure propagation area. FTNIR analysis confirmed this observation irrespective of the surface treatment [21]. Thus, both SEM observation and FTNIR analysis confirm that the failure in our systems takes place at the interphase/metal interface.

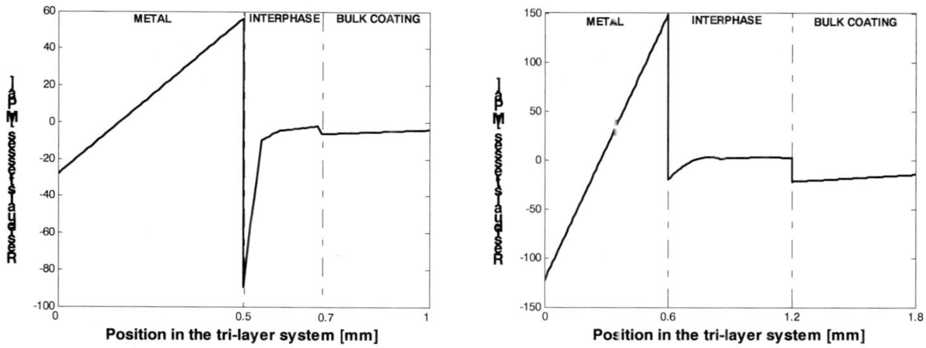

Degreased aluminium *Chemically etched titanium*

Fig. 4: Residual stress profiles in tri-layer systems for both degreased aluminium and chemically etched titanium.

Lastly, the ultimate load obtained for three different aluminium surface treatments was plotted against calculated residual stresses at the interphase/metal interface (Fig. 6). It can be seen that the work of adhesion determined using an ultimate parameter depends on residual stresses. Indeed, when residual stresses at the interphase/metal interface increase, the ultimate load decreases. From ultimate load values obtained (Table 1) and by using residual stress profiles (fig. 4) for both chemically etched titanium and degreased aluminum, the critical energy release rate ($G_{Ic}^{flexure}$) to initiate the failure (Table 1) was determined. If both the interphase and residual stresses were not considered, it was observed that the critical energy release rate values obtained by three point flexure ($G_{Ic}^{flexure}$) or tapered double cantilever beam (G_{Ic}^{TDCB}) tests, where an interfacial failure was always observed, were in good agreement. This confirms the validity of our assumptions. Without the residual stress consideration, $G_{Ic}^{flexure}$ values were equivalent for both bi-layer and tri-layer systems. It can be assumed that the gradient of Young's modulus within the interphase can also be neglected irrespective of the substrate nature. However, when residual stresses were considered, $G_{Ic}^{flexure}$ values obtained were higher than those calculated without residual stresses. Therefore, residual stresses act as a potential deformation energy in bonded or painted systems. Obviously as soon as this potential deformation energy was higher than the intrinsic adhesion energy of the coating, a spontaneous coating failure will occur.

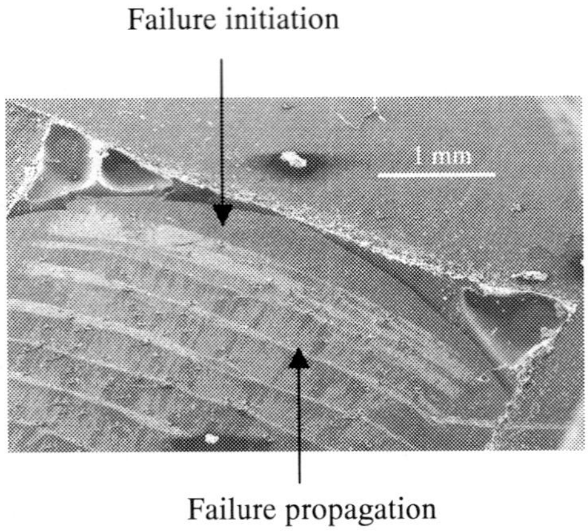

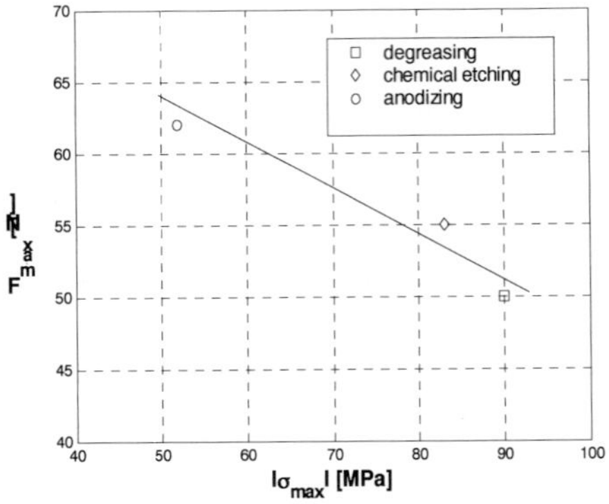

Fig. 5: SEM observation of the failed chemically etched substrate.

Fig. 6: Ultimate load (F_{max}) as a function of the maximal interphase/metal interface residual stress intensity (σ_{max}) for a coating thickness of 0.5 mm.

When both residual stresses and interphase properties were considered, $G_{Ic}^{flexure}$ values for bi-layer or tri-layer system were different in the case of titanium substrate, but about the same for aluminum substrate. This means that interphase thickness has to be considered when there is a thick interphase. Overall properties (thickness, Young's modulus gradient, residual stresses) of the interphase have to be considered as soon as the interphase thickness cannot be neglected when compared to the entire specimen thickness.

Table 1. Ultimate load, $G_{Ic}^{flexure}$ and G_{Ic}^{TDCB} values for both aluminum and titanium substrates.

Degreased aluminum			
Calculation assumptions	$G_{Ic}^{flexure}$ [J/m²]	G_{Ic}^{TDCB} [J/m²]	F_{max} [N]
Bi-layer system	41 ± 14%	41 ± 10%	20 ± 8%
Tri-layer system	41 ± 14%		
Bi-layer system + residual stresses	66 ± 9%		
Tri-layer system + residual stresses	66 ± 9%		
Chemically etched titanium			
Calculation assumptions	$G_{Ic}^{flexure}$ [J/m²]	G_{Ic}^{TDCB} [J/m²]	F_{max} [N]
Bi-layer system	358 ± 19%	224 ± 51%	25 ± 12%
Tri-layer system	359 ± 19%		
Bi-layer system + residual stresses	520 ± 15%		
Tri-layer system + residual stresses	468 ± 16%		

CONCLUSION

In this work, a tri-layer model (bulk coating/interphase/metal) was considered to study the interfacial failure in the case of the three-point flexure test taking into account the mechanical properties (residual stress profile and Young's modulus gradient) of the formed interphase. The work of adhesion was determined in terms of an energy parameter: the critical strain energy release rate noted $G_c^{flexure}$. For a bi-layer system (without interphase and without residual stresses), the critical energy release rate values obtained by three-point flexure or tapered double cantilever beam tests were in good agreement irrespective of the metallic substrate. This confirms the validity of our assumptions. The $G_{Ic}^{flexure}$ results obtained for aluminum and titanium substrates show that interphase mechanical properties have to be considered as soon as the interphase thickness cannot be neglected. Indeed, residual stresses act as a deformation potential energy within coated or bonded systems. Thus, it was possible to determine from a three-point flexure test an intrinsic parameter representing the work of adhesion.

REFERENCES

1. Scafidi, P. and Ignat, M. (1998) *J. Adhesion Sci. Technol.* **12**, 1219.
2. Orsini, H. and Schmit, F. (1993) *J. Adhesion* **43**, 55.
3. Thouless, M.D. and Jensen, H.M. (1994) *J. Adhesion Sci. Technol.* **8**, 579.
4. Mittal, K.L. (1978). In: *Adhesion Measurement of Thin Films, Thick Films and Bulk Coatings*, pp. 5-17, K.L. Mittal (Ed.), ASTM, Philadelphia.
5. Roche, A.A., Dole, P. and Bouzziri, M. (1994) *J. Adhesion Sci. Technol.* **8**, 587.
6. Schön, J., Nyman, T., Blom, A., and Ansell, H. (2000) *Composites Science Technol.* **60**, 173.
7. Todo, M., Jar, P.Y.B. and Takahashi, K. (2000) *Composites Science Technol.* **60**, 263.
8. Williams, J.G., (1980). *Fracture Mechanics of Polymers*, Ellis Horwood, Chichester.
9. Scott, J.M. and Phillips, D.C., (1975) *J. Mater. Sci.* **10**, 551.
10. Meiller, M., Roche, A.A. and Sautereau, H. (1999) *J. Adhesion Sci. Technol.* **13**, 773.
11. Kinloch, A.J. (1987). *Adhesion and Adhesives*, Chapman and Hall, London.
12. Trantina, G.G. (1972) *J. Composite Mater.* **6**, 192.
13. Mostovoy, S. and Ripling, E.J. (1966) *J. Appl. Polym. Sci.* **10**, 1351.
14. Nairn, J.A. (2000) *Int. J. Adhesion Adhesives* **20**, 59.
15. Bouchet, J., Roche, A.A. and Hamelin, P. (1999) *Thin Solid Films* **355-356**, 270.
16. Bentadjine, S., Roche, A.A. and Bouchet, J. (2001). In: *Adhesion Aspects of Thin Films*, Volume 1, pp. 239-260, Mittal, K.L. (Ed.), VSP, Utrecht.
17. Bentadjine, S., Roche, A.A., Massardier, V. and Petiaud, R. (2001) *Polymer* **42**, 6271.
18. Roche, A.A., Behme, A.K. and Solomon, J.S. (1982) *Int. J. Adhesion Adhesives* **2**, 249.
19. Roche, A.A., Gaillard, F., Romand, M. and VonFahnestock, M. (1987) *J. Adhesion Sci. Technol.* **1**, 145.
20. Rives, B. (1999). Ph.D. thesis, Université Paul Sabatier de Toulouse III, France.
21. Bouchet, J., Roche, A.A. and Jacquelin, E. (2001) *J. Adhesion Sci. Technol.* **15**, 321.
22. Bouchet, J., Roche, A.A., Jacquelin, E and Scherer, G.W. (2001). In: *Adhesion Aspects of Thin Films*, Volume 1, pp. 217-237, Mittal, K.L. (Ed.), VSP, Utrecht.
23. Bouchet, J. and Roche, A.A. (2002). *J. Adhesion* **78**, 799.
24. Bouchet, J., Roche, A.A. and Jacquelin, E. (2001) *J. Adhesion Sci. Technol.* **15**, 345.

The Application of Fracture Mechanics to Polymers, Adhesives and Composites
D.R. Moore (Ed)

SCRATCH RESISTANCE AND EMBRITTLEMENT OF COATED POLYMERIC SYSTEMS

F. RAMSTEINER, T. JAWOREK, M. WEBER, St. FORSTER

INTRODUCTION

To improve the surface properties of polymeric items they are very often covered with a hard coating. However, this procedure can entail embrittlement of the system. This premature damaging is likely to be caused by the flaws initiated in the brittle coating and the crack propagating into the otherwise unnotched tough polymeric substrate. The co-operative sequence of events depends on the brittleness of the coating, the toughness of the polymeric substrate, and the adhesion between the coating and the substrate. The toughness of the coating on the substrate can be characterised in a tensile test by the critical strain at which the first cracks are observed under the microscope. The higher the critical strain the tougher is the coating. It is not necessary to test the isolated film because the mechanics of thin films are different from their constrained situation on a bulk polymer and the latter feature is important in applications. The scratch resistance was measured [1] by the needle test in which needles are moved with different angles and normal forces on the surface. The damage of the coating was judged in the REM and graded from 0 to 5. The results were plotted in a two dimensional failure map given by the angle of the needle and the normal force as described for pure polymers by Briscoe et al.[2]. In accordance with the main intention of this work which is to understand the embrittlement of the coated system, only that aspect is considered in this paper.

MEASUREMENTS AND DISCUSSION

Figure 1 shows the force deflection record in three point bending Charpy impact test at -20°C of an uncoated and of two coated compression moulded specimens of an ASA/PC blend. The two different thickness coatings are on the tensile side of the specimens. The uncoated specimen is very tough and it does not break in the test. However, a 116 μm thin coating film reduces the elongation drastically so that the specimen breaks near the peak force. If the coating is 158μm thick, the specimen breaks at a still lower force and deflection. The coated specimens appear to embrittle with increasing thickness of the coating. The initial rise of the force is however independent of the coating, including the impact oscillations since the thickness of the coating is too thin to influence the deformation curve at the beginning.

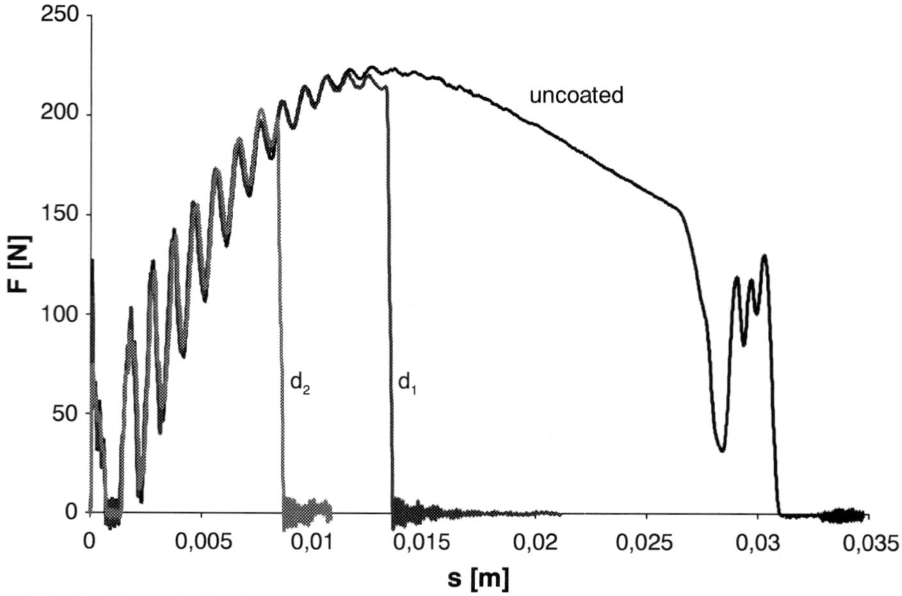

Fig. 1. Force deflection record of an uncoated and of two coated ASA/PC specimens (-20°C) in the three point bending impact test (thickness of the coating: d_1= 116 μm, d_2 = 158μm)

The very often observed generally decreasing toughness with increasing thickness of the brittle coating is demonstrated in Fig.2 also for a ABS material [3], coated with its brittle SAN matrix.

If the flaw in the coating is regarded as an initial notch for the specimen, the reduction in toughness is not surprising. According to linear elastic fracture mechanics a decrease in failure strength is expected with increasing notch length according to

$$P_{max} \approx K_{IC}/f\sqrt{a} \qquad (1)$$

with P_{max} = peak force, K_{IC} = critical stress intensity factor, a = initial crack length, and f is a geometrical factor which depends on the geometry of the specimen [4].

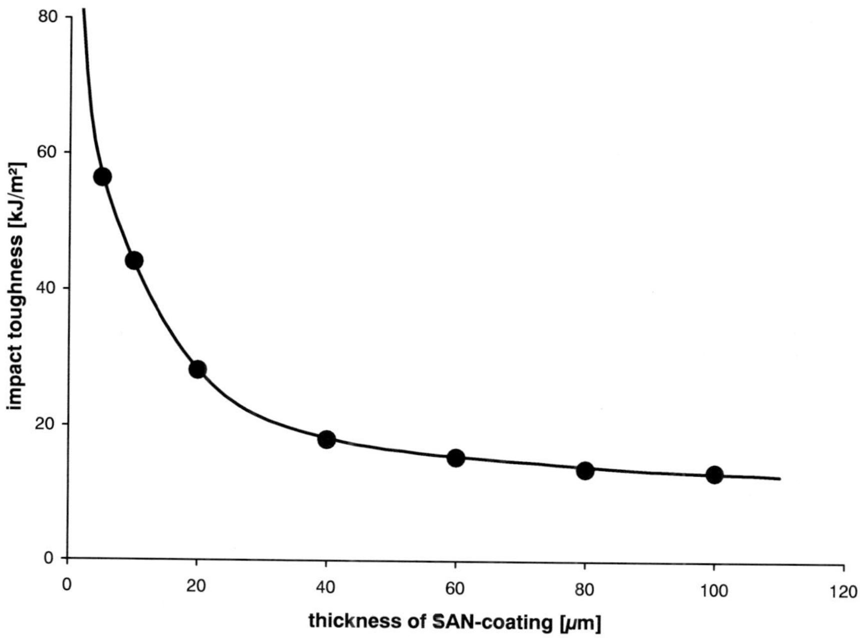

Fig. 2. Impact toughness of ABS as a function of the thickness of a brittle SAN film calendared on the surface of the ABS-specimen [3].

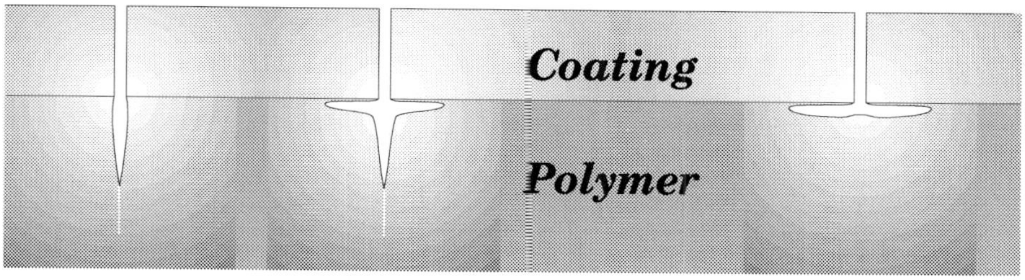

a b c

Fig .3. Models for crack transition from a coating to the substrate

In the coated systems the initial short flaw in the coating cannot act as a normal notch in a polymer because the crack propagation into the substrate is modified by the interface [5] as is shown schematically in Fig. 3. Only if the coating is fixed very strongly to the substrate can the initial flaw propagate into the polymeric substrate. If however the coating delaminates during stressing the tip of the flaw blunts and can no longer work as a stress intensifier. The initial stress initiating flaw does not act as a notch (Fig. 3c) in the direction of the original growth. In the case of medium bonding between coating and substrate, both features can happen, depending on the local situation near the interface. Additionally, it must be kept in mind that in each of these cases the plastic deformation of the polymer near a crack tip at the interface is restricted by the hard coating in contrast to a free surface as long as the yield zone at the crack tip is influenced by the coating.

The consequences of the different adhesion between coating and substrate is shown in Figs. 4 and 5. In both cases the same coating was used, but the polymeric substrate was changed. The three point bending impact measurements were performed at room temperature. In the case of the ASA/PC blend (Fig. 4), the mechanically deeply notched specimens show an increase of the peak force with the roughly

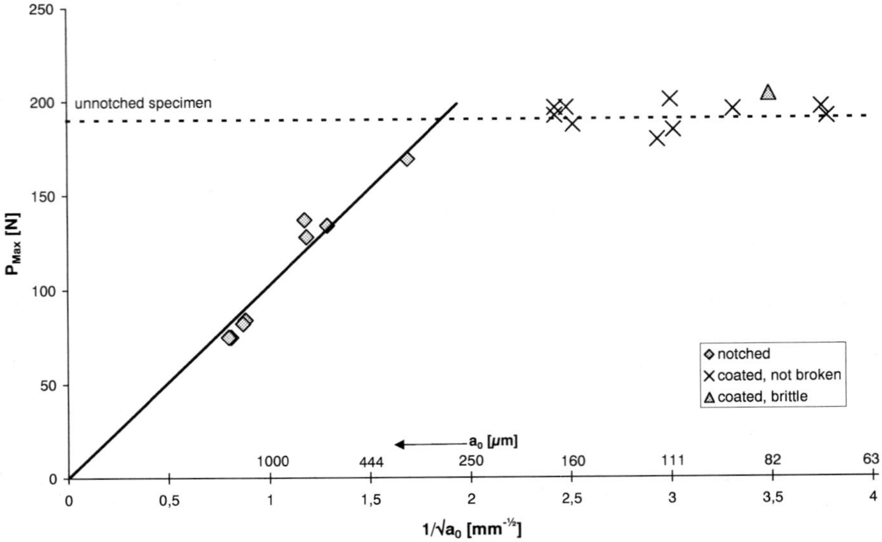

Fig. 4. Maximum force in three point bending test of coated ASA/PC as a function of the inverse root of the coating thickness and the notch length. The dotted horizontal line is given by the yield strength of the unnotched specimen.

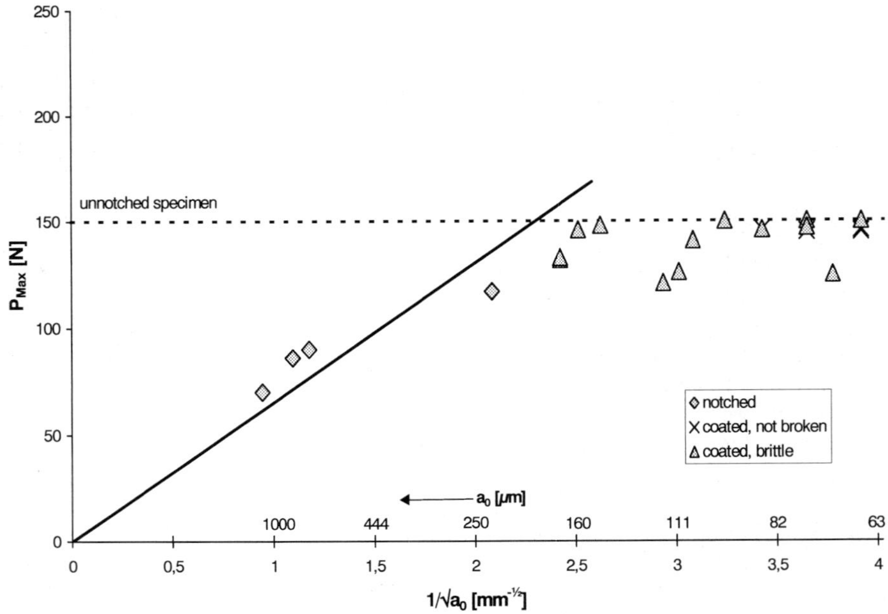

Fig. 5. Maximum force in three point bending test of coated triblend as a function of the inverse root of the coating thickness and the notch length. The dotted horizontal line is given by the yield strength of the unnotched specimen.

inverse root of the initial notch length. According to eq.(1) the dependence of the root is strongly valid only if the factor f is included. In the coated systems with the assumption, that the initial notch length of the system is given by the thickness of the coating, the yield strength of the unnotched uncoated specimen is attained. The coating has obviously no influence on the yield strength of the pure substrate. In this case, the coating has delaminated as is shown in Fig. 6 with light microscopy images on the top (left hand side) and the side (right hand side) of the specimen. Clearly many flaws were generated in the coating but were deflected in the interface. These debonded parts shine brightly due to light scattering and no cracks propagated into the substrate.

The situation is different with the same coating on a triblend, which is a compound of PA and ABS (Fig. 5). Again the maximum bending force of the mechanical deeply notched specimens is likely to increase roughly with the inverse of the root of the initial notch length. However, the coated systems with different thickness of the coating films reach the yield strength of the unnotched specimen only with coating films thinner than 100μm. Otherwise the coating reduces the yield or peak stress. In this system the coating has not delaminated as is shown in the microscope images of the side (right hand side) and on top of the coating (left hand side)

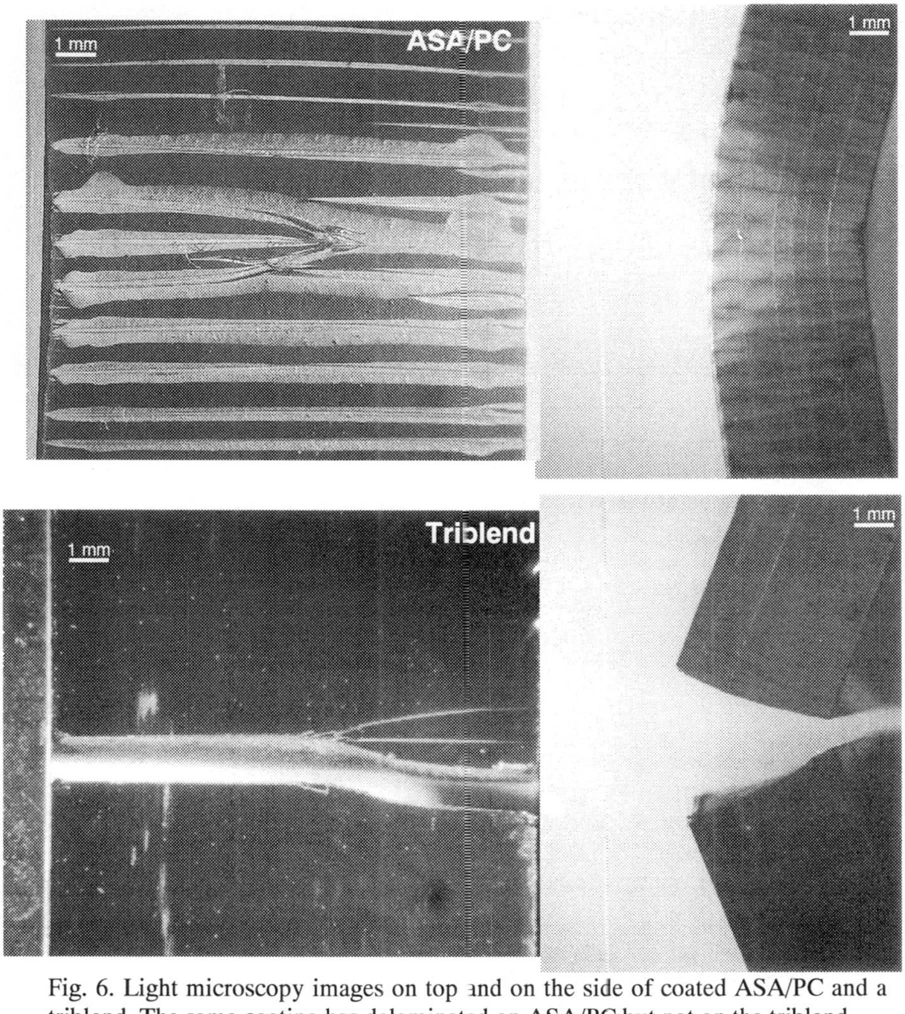

Fig. 6. Light microscopy images on top and on the side of coated ASA/PC and a triblend. The same coating has delaminated on ASA/PC but not on the triblend

of this specimen in Fig. 6. There the coating with its flaws has initiated premature fracture. The thickness of the coating is not expected to equal the length of initial notch. The reason is that the substrate is tough and linear fracture mechanics is not applicable. Additionally, the tip of the small crack is not blunted as in the bulk at least as far as its yield zone is restricted by the stiff coating. Anyway, fracture mechanics offers a better understanding of the embrittlement.

High scratch resistance is likely to be preserved when a thin viscoelastic interface is inserted between the substrate and the hard coating so that the tip of the flaw is blunted [1]. Another step in blunting the propagating crack tip is to orient the molecules in the substrate perpendicular to the crack propagation. In Fig. 7. crack propagation is shown for an injection moulded and an annealed substrate of an ASA/PC blend in the three point bending test. Whereas in the annealed specimen (Fig.7b) the crack propagates from the coating straight into the substrate, in the injection moulded specimen (Fig.7a) the crack deviates due to the oriented molecules beneath the coating perpendicular to the original direction and the system is tough.

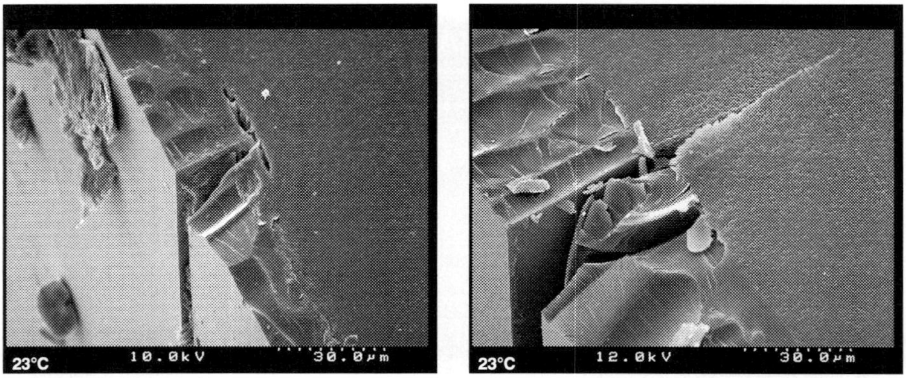

a b

Fig. 7. REM images of sheets cut parallel to the tensile direction on the bent side of the specimen showing crack propagation into the substrate near the interface in an injection moulded (a) and an annealed specimen (b) of ASA/PC.

ACKNOWLEDGEMENT

The contributions by Mr. Schill (REM results) are gratefully acknowledged.

REFERENCES

1. Ramsteiner, F., Forster, St., Polymer Testing, accepted for publication
2. Briscoe. B.J., Pellilo, E., Ragazzi, F., Sinha, S.K. (1998) Polymer **39**, 2161
3. Wiebusch, W. (1983) internal report

4. Moore, D.R., Pavan, A., Williams, J.G. (2001) Fracture Mechanics Testing Methods for Polymers Adhesives and Composites, ESIS Publication 28 Amsterdam, London, New York, Oxford, Paris, Shannon, Tokyo
ISBN 00804368897, Elsevier
5. Cook, L., Gordon, J.E. (1964) Proc. Roy. Soc., London, **A282**, 508

The Application of Fracture Mechanics to Polymers, Adhesives and Composites
D.R. Moore (Ed)

EVALUATION OF THE INTERFACIAL STRENGTH IN POLYMERIC SYSTEMS

F. RAMSTEINER

INTRODUCTION

The strength of interfaces in polymeric systems are of growing importance in the development of new products with new properties, since more and more materials are styled by blending of standard polymers to exploit synergistic effects. Most of them are incompatible and thus internal interfaces are formed in the mixtures. In many cases, the strength of the interfaces in those composites determines their mechanical behaviour. Additionally, failure at internal interfaces in homopolymers must be regarded, e.g. after welding or in semi-crystalline polymers at the spherulite borders. Fracture mechanics is a very suitable method to study this crack growth along internal interfaces starting at the tip of an initial notch in the interface. Thus fracture mechanics was used in the past by many authors.
This paper shows some of our applications of fracture mechanics for

- internal failure in homopolymers,
- welding experiments with polystyrene,
- characterisation of the influence of powder on the welding process,
- measurement of the interfacial strength between different polymers, and
- elaboration of the efficiency of compatibilizers in improving the interfacial strength.

EXPERIMENTAL METHODS

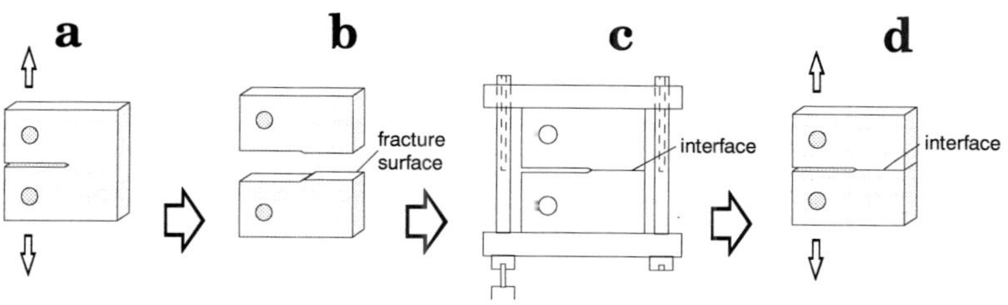

Fig.1 Schematic procedure to measure interfacial strength with CT-specimens
CT-specimen, b) semi specimens, c) annealing , d) tensile test

Brown [1] e.g. in his studies of the strength of interfaces in polymeric materials pressed a razor blade into the interface of a double cantilever beam for crack initiation. Jud, Kausch and Williams [2] introduced the classical CT-specimen into their pioneering work on crack healing in PMMA . CT-specimens were used also in this paper to study the strength of interfaces in polymeric systems. The procedure is shown in fig.1 [3]. CT specimens (fig.1a) were broken or two halves of the CT specimens (fig.1b) were prepared. These two parts were brought in contact at higher temperatures using some normal forces to weld them together (fig.1c). During this welding process the notch is protected from annealing by an Al foil or restored after welding by milling a notch along the interface. The strength of the interface ahead of the initial notch is measured by the residual strength after welding and/or by following the crack growth in the interface (fig.1d). In linear fracture mechanics failure is characterised by the critical stress intensity factor [4]. However, if the CT-specimens have the same geometry with the exception of their thickness, failure is characterised already by the failure strength.

EXPERIMENTAL RESULTS

Failure at the spherulite boundaries

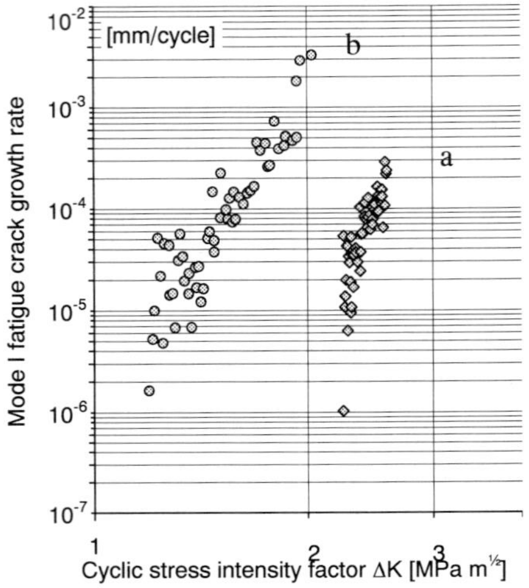

Fig.2 Velocity of crack propagation in PP as a function of the amplitude of the stress intensity factor in fatigue (10 Hz)
a) after fast cooling,
b) after annealing for 2hrs at 130°C

a

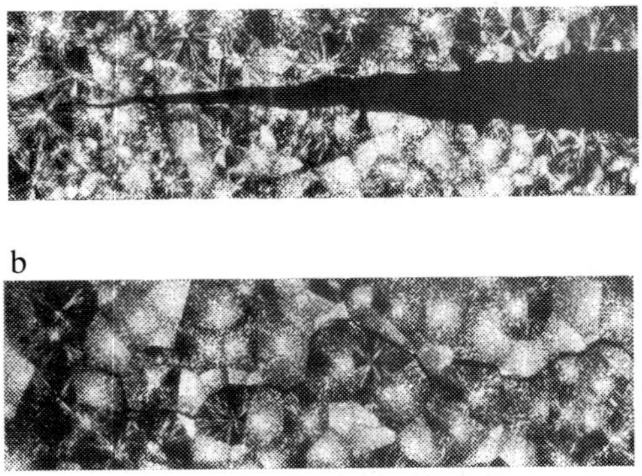

b

Fig.3 Transmission images in polarised light of crack propagation in PP
a) after fast cooling,
b) after annealing for 2 hrs at 130°C

Fig.2 shows for a fatigue test the plot of the crack growth per cycle da/dN in PP as a function of the amplitude of the stress intensity factor. One of the specimens was fast cooled from the melt and the other one annealed for 2 hrs at 130°C. CT specimens were used according to the ESIS protocol for fatigue [4] at 10 Hz. Crack propagation was recorded by measuring the change of the compliance with increasing crack length. The crack growth is faster in the annealed than in the fast cooled material. The reason can be seen in fig.3 where the side views of thin sheets normal to the tensile direction are reproduced. In the fast cooled specimen the crack propagates ahead of the initial notch more or less along the direction perpendicular to the tensile direction. However in the annealed specimen the crack grows along the spherulite borders . Obviously the number of the stress transferring tie molecules between the spherulites has been reduced by increased crystallisation during annealing. This reduced bonding strength is the reason for the occasional observed embrittlement of semi-crystalline polymers after annealing.

Healing of cracks

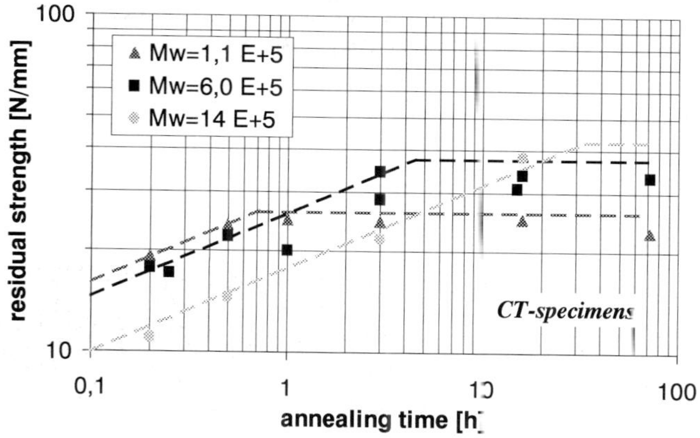

Fig.4 Residual strength (N/mm) of PS specimens after healing of fracture cracks at 120°C as a function of annealing time

In injection moulding very often two streams of polymer melts flow together e.g. after an obstacle in the cavity mould forming a contact plane, the so called welding line. This internal contact plane must be at least partly strengthened by inter-diffusion of molecules across this internal interface to avoid weakness. This welding process depends on the temperature and residence time above the melting temperature. The consequence of interdiffusion for strength can be observed by healing of cracks as proposed by Jud et al [2]. Thus CT specimens are broken according to fig.1, brought together and annealed for various times at a given temperature. The residual stress intensity factor after annealing indicates the increasing strengthening by interdiffusion during annealing. If the geometry of the various specimens, especially the notch length are kept constant, then already the fracture stress characterises the progress in the healing process. In fig. 4 the residual stress is plotted as a function of annealing time at 120°C for PS with different molecular weights. The upper horizontal dashed lines represent the failure strength of the original CT specimens The data reflects the decreasing interdiffusion rate with increasing molecular weight of polystyrene. The increase in strength at the short annealing times is given in fracture mechanics by the well known $t^{1/4}$ relation [2,5,6]. The final strength increases with the molecular weight. Thus low molecular weight favours fast increase of strength at short times, high molecular weight products need long annealing times for the high strength. This method in combination with optical or spectroscopic methods allows the study of the interdiffusion processes [5,6] in more detail.

Welding

The effect of coating agents on granules or prefoamed beads is to avoid welding between them during their transport at lower temperatures and to reduce electrical charging [3]. However the coating agent should not hinder welding of the beads at higher temperatures during the foaming process. The details of this welding and especially the effect of the coating agent can be studied

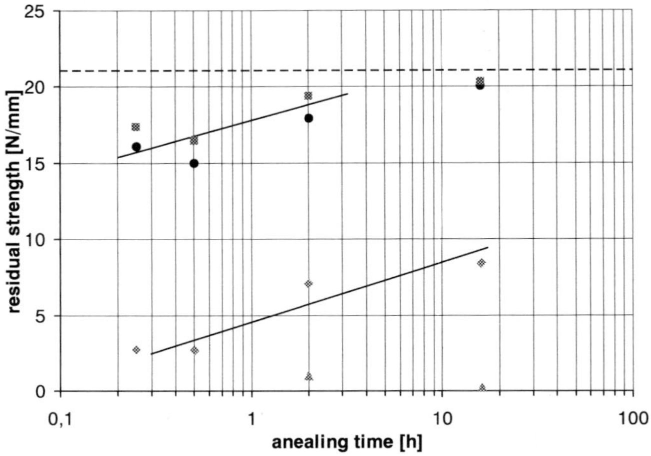

Fig.5 Residual strength (N/mm) of PS specimens after annealin; at 120°C of the fracture surfaces with different coating agents
● Pure PS ◆ Zinc stearate ■ Glycerine monostearate
• Silicidacid

analogue to the healing procedure. Thus it is started with CT-specimens, they are broken, the fracture surfaces are covered with a given amount of the coating agent, the two parts of the specimen are brought in contact, and then each specimen is annealed for various times at a given temperature. The residual strength after this test reflects the extent of welding. If the geometry of the specimen is kept constant then the strength represents the critical stress intensity factor. In fig. 5 the residual strengths of PS-specimens with different coating agents is plotted as a function of the annealing time at 120°C [3]. For comparison the specimen without a coating was also tested. If coating agents are used with a softening temperature (melting temperature, glass transition temperature) higher than the welding temperature, the agent restricts the interdiffusion, the small particles act like distance holders, welding is hindered. Silicidacid is typical of this kind of an agent. If the agent melts below the welding temperature, the molecules can interdiffuse across the interface, the strength at the interface increases. An example for this situation is zinc stearate. In the case where low molecular agents are able to plastisize the material, an accelerated welding is likely to occur as shown in fig.5 for Glycerine monostearate.

Interfaces in blends

To study the interfacial strength between different polymers, two parts of the pure particular polymers are welded together within a CT-specimen. The residual strength is measured after annealing at 160 for 2 hrs. The results are visualised in fig.6. Within the circle for each homopolymer the original strength of the CT specimens and the interfacial strength after fracture and annealing are given for comparison [3]. In the small circles between the polymers the interfacial strengths between this pair after annealing is given. In those cases where the interfacial strength is weak (PS/SAN), because the individual polymers are highly incompatible, the crack propagates along the interface at low forces, welding is not possible. If, however, the compatibility is sufficiently high (SAN/PMMA), interfacial strength is increased. In combination with other methods more details of the interdiffusion zone can be elaborated [7]. In those cases with mechanically different polymers the crack can deviate from the interface [3,6,8]. The deformation mode is no longer pure Mode I, shear Mode II is

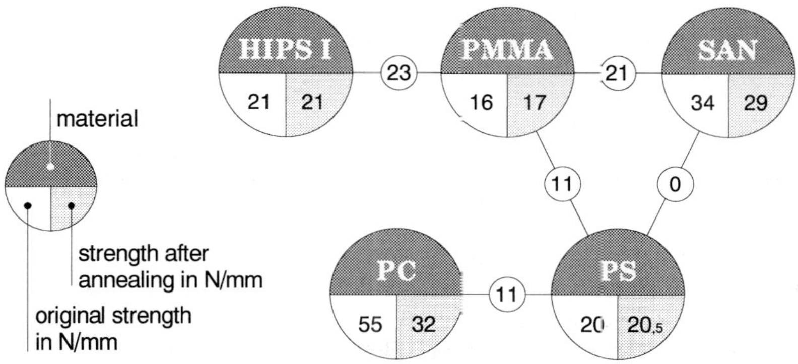

Fig.6 Interfacial strength (N/mm) between different polymers after annealing for 2hrs at 160°C

superposed [8,9]. By using asymmetric CT-specimens, the crack can be forced into the interface or even into the opposite material [3]. By using highly crystalline polymers the welding temperature must be higher than the melting point, otherwise interdiffusion is restricted by the crystals and compatibility cannot be assessed from this kind of measurement [10]. Oligomers seem to segregate preferentially in the interface having a detrimental effect on interface toughness [11]. The method is also suitable to study compatibility on changing the ratio of the monomers in copolymers [11].

Effect of compatibilizer

To improve the strength between an immiscible pair of polymers, e.g. in blends, compatibilizers are used. Very often, they consists of diblock-copolymers, each of the ends is compatible with one the polymers. By bridging in this way the interface strength is increased (fig.7). An experimental example is given in fig.8 with a 0.5mm thin foil of a PS/hydrogenated PBu compatibilizer between PE and HIPS. In this case it must be decided by optical methods, whether fracture occurs at a still weak interface between the compatibilizer and one of the polymers or inside the compatibilizer in the case of good compatibility.

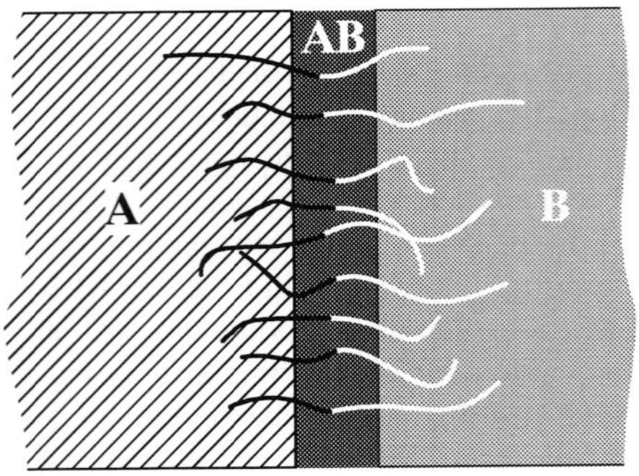

Fig.7 Schematic bonding of polymer A and B by AB- blockcopolymer-chains in the interface

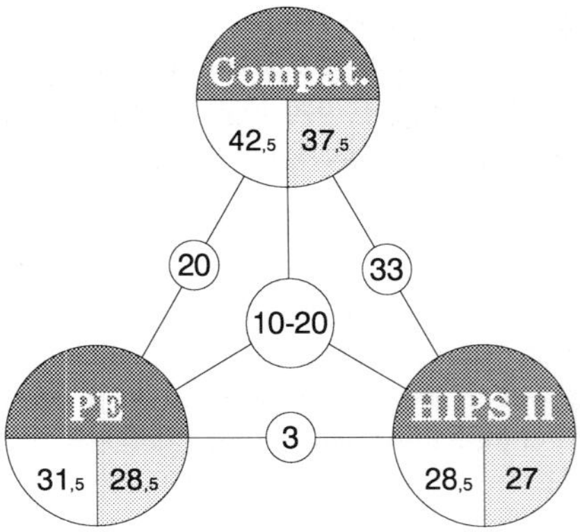

Fig.8 Influence of compatibilizer on the interfacial strength (N/mm) between HIPS and PE [3]

Within the circles for each of the materials in fig.8 the left hand side is the tensile strength of the CT-specimen of the pure polymer, the right hand side gives the strength of the broken and afterwards for 2 hrs at 120°C annealed CT specimen. The values in the circles between the materials are respectively the strengths of the interfaces. As expected, there is nearly no adhesion between the incompatible PE and HIPS. The situation is improved between PE and HIPS with the compatibilizer. If the compatibilizers is added in the interface between PE and HIPS, then the interfacial strength increases to 10 –20 N/mm, and depends to some extent on the position of the initial notch in relation to the thickness of the compatibilizer layer. Details of the influence of length and concentration of compatibilizers were studied with a molecular layer between the polymers by Creton and Kramer [12] in this way.

ACKNOWLEDGEMENT

The contributions by Mr. Forster and Mr. Schuster are gratefully acknowledged.

REFERENCES

1. Brown, H.R. (1990) J. Mater. Science **25**, 2791
2. Jud, K., Kausch, H.H., Williams, J.G. (1981) J. Mater. Science **16**, 204
3. Ramsteiner, F. (1996), Polymer Testing **15**, 573
4. ESIS Protocol in Moore D.R., Pavan, A., Williams, J.G. (2001) Fracture Mechanics Testing methods for Polymers Adhesives and Composites, Elsevier Science
5. Kausch, H.H., (1991) Macrom. Chem., Macromol. Symp. **41**, 1
6. Wool, R.P.(1995) Polymer Interfaces, Hanser Publishers, Munich Vienna New York
7. Jabbari, E., Peppas, N.A.(1994) J. Mater. Sci. **29**, 3969
8. Comninou. M. J. (1977) Appl. Mechanics **44**, 631

9. Erdgogan, F. (1965) J. Appl. Mechanics **87**, 403
10. Xue, Y.Q., Tervoort, T.A., Lemstra, P.J. (1998) Macromolecules **31**, 3075
11. Janarthan,V., Stein, R.S., Garrett, P.D. (1994) Macromolecules **27**, 4855
11. Creton, C., Kramer E.J., Hui, C-Y. Brown. H.R. (1992) Macromolecules, **25**, 3075

The Application of Fracture Mechanics to Polymers, Adhesives and Composites
D.R. Moore (Ed)

THE USE OF FRACTURE MECHANICS TECHNIQUES TO PREDICT THE SERVICE LIFE OF ADHESIVE JOINTS

A.J. KINLOCH and A.C. TAYLOR

INTRODUCTION

The use of adhesive bonding in industry has greatly increased in recent years. However, the use of adhesives in truly structural applications is still often limited. This is mainly due to a lack of confidence in the performance of adhesive joints, especially as the mechanical performance of the joints may deteriorate with exposure to moisture and/or cyclic-fatigue loading. Thus, the ability to quantitatively describe this reduction in performance and to predict the lifetime of bonded joints would be a powerful tool, enabling manufacturers to make wider and more efficient use of adhesive bonding. This can be achieved using a linear-elastic fracture-mechanics (LEFM) approach.

This fracture-mechanics approach to lifetime prediction described here assumes that the joint is cracked initially, i.e. that there is no initiation phase before crack propagation. As the joint life is dominated by the propagation of cracks, rather than their initiation, the predictions provide a 'lower bound' value for the joint lifetime.

PREDICTION METHODOLOGY

The first step in the prediction of the lifetime of the bonded joints is to perform fracture mechanics tests in cyclic fatigue on the adhesive/substrate system of interest. An expression is used to describe these data, i.e. the relationship between the rate of crack growth, da/dN, and the maximum strain-energy release-rate, G_{max}. Secondly, the variation of G_{max} with crack length in the joint or component of interest is theoretically modelled, using either an analytical or a finite-element approach. Finally, these data are combined and the resulting expression is integrated to predict the long-term fatigue life of the joint or component.

MODELLING THE FRACTURE MECHANICS DATA

Introduction

The cyclic-fatigue data can be obtained using any standard LEFM test geometry, e.g. tapered double-cantilever beam (TDCB), double-cantilever beam (DCB) or compact tension (CT) adhesive-joint specimens. These tests enable the measurement of the rate of fatigue crack growth, da/dN, versus the maximum strain-energy release-rate, G_{max}, applied during the fatigue cycle. The fracture-mechanics tests are conducted under the same test conditions as the joints whose service-life is to be predicted. It is also important to ensure that the LEFM fatigue test specimens exhibit the same failure path (i.e. cohesively through the adhesive layer or interfacially between the adhesive and substrate) as the joint or component of interest [1].

Modelling the Fatigue Data

The LEFM data obtained relate the rate of fatigue crack growth, da/dN, to the maximum strain-energy release-rate, G_{max}, applied during a fatigue cycle, see Figure 1 for example. This relationship is of a sigmoidal form, with a lower-bound at the fatigue threshold, G_{th}, where the crack growth rate is negligible ('Region I' in Figure 1) and an upper-bound occurs at the adhesive fracture energy, G_c, measured at a constant displacement-rate ('Region III' in Figure 1). This relationship can be described using a modified form of the Paris Law [2, 3]:

$$\frac{da}{dN} = DG_{\max}^{n}\left(\frac{1-\left(\frac{G_{th}}{G_{\max}}\right)^{n_1}}{1-\left(\frac{G_{\max}}{G_c}\right)^{n_2}}\right) \tag{1}$$

where G_{th} and G_c are the values of the fatigue threshold and constant displacement-rate adhesive fracture energy respectively, and the values of the fatigue coefficients D, n, n_1 and n_2 are calculated from the experimental data. Initially, experimental data from the linear 'Region II', see Figure 1, are fitted with a linear slope of gradient n, and intercept D. The values of n_1 and n_2 represent the acuity of the change in gradient, as may be seen from Figure 1, due to the transition in the fatigue data from the linear 'Region II' to the threshold 'Region I'; and also due to the transition in the data from the linear 'Region II' to the near fast-fracture 'Region III'. The values for n_1 and n_2 are constrained in their range of possible values, as suggested by Taylor [4], such that $0.1 < n_1, n_2 \leq 10$. (If the values of n_1 and n_2 are not constrained in the range that they can take, then the slope of the 'Region II' relationship deviates from that already calculated.) Further, the value of n_2 is generally chosen to be equal to n_1, since it is not possible to obtain data from the near fast-fracture region, as the rate of crack growth is too large to be measured accurately.

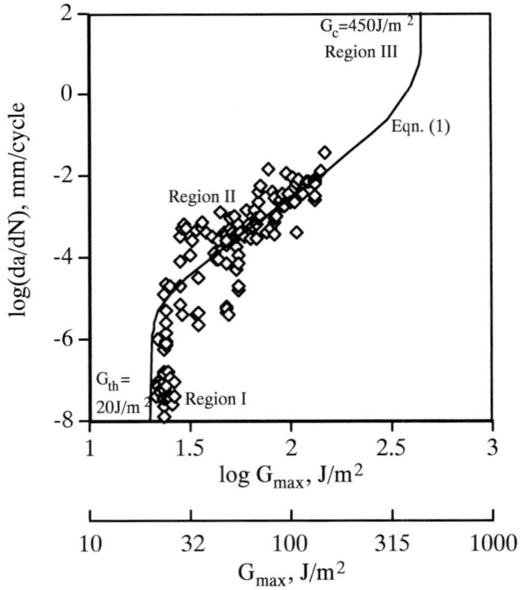

Fig. 1: Logarithmic crack growth rate per cycle, da/dN, versus logarithmic, and linear, G_{max} for cyclic-fatigue fracture-mechanics tests. The relationship from Equation (1) is shown.

SERVICE LIFE PREDICTION

Introduction

The second stage of the lifetime prediction methodology is to describe the variation in the maximum strain-energy release-rate in a fatigue cycle, G_{max}, with the crack length and the applied load in the joint of interest. This can be done using an analytical or a finite-element approach. These data are then combined with the fracture-mechanics data described using Equation (1), and the resulting expression is integrated to give the theoretical service life. This process will now be illustrated using two common joint geometries.

The Single-Lap Joint - Analytical Approach

Various models are available to describe the variation of the maximum applied strain-energy release-rate with crack length for the single-lap joint, e.g. [5, 6]. For example, Kinloch and Osiyemi [3] derived the expression:

$$G_{max} = \frac{12}{E_s h^3}\left(\frac{T_{max}(h+t_a)}{2}\right)^2\left(1+\left(\frac{12T_{max}(1-\upsilon^2)}{E_s h^3}\right)^{1/2}(c-a)\right)^{-2} \qquad (2)$$

where T_{max} is the maximum load per unit width in a fatigue cycle, E_s is the substrate modulus, h is the substrate thickness, t_a is the adhesive layer thickness, υ is the adhesive Poisson's ratio, c is half the overlap length and a is the crack length.

The integration limits are the initial flaw size, a_o, and the final crack length, a_f. The initial flaw size may be deduced from the Griffith equation:

$$a_o = \frac{E_a G_c}{\pi \sigma_a^2} \qquad (3)$$

where E_a, G_c and σ_a are the adhesive modulus, fracture energy and tensile strength respectively. Note that the choice of the value of a_o does not have a significant effect on the predicted lifetimes [7]. The final crack length, a_f, where rapid crack growth and failure occurs, can be obtained from Equation (2), when G_{max} reaches the adhesive fracture energy, G_c. Note that as cracks grow from both ends of the overlap, $a_f \leq c$, where c is half the overlap length.

Lap-joint specimens were manufactured according to the method described in ASTM D1002 [8], using 2 mm thick mild-steel substrates. Fatigue tests were conducted at a range of maximum loads, using a load ratio, T_{min}/T_{max}, of 0.5, with a sinusoidal loading waveform at a frequency of 5 Hz. Maximum nominal stresses from 30 to 90 % of the average initial value, obtained from the constant displacement-rate tests were applied. The number of cycles, N_f, to failure were recorded. The tests were conducted with the joints immersed in distilled water at 28±1°C.

The lifetimes predicted using the analytical method and the experimental results for the single-lap joints are shown in Figure 2. The agreement between the analytical method and the experimental results is only moderate. The fatigue life is underestimated, and the predicted threshold of approximately 20 % of the nominal failure stress is lower than the 30 % observed experimentally. Thus, the prediction is a conservative one. This analytical method of failure prediction has been shown to fit the experimental data very well for some adhesive/substrate combinations, but poorer matches have also been found [1]. (Although, invariably the threshold values of the load per unit width, below which no significant fatigue crack growth occurs in the single-lap joint, are predicted with a relatively high degree of accuracy.)

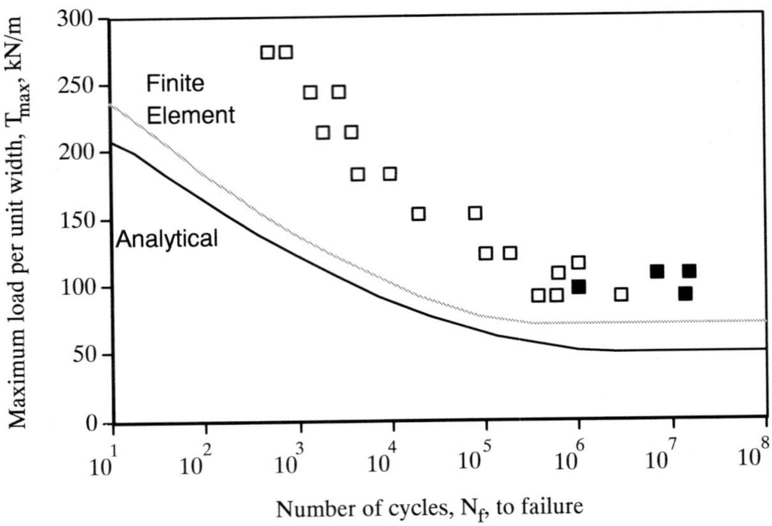

Fig. 2: The number, N_f, of cycles to failure for the single-lap joints as a function of the maximum load, T_{max}, per unit width applied in a fatigue cycle. The points represent the experimental data, the open symbols indicating joint failure and the filled symbols indicating that the joint did not fail. The lines are the predicted lifetimes using the finite-element and analytical models, using a_o = 85 μm.

The Single-Lap Joint - Finite-Element Approach

The finite-element model of the single-lap joint was considered as consisting of two isotropic, homogeneous and linear-elastic materials joined together along a common interface, and containing two interfacial cracks. These crack locations were the most critical locations, and where failure was observed experimentally [4]. It should be noted that the analytical model described above cannot distinguish between crack growth cohesively through the adhesive layer and crack growth along the adhesive/substrate interface. The J-integral method was used to calculate values of G_{max} as a function of the crack length, a, in the joint and the applied load, T_{max}, per unit width [7].

As G_{max} is a function of both the crack length, a, and the applied load, T_{max}, per unit width, the function to describe G_{max} was formulated in two parts. Firstly, values of G_{max} were plotted against the values of T^2_{max} for each crack length. Secondly, the gradient of each line, β, (where $\beta = G_{max} / T^2_{max}$) was then plotted against the crack length. The relationship between the slope, β and the crack length may be described using an exponential equation [9]. This model can be combined with the fracture-mechanics data, and the predicted number of cycles, N_f, to failure can be deduced by integrating Equation (1) between the limits a_o and a_f.

The predictions using the finite-element approach to model the variation of G_{max} with crack length lie closer to the experimental data than those using the analytical method, as shown in Figure 2. The predictions are good, and are conservative as discussed above.

The Bonded Component

The prediction methodology can also be used for more complex joint geometries. An example component was designed using an end-loaded 250 mm long top-hat' section made from 2 mm thick mild-steel bonded to a baseplate, see [7] for a full description. A reinforcing plate was bonded under the 'top-hat' section to prevent plastic deformation. The adhesively-bonded flange-width, c, was 15 mm. The bonded component was tested in fatigue using a constant maximum displacement of 1.2 mm, a frequency of 5 Hz, and a displacement ratio, u_{min}/u_{max}, of 0.5. The tests were undertaken in a 'dry' environment of 23±1°C and approximately 55% relative humidity. The resulting fatigue crack, which propagated through the adhesive layer of each flange, was monitored optically using a pair of travelling microscopes. The rate of crack growth per cycle, da/dN, in each bonded flange of the component was calculated from the measured crack length data.

A three-dimensional finite-element model of the bonded component was developed. The steel and adhesive were assumed to undergo elastic deformation only, as no plastic deformation was observed experimentally. Cracks were placed at the interface between the adhesive and the baseplate, corresponding to the locus of failure observed experimentally. This locus of failure, at the interface, was also observed for the fracture-mechanics fatigue tests. From the finite-element model, the theoretical relationship between the strain-energy release-rate, G_{max}, and crack length can be obtained. The relationship between da/dN and G_{max} was obtained from the fracture-mechanics fatigue data, as discussed above, as the locus of failure was the same in both cases. The predicted rate of crack growth per cycle, for a given crack length, was calculated using the modified Paris law, Equation (1), and the value of G_{max} from the finite-element model. The experimental results and the predictions are shown in Figure 3. The agreement between the predicted values and the experimental data is very good.

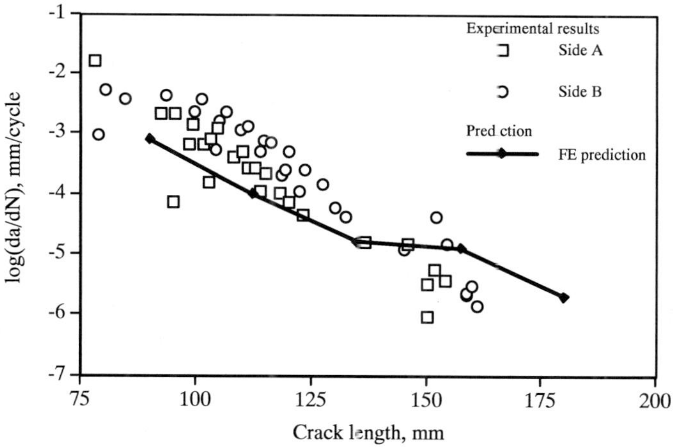

Fig. 3: Logarithmic rate of crack growth per cycle, da/dN, versus the length, a, of the propagating cyclic-fatigue crack for the bonded 'top-hat' component. The open points represent the experimental data, whilst the solid line is the predicted crack growth rate.

CONCLUSIONS

A fracture-mechanics approach has been used to predict the lifetime of bonded joints and components. The fracture-mechanics fatigue data are modelled using a modified Paris law relationship. An analytical or finite-element model is used to evaluate the variation of G_{max} with crack length. Substitution of the expression for G_{max} and integration allows the number of cycles to failure to be predicted. The method was illustrated by predicting the lifetimes of single-lap joints using an analytical and a finite-element approach, and of a bonded 'top-hat' beam component. The method gave good predictions of the fatigue life in all cases.

ACKNOWLEDGEMENTS

Dr. Ambrose Taylor would like to thank the Royal Academy of Engineering for a Post-Doctoral Research Fellowship. The authors wish to thank the EPSRC for financial support through an Engineering Platform Grant.

REFERENCES

1. Hadavinia, H., Kinloch, A.J., Little, M.S.G. and Taylor, A.C. *Int. J. Adhesion Adhesives* (To be published).
2. Martin, R.H. and Murri, G.B. *ASTM STP* (1990) **1059**, 251.
3. Kinloch, A.J. and Osiyemi, S.O. *J. Adhesion* (1993) **43**, 79.
4. Taylor, A.C. (1997) PhD Thesis, Imperial College London.
5. Hu, G. *Composites Eng.* (1995) **5**, 1043.
6. Papini, M., Fernlund, G. and Spelt, J.K. *Int. J. Adhesion Adhesives* (1994) **14**, 5.
7. Curley, A.J., Hadavinia, H., Kinloch, A.J. and Taylor, A.C. *Int. J. Frac.* (2000) **103**, 41.
8. American Society for Testing and Materials, *ASTM Standard* (1983) **D1002**.
9. Curley, A.J., Jethwa, J.K., Kinloch, A.J. and Taylor, A.C. *J. Adhesion* (1998) **66**, 39.

The Application of Fracture Mechanics to Polymers, Adhesives and Composites
D.R. Moore (Ed)

RESIDUAL STRESS EFFECTS IN FRACTURE OF COMPOSITES AND ADHESIVES

JOHN A. NAIRN

INTRODUCTION

Composites and structures, such as adhesive joints, are comprised of multiple phases having different thermal expansion coefficients. Because manufacturing methods typically include elevated temperature steps, when the final structure is cooled to use temperature, residual stresses occur. Residual stresses are not necessarily a problem; in some situations they can be beneficial. In other situations, however, residual stresses have an adverse effect on failure properties. Residual stresses may cause a composite to fail sooner than expected, to have greater susceptibility to solvents, or to have reduced durability due to accelerated aging or fatigue damage mechanisms. Thus an important question in all design and analysis with composite materials and adhesive joints is how do residual stresses affect failure properties?

This chapter considers fracture mechanics analysis of composites and adhesive joints with residual thermal stresses. The methods here are restricted to structures with linear thermoelastic phases subjected to a uniform change in temperature. Application of fracture mechanics requires evaluation of the energy release rate, G, for propagation of cracks in the presence of residual stresses. In many important problems, the effect of residual stresses on G can be evaluated without the need for any thermoelasticity analysis for residual stresses. This general result makes it possible to easily include residual stress effects in models that previously incorrectly ignored them. The general theory for fracture of structures with residual stresses is given in the theory section. The results section gives some examples of residual stress effects on failure in composites and adhesive joints.

THEORY OF RESIDUAL STRESS EFFECTS

Consider an arbitrary structure subjected to a uniform temperature change of ΔT and to any mixed traction and displacement boundary conditions as illustrated in Fig. 1. This structure may contain cracks within phases or cracks spanning phases. The goal of any fracture mechanics analysis is to derive the energy release rate for crack growth, which can be calculated from global thermoelastic potential energy using

$$G = -\frac{d\Pi}{dA} = \frac{d(W-U)}{dA} \tag{1}$$

where Π is the thermoelastic potential energy, W is external work, U is thermoelastic internal energy, and A is crack surface area. By partitioning total stresses into mechanical and residual stresses (σ^m and σ^r) and making use of virtual work methods and the divergence theorem, G can be expressed in a useful and general form as (see Refs. [1] and [2] for details):

$$G = G_{mech} + \frac{V\Delta T}{2}\left(2\frac{d\langle\sigma^m \cdot \alpha\rangle}{dA} + \frac{d\langle\sigma^r \cdot \alpha\rangle}{dA}\right) \tag{2}$$

where V is total volume and G_{mech} is the energy release rate in the absence of residual stresses:

$$G_{mech} = \frac{d}{dA}\left(\frac{1}{2}\int_{S_T} T^0 \cdot u^m dS - \frac{1}{2}\int_{S_u} T^m \cdot u^0 dS\right) \tag{3}$$

The angle brackets indicate volume-averaged quantities; here they are average mechanical or residual stresses weighted by the phase-dependent thermal expansion tensor $\boldsymbol{\alpha}$. $\boldsymbol{T}$ and $\boldsymbol{u}$ in Eq. (3) are surface tractions and displacements; superscript m indicates the mechanical component of the boundary terms. These results assume traction free crack surfaces and prefect interfaces. The extension of traction loaded cracks and imperfect interfaces is given in Ref. [2].

The first term in Eq. (2) is G in the absence of residual stresses; thus the other terms are an exact expression of the residual stress effect in composite fracture. An important special case of Eq. (2) is to consider mode I fracture. During pure mode I crack growth (and similarly during pure mode II or mode III crack growth) the mode I energy release rate must be proportional to mode I stress intensity factor squared, K_I^2. Furthermore, for linear-elastic materials in which all applied tractions and displacements are scaled by a factor P and the temperature difference ΔT is scaled by a factor T^*, K_I must scale by a linear combination of P and T^*. In other words, the mode I energy release rate must have the form

$$G_I \propto K_I^2 = \left(c_1 P + c_2 T^*\right)^2 \tag{4}$$

where c_1 and c_2 are constants that depend on the specific problem. Comparing Eq. (4) to Eq. (2) and realizing that σ^m is proportional to P while σ^r is proportional to T^*, it is possible to eliminate one term and simplify G_I to [1,2]

$$G_I = G_{mech}\left(1 + \frac{V\Delta T}{2G_{mech}}\frac{d\langle\sigma^m \cdot \alpha\rangle}{dA}\right)^2 \tag{5}$$

Notice that G_I depends only on mechanical stresses. Thus once the mode I fracture problem is solved ignoring residual stresses, the energy release rate in the presence of residual stresses can be determined exactly without any need for thermoelasticity analysis of the structure.

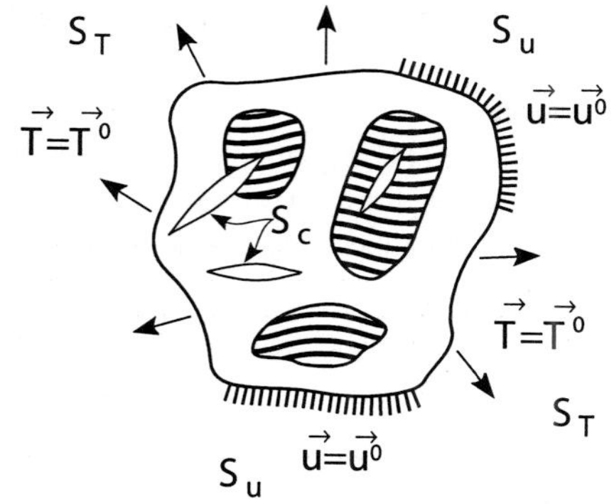

Fig. 1: An arbitrary multiphase composite subjected to tractions $\mathbf{T}^0$ on surface S_T and displacement boundary conditions $\mathbf{u}^0$ on surface S_u and containing crack surfaces (S_c).

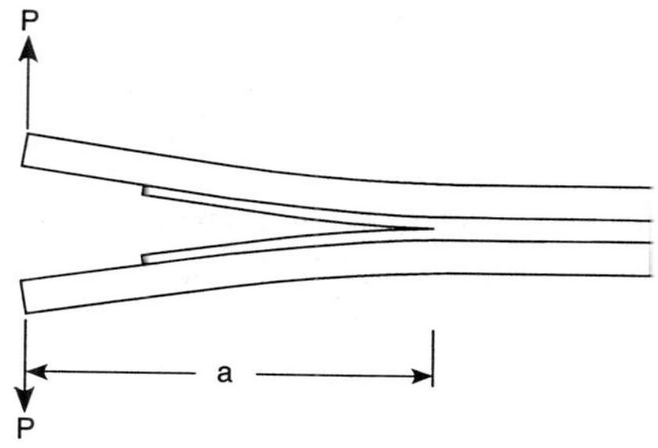

Fig. 2: A DCB specimen used to measure mode I toughness of adhesive joints. P is the applied load and a is crack length.

The general result in Eq. (5) has many applications for including residual stress effects in the analysis of fracture of composites and structures. It is analogous to the well-known Levin [3] analysis that derived a method to exactly calculate effective thermal expansion coefficient of a composite from knowledge of mechanical stresses alone. The Levin analysis used virtual work methods and did not need any thermoelasticity results. Similarly, Eq. (5) was derived with virtual work methods an only needs a mechanical stress analysis. Both analyses are limited to linear-elastic materials; Eq. (5) is further limited to pure mode fracture (pure mode I, II, or III), but many important failure problems are pure mode fracture problems.

RESIDUAL STRESS EFFECT RESULTS

Figure 2 shows a double cantilever beam specimen (DCB) used to measure the fracture toughness of adhesive joints [4]. Assuming the crack runs down the middle of the adhesive as pure mode I fracture, the energy release rate in the presence of residual stresses can be evaluated using Eq. (5) and corrected beam theory to be [5]

$$G_I = (C_m\,(a + 1.15\Delta_W)P + C_r\,\Delta T)^2 \tag{6}$$

where the mechanical and residual constants are given by

$$C_m = \sqrt{\frac{C_K^*}{B}} \quad \text{and} \quad C_r = \frac{\alpha_K^*}{\sqrt{C_K^* B}} \tag{7}$$

Here C_K^* and α_K^* are the simple beam theory compliance and thermal curvature coefficient of one arm of specimen [5], and B is width of the arms. The mechanical term includes an *effective* crack length ($a + 1.15\Delta_W$) as discussed elsewhere for corrected beam theory of DCB specimens [5,6]; the coefficient 1.15 on the correction term Δ_W [6] was determined numerically by finite element analysis of numerous DCB specimens [5]. The thermal term does not require correction because it is accurate with simple beam theory alone [5]. Comparison to finite element analysis shows that Eq. (6) is accurate within 1% for all adhesive DCB geometries [5]. It is thus easy to correct adhesive tests for residual stress effects. Notice that the residual stress effect in G_I is proportional to α_K^*, which means the residual stress effect is caused by a tendency of the arms of the specimen to curve due to residual stresses.

Many fracture studies on adhesives have ignored residual stresses [4]. Equation (6) can be used to calculate the errors of such an approach [5]. Assume that some particular adhesive has a *true* fracture toughness of G_{Ic}. If an adhesive DCB specimen with this adhesive is tested, it will fail when $G_I = G_{Ic}$, which, by Eq. (6), happens when the load reaches

$$P = \frac{\sqrt{G_{Ic}} - C_r\Delta T}{C_m(a + 1.15\Delta_W)} \tag{8}$$

If this observed failure load is then used in an analysis that ignores residual stresses (*i.e.*, Eq. (6) with ΔT=0), the calculated or apparent toughness is

$$G_{Ic}^{app} = \left(\sqrt{G_{Ic}} - C_r\Delta T\right)^2 \tag{9}$$

The difference between G_{Ic} and G_{Ic}^{app} is the error caused by ignoring residual stresses. In typical polymer adhesives between metal adherends, $C_r\ \Delta T$ is negative which causes G_{Ic}^{app} to be *higher* that the actual toughness. In other words, the presence of residual stresses causes the adhesive to appear tougher than it does without residual stresses. Some sample calculations of errors due to ignoring residual stresses are plotted in Fig. 3 as the percentage error in G_{Ic}^{app} as a function of modulus ratio between adherend and adhesive, R, for various adherend to adhesive thickness ratios, λ [5]. The errors are large for low R and λ and decrease as either R or λ increases. The dashed vertical line shows a typical R value for aluminum-epoxy specimens. The aluminum-epoxy errors exceed 1% even with a very thin adhesive (λ=64) and exceed 40% for a thick adhesive (λ=2).

A similar analysis [5] can be done for laminate double cantilever beam specimens used to measure delamination toughness. The residual stress effect in DCB specimens can be eliminated by constructing doubly-symmetric double cantilever beam specimens in which each arm of the specimen is a symmetric laminate and thus has $\alpha_\kappa^* = 0$. If the arms are not symmetric laminates, however, the residual stress effect can be very large. For example, the errors in G_{Ic}^{app} for a variety of laminates were calculated to range from -55% to +76% [5]. In other words, the errors caused by ignoring residual stresses can be large and can cause the G_{Ic}^{app} result to be either significantly too high or significantly too low.

Additional examples of residual stress effects in fracture of composite or structures include matrix microcracking in laminates [7,8], cracking of paints or coatings [9], and interfacial crack growth such as in the microbond or pull-out specimens [10-12]. In matrix microcracking of cross-ply laminates, the residual stress effect induces tensile stresses in the 90° plies that promote microcracking. The general methods above can include residual stresses in the energy release rate for microcracking; proper analysis of microcracking experiments requires inclusion of the residual stress term. A master plot method is available which make it possible to determine both toughness and residual stress effects from experimental cracking experiments [7,8]. In other words, the value of ΔT in the analysis does not have to be measured or be assumed to be due to linear thermoelastic phases; it can be determined from fracture experiments as an effective ΔT for residual stresses. Cracking of coatings is analogous to

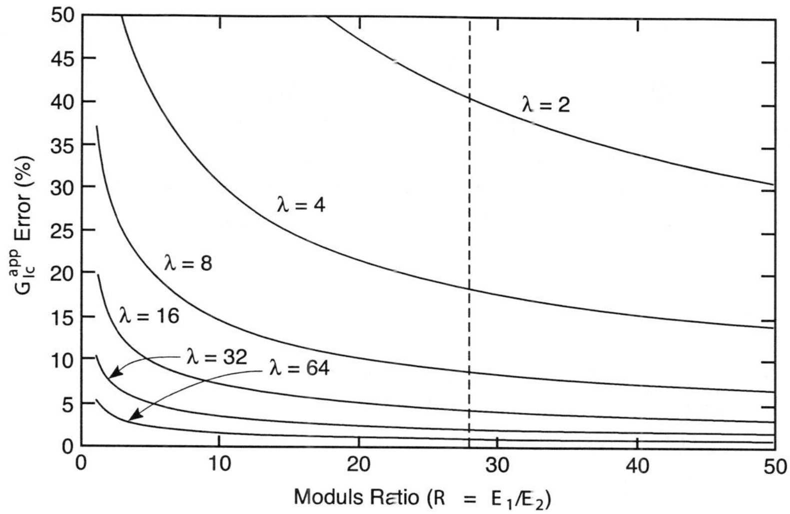

Fig. 3: The percentage error in the apparent toughness G_{Ic}^{app} in adhesive DCB specimens as a function of adherend to adhesive modulus ratio, *R*, for various values of adherend to adhesive thickness, λ. The calculations assumed G_{Ic}=200 J/m^2, $\Delta\alpha$ =-40 X10^{-6} K^{-1}, and ΔT = -100°C. The vertical dashed line corresponds to *R*=28, which is a typical value for aluminum-epoxy specimens.

microcracking of laminates [9]. An additional source of residual stresses in coatings might arise from chemical shrinkage as the coating cures or dries. Such residual stresses can be included in a thermal stress analysis by replacing thermal strain terms like $\Delta\alpha\Delta T$ by an effective residual strain term due to both chemical and thermal residual stresses [9].

In the microbond specimen [13], a droplet of matrix is deposited on a fiber and an interfacial crack is induced by pulling the fiber while restraining the matrix. In this specimen, all loading is on the top of the droplet while the bottom of the droplet is stress free. A global energy analysis shows that the bulk of the energy released due to crack growth is due to the release of residual stress energy as the droplet and matrix become debonded [10]. Although many interpretations of microbond results have ignored residual stresses, such models have serious errors. By including residual stresses and interfacial friction effects, the microbond specimen can become a fracture mechanics test for determining interfacial mode II fracture toughness [10,14]. For example, it can be used to assess the role of physical aging and residual stress relaxation on interfacial properties [11]. Similar fracture mechanics methods can be used to include residual stress effects in the analysis of single-fiber, pull-out experiments [10.12].

RECOMMENDATIONS

Residual stresses are always present in composites and structures such as adhesive joints. Proper analysis of fracture experiments with such materials must therefore always include residual stress effects. In some geometries, such as tapered DCB specimens with very large R and λ, residual stresses can be demonstrated to be small and can be ignored [15]. In many important specimens, however, residual stresses are non-negligible and can even be the dominant effect causing fracture. The methods above give general tools for including residual stresses. These tools assume linear thermoelastic phases with temperature-independent properties. The general equations can be extended to more general phases and to other sources of residual stresses by replacing ΔT by an effective value that gives the actual level of residual stresses. The use of an effective ΔT usually means the fracture experiments must be coupled with additional experiments that measure the level of residual stresses in the specimen.

REFERENCES

1. J. A. Nairn (1997) *J. Appl. Mech.* **64**, 804-810.
2. J. A. Nairn (2000) *Int. J. Fract.* **105**, 243-271.
3. V. M. Levin (1967) *Mechanics of Solids*, **2**, 58-61.
4. S. Mostovoy and E. J. Ripling (1966) *J. Appl. Polym. Sci.* **10**, 1351-1374.
5. J. A. Nairn (1999) *Int. J. Adhesion & Adhesives* **20**, 59-70.
6. J. G. Williams (1995) *Proc. Int'l Mechanical Engineering Congress*, San Francisco, USA, November 12-17.
7. J. A. Nairn and S. Hu (1994) in *Damage Mechanics of Composite Materials*, eds., Ramesh Talreja, 187-243.
8. J. A. Nairn (2000) in *Polymer Matrix Composites*, eds., R. Talreja and J.-Å. E. Manson, 403-432.
9. S. R. Kim and J. A. Nairn (2000) *Engr. Fract. Mech.* **65**, 573-593.
10. J. A. Nairn (2000) *Adv. Comp. Letts.* **9**, 373-383.
11. S. Zhandarov, E. Pisanova, E. Mader, and J. A. Nairn (2000) *J. Adhesion Sci. & Tech.* **15**, 205-222.
12. D.-A. Mendels, Y. Leterrier, J.-A. E. Månson, and J. A. Nairn (2002) *J. Comp. Mat.* in press.
13. B. Miller, P. Muri, and L. Rebenfeld (1987) *Comp. Sci. & Tech.* **28**, 17-32.
14. C. H. Liu and J. A. Nairn (1999) *Int. J. Adhesion & Adhesive*. **19**, 59-70.
15. B R. K. Blackmand, H. Hadavinia, A. J. Kinloch, M. Paraschi, and J. G. Williams (2002) *Eng. Fract. Mech.* **70**, 233-248.

Application of Fracture Mechanics to Composites

The Application of Fracture Mechanics to Polymers, Adhesives and Composites
D.R. Moore (Ed)

TOUGHNESS COMPARISONS WITH AND WITHOUT LINEAR ELASTIC FRACTURE MECHANICS

D.R. MOORE

INTRODUCTION

Pendulum impact tests for the measurement of toughness have been in common use in industry for many years. Specimen geometry may or may not include a notch and is based on three-point flexure (Charpy test) and clamped cantilever (Izod test). Toughness is expressed as an absorbed energy to fracture per ligament area and in the ISO or ASTM versions of these methods there is no instrumentation of the procedure i.e.there is an absence of accompanying force-displacement signals associated with data collection.

The use of these methods is so deep rooted in industrial practice and has been for so many decades that it seems inconceivable that the procedures have weaknesses. However, it is apparent that for certain classes of material the methods are flawed. In particular, their application to low toughness systems can lead to misleading assessments of toughness.

This article will present data for some phenolic compounds that can be toughened with a thermoplastic additive and that can incorporate glass fibre reinforcement. Results will be presented for notched Izod pendulum data that significantly misrepresent the toughness characteristics for these materials. Linear elastic fracture mechanics (LEFM) results will then be given to both remedy and explain these phenomena.

MATERIALS

All materials employed in this study are based on a phenol formaldehyde resin with a hexamethylene tetramine crosslinking agent. Two resin formulations are used; one without a toughening additive and another with a poly ether sulphone (PES) copolymer toughening material that is added at 35% w/w. Details of the structure-property characteristics of resins, including those incorporating a toughening agent, are discussed elsewhere [1]. The main purpose of the addition of the PES copolymer is to improve significantly the toughness of the resin formulation

These resin systems were used as the basis for the preparation of some discontinuous fibre reinforced composites where a glass fibre concentration in the range (20-60)% w/w was used. The glass fibre reinforcement was in the form of a 3mm chopped glass fibre. Plaque mouldings were prepared in these materials of dimensions 150 mm x 100 mm x 3 mm (nominal). This was achieved by first dissolving the components of the formulations in methyl chloride. The glass fibres were then added to this solution and the solvent evaporated. The mixture was then heated to 60 °C and then compression moulded at 150 °C. The compression moulding involved a combination of pressure on the mixture that was also in a vacuum bag. Compression and heat was applied for 1 hour. These plaque mouldings exhibited macroscopic isotropy as measured by modulus at mutually perpendicular directions within the plane of the mouldings [2].

NOTCHED IZOD RESULTS

The configuration of the notched Izod test is shown in Figure 1 and includes specimen dimensions. The standard notch tip radius for the specimen is 250 μm.

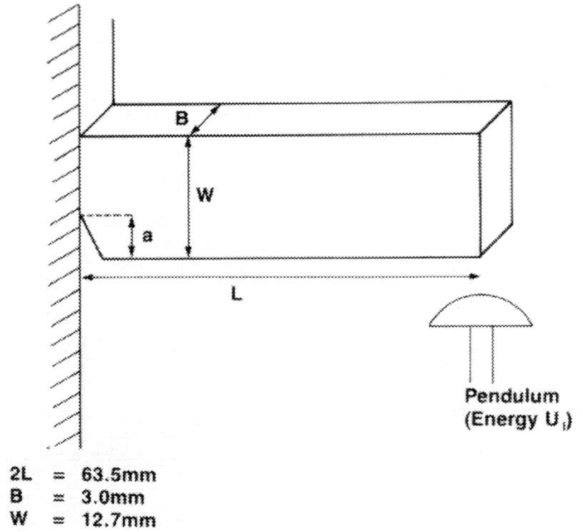

Figure 1 Configuration and specimen details for the notched Izod test.

A number of compounds at different glass fibre concentrations were tested in the notched Izod configuration. Two series of samples were used; one based on a standard untoughened phenolic resin system, i.e. without the addition of the PES copolymer and a second with addition of 35% w/w of the PES copolymer to the phenolic resin (toughened system). The notched Izod results at 23 °C are shown in Figure 2.
It is quite apparent that there is little difference in Izod toughness for the compounds based on the two resin systems. Apparently, the addition of the thermoplastic additive is offering no enhancement of toughness! This would be considerably at variance with other studies of toughening [3]. There is therefore a strong hint that these Izod results might be misleading.

TOUGHNESS FROM LINEAR ELASTIC FRACTURE MECHANICS

Toughness can also be measured using a linear elastic fracture mechanics approach for initiation and propagation of a crack. In general, the protocols established by the European Structural Integrity Society (ESIS) are adopted [4]. Initiation toughness is measured with a single edge notch beam (SENB) geometry, whilst a compact tension (CT) geometry is used for measuring both initiation and propagation toughness, where results are then presented as plots of toughness versus crack length ("R-curves"). Figure 4 illustrates such a curve for a toughened resin system. Modulus and yield strength are also measured for these materials. Modulus is calculated from the measurement of beam stiffness in three-point bending at 23 °C and 1 mm/min. Yield strength is measured by uniaxial compression on square platelet type specimens [5] at 1 mm/min at 23 °C.

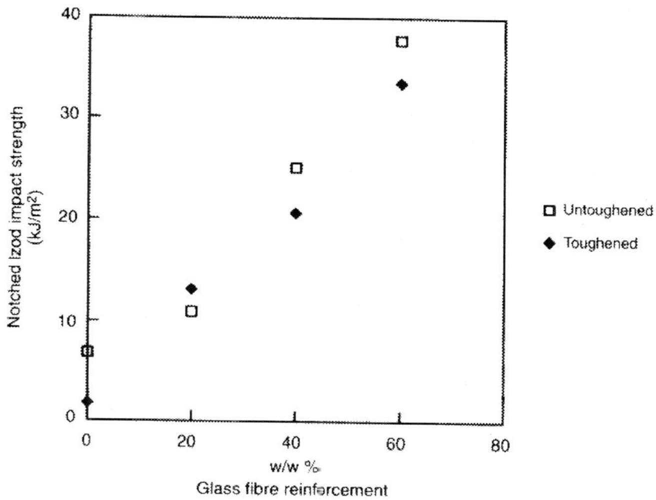

Figure 2 Notched Izod impact data for toughened (with PES copolymer in the phenolic resin system) and untoughened resin.

There are a number of difficulties to overcome in the fracture toughness measurements.

(i) For the composite samples, it was difficult to detect crack initiation from the measured force-displacement curves. Instrumentation of the specimens was used to remedy this problem.

(ii) For the resin samples, introduction of the sharpest notch (tip radius 1 μm) was not commensurate with a natural crack. Consequently initiation of a crack was from an effectively blunt crack followed by crack sharpening as the crack grew. Therefore, only fracture toughness measurements after some crack propagation related to valid plane strain fracture toughness. Again instrumentation of the specimen was required.

(iii) The majority of fracture toughness measurements were made at test speeds of 1 mm/min. Measurement at impact speeds commensurate with the Izod test (about 2 m/s) could not be made. This was not a problem because comparison of results at common test speeds was not required, merely comparison of toughness between "toughened" and "untoughened" resin samples.

Full details of the fracture toughness measurements are given elsewhere [6]. Figure 3 shows the fracture toughness results plotted as G_c versus glass fibre content for toughened and untoughened resin systems.

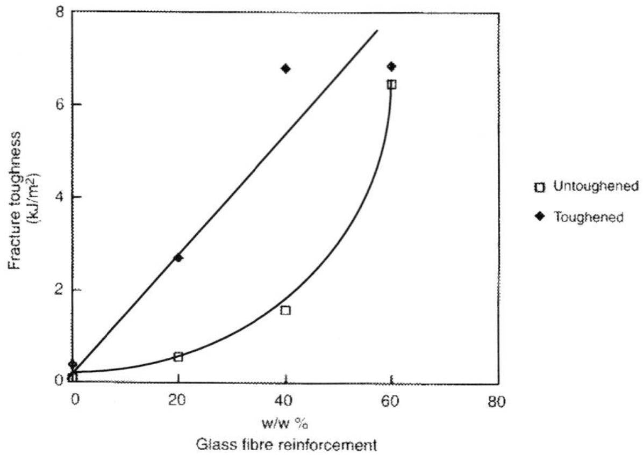

Figure 3 G_c plotted against glass fibre content for toughened and untoughened resin systems

The data in Figure 3 show a clear toughness benefit by the addition of the thermoplastic additive to the resin system. However, the trends lines associated with each set of data should not be taken too seriously, particularly for relatively large fibre contents. Nevertheless, the results are in complete contrast to the notched Izod data. The question is which set of data is correct and why the other is misleading.

AN EXPLANATION.

Both Izod and LEFM tests involved the fracture of a notched specimen. The conceptual idea in testing a notched specimen is to consider the notch as being equivalent to a crack. A pragmatic criterion for equating notches and natural cracks is that the plastic zone (radius r_p) around the crack is an order of magnitude larger than the tip radius (μ) of the notch:

$$\mu << r_p \qquad (1)$$

The plane strain plastic zone radius can be calculated from knowledge of fracture toughness (K_c) and yield strength (σ_y):

$$r_p = \frac{1}{6\pi}\left(\frac{K_c}{\sigma_y}\right)^2 \qquad (2)$$

Values of these properties were obtained and enabled the plastic zone radius to be calculated. Table 1 shows data for the unreinforced resin systems and the composites based on 60% w/w glass fibre reinforcement. It is assumed that a yielding process can occur in a highly reinforced glass composite and that LEFM can also be applied to such a material.

Material	K_c (MPam ^{½})	σ_y (MPa)	r_p (μm)
Toughened phenolic resin	0.5	220	0.27
Untoughened phenolic resin with 60 % w/w glass	9.0	673	9

Table 1 Property values and plastic zone radii for selected compounds

These data can now be applied to the Izod results. The notch tip radius in the Izod specimen is 250 μm. For such a notch to act like a sharp crack, the criterion in equation (1) requires that μ/r has a value of 0.1 or less. The data in Table 1 indicate that this ratio will be 926 for the resin system and 27 for the 60% w/w glass composite. Therefore the Izod notch is far too blunt if it is to simulate a natural crack for these materials. The consequence of the notch being blunt leads to additional energy absorption at the initiation of crack growth. Therefore, the Izod energy data are too large.

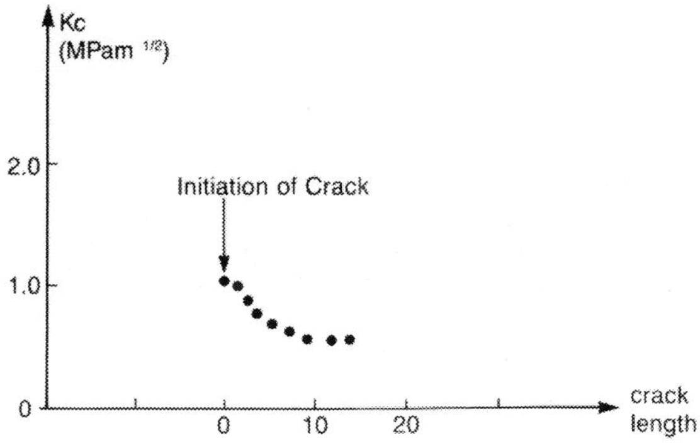

Figure 4 Fracture toughness versus crack length for toughened phenolic resin system for a compact tension test geometry [6]

The influence that a blunt crack can have on the measured toughness was investigated for the toughened resin system, without the presence of glass. Figure 4 shows fracture toughness (Kc) plotted against crack length for a compact tension (CT) specimen. The measured fracture toughness at initiation is too large because the tip radius of the notch is too large. The initial crack tip radius for this CT specimen is 1 μm and therefore μ/r_p is 3.7 (instead of the required 0.1). However, as the crack grows the notch sharpens and becomes a natural crack. In turn, the fracture toughness reduces. This is the process that occurs with the notched Izod tests. However, the bluntness of the Izod notch is considerably higher, since μ/r_p lies between 27 and 926 in these data. Therefore, the notched Izod data are considerable at variance with a valid measurement of toughness. Moreover, since μ/r_p is changing from sample to sample, there can be no consistency in the assessment of toughness by the notched Izod test, as indeed observed by the data in Figure 2.

CONCLUDING COMMENTS

This article has demonstrated that toughness assessment by a standard notched Izod test can be considerably misleading. Moreover, that application of LEFM methods can remedy the problem and provide an explanation of why there are problems with the Izod test. It is not to be implied that standard pendulum impact tests inevitably lead to a flawed assessment of toughness. The issue seems to relate to the level of toughness being assessed. If the toughness is relatively large, then there is unlikely to be an issue about the notch acting as a natural crack. However, when toughness is low, the Izod notch is likely to be too blunt and this will encourage other energy absorbing mechanisms to be associated with crack growth.

REFERENCES

1 Moore D.R., Colclough W.G., Carter J. T., McGrail P. T., Choate M. A., (1990) SAE Tech. Paper series 900069, Michigan.
2 Williams J. G., Pavan, A., *Impact and Dynamic Fracture of Polymers and Composites.* pp 383-400 ESIS Pub 19, ISBN 0852989466 (1995) MEP London
3 Carter J., Choate M., Colclough W., McGrail P. T., Moore D. R., SAE Tech. Paper Series 890251, Michigan, 1989, p 91
4 Moore D.R., Pavan A, Williams J.G. ed *Fracture Mechanics Testing Methods for Polymers Adhesives and Composites* ESIS Publication 28, Elsevier 2001.
5 Davies M., Moore D. R., Comp. Sci. & Tech., 40, 1991, p 131
6 Lowe A. C., Moore D. R., Rutter P. M., *Impact and Dynamic Fracture of Polymers and Composites* ESIS pub 19 (1995) p 383 ISBN 85298 946 6 MEP, London

The Application of Fracture Mechanics to Polymers, Adhesives and Composites
D.R. Moore (Ed)

FINITE FRACTURE MECHANICS OF MATRIX MICROCRACKING IN COMPOSITES

JOHN A. NAIRN

INTRODUCTION

The first form of damage in composite laminates is often matrix microcracks that are defined as intralaminar or ply cracks across the thickness of the ply and parallel to the fibers in that ply (see reviews in [1,2]). Tensile loading, fatigue loading, environment, and thermal cycling can all lead to microcrack formation. Microcracks can form in any ply that has a significant component of the applied load transverse to the fibers in that ply. Microcracks lead to degradation in properties of the laminate including changes in effective moduli, Poisson ratios, and thermal expansion coefficients [3]. Although these changes are sometimes small, microcracks can nucleate other forms of damage. For examples, microcracks can lead to delaminations, cause fiber breaks, and provide pathways for entry of corrosive liquids. An important issue in design of composite laminates is to be able to predict the initiation and development of microcracking damage following complex loading conditions. This chapter describes a fracture mechanics approach to the microcracking problem.

A complicating feature of composite fracture mechanics analysis is that laminates often fail by a series of fracture events instead of by continuous crack growth. Microcracking is a prime example. When cross-ply laminates are loaded in tension, the microcracking process is a series of events in which a single microcrack forms and instantaneously (on observation time scale) propagates until it fills the entire cross-sectional area of the ply. Conventional fracture mechanics deals with predicting the propagation of an existing crack. One could imagine analyzing microcrack propagation within a ply by standard methods, but there is little incentive to tackle this problem. The analysis could not be compared to experimental results for events and the analysis of a single crack does not answer the problem of predicting the extent or number of microcracks that form under various loading histories. Some microcracking models have abandoned fracture mechanics and used critical stress criteria instead; these models do not work well [4]. A better approach is to extend fracture mechanics methods to handle fracture events. Hashin has coined the term “finite fracture mechanics” to describe prediction of fracture events by comparing the total energy released due to a finite amount of crack area to an event toughness [5]. A finite fracture mechanics model for matrix microcracking can correlate a large body of experimental observations and can predict the extent of microcracking damage under various loading conditions [1,2].

FINTE FRACTURE MECHANICS PRINCIPLES

The development of a finite amount of fracture area, ΔA, must conserve energy. By the first law of thermodynamics, energy balance for an elastic material can be expressed as

$$\frac{\Delta w}{\Delta A} - \frac{\Delta U}{\Delta A} = 2\gamma + \frac{\Delta K}{\Delta A} \tag{1}$$

where w is external work, U is internal energy, γ is surface energy, and K is kinetic energy. Conventional fracture mechanics deals with infinitesimal static crack growth for which $\Delta A \rightarrow da$ and $\Delta K \rightarrow 0$. Crack growth occurs when energy release rate, G, is equal to the critical energy release rate, G_c:

$$G = \frac{dw}{dA} - \frac{dU}{dA} = 2\gamma = G_c \tag{2}$$

G_c is used in place of 2γ because experimental observations show that energy released during crack growth is always much larger than the thermodynamic surface energy (2γ). In other words, G_c is an *effective* material property that accounts for crack-tip energy dissipation not included in a linear-elastic stress analysis of crack-tip stresses. The logical extension to finite fracture mechanics is to assume a fracture event occurs when

$$G^* = \frac{\Delta w}{\Delta A} - \frac{\Delta U}{\Delta A} = 2\gamma + \frac{\Delta K}{\Delta A} = G_c^* \tag{3}$$

Here G^* is a finite energy release rate and G_c^* is an *effective* material property or event toughness that accounts for energy dissipation and kinetic energy effects not included in static linear-elastic analysis of work and energy before and after a fracture event. Conventional fracture mechanics works well provided G_c is found to be independent of specimen geometry. Similarly, finite fracture mechanics works well provided G_c^* is independent of specimen geometry and current damage state and provided initiation of the event is facilitated by conditions such as existing flaws or stress concentrations. The appearance of ΔK in G_c^* might render some fracture events unsuitable for finite fracture mechanics. ΔK can be included in an *effective* toughness, however, provided either it is small ($\Delta K << G^*$) or it is a constant that is characteristic of a particular fracture event [6].

APPLICATION TO MICROCRACKING

To verify the use of finite-fracture mechanics for microcracking, predictions can be compared to experiments. The procedure is to evaluate G^* and then predict microcrack formation by assuming the next crack forms when $G^* = G_c^*$. A unique feature of finite fracture mechanics is that G^* is different for load-control *vs.* displacement-control experiments [2,6]. Most static experiments use displacement control, but fatigue or thermal cycling experiments use load control. Both conditions must be analyzed. By using global energy methods, which include residual stresses effects [7], G^* for formation of a new microcrack in a $[0_n/90_m]_s$ laminate midway between two existing microcracks separated by $2mt_{ply}\rho_i$, where t_{ply} is thickness of a single ply, can be expressed exactly as [2]

$$G^* = \frac{\sigma_{90}^2 \rho_i B}{2k_m^2}\left(\frac{1}{E_A(\rho_i/2)} - \frac{1}{E_A(\rho_i)}\right) \qquad \text{(load control)} \qquad (4)$$

$$G^* = \frac{E_{A,N}^* E_{A,N+1}^*}{E_{A,0}^{*\,2}} \frac{\sigma_{90}^2 \rho_i B}{2k_m^2}\left(\frac{1}{E_A(\rho_i/2)} - \frac{1}{E_A(\rho_i)}\right) \qquad \text{(displacement control)} \qquad (5)$$

Here $\sigma_{90} = k_m\sigma_{app} + k_{th}\Delta T$ is the stress transverse to the fibers in the 90° plies in the absence of damage (it is linearly related to applied load, σ_{app}, and temperature difference, ΔT, by laminate-dependent mechanical and thermal stiffnesses, k_m and k_{th}), B is laminate thickness, and $E_A(\rho)$ is the effective axial modulus of a microcracking interval of dimensionless length ρ. In the displacement-control equation, $E_{A,N}^*$ is the axial modulus of a damaged laminate with N microcracks. Analogous, but different, results can be derived for microcracking of surface plies in $[90_m/0_n]_s$ laminates [2].

Figure 1 compares predictions to experiments for three different cross-ply laminates. All results are fit well with $G_c^* = 230 \pm 30$ J/m^2. The analysis for these experiments required input of $E_A(\rho)$ and ρ_i. $E_A(\rho)$ can be found by various stress-analysis methods [1]. The analysis for Fig. 1 used a complementary energy solution [8]. Simpler, albeit less accurate, shear lag methods are available [9]. The complementary energy solution loses accuracy when the ratio of the 0° to 90° thicknesses increases [10]; for these geometries $E_A(\rho)$ can be found by potential energy solutions or by numerical methods [2,10]. The term ρ_i is the dimensionless length of the microcracking interval where the next crack forms. This term is not the *average* crack spacing because experimental observations and theory both reveal that larger crack intervals are more likely to crack than smaller crack intervals. Unless experiments are coupled with direct observation of each microcracking event, the recommended analysis method is to replace ρ_i by $f\langle\rho\rangle$ where f is a parameter greater than 1 and $\langle\rho\rangle$ is the average microcrack spacing [2,4]. In a deterministic microcracking model where the crack always forms in the largest crack interval, f would oscillate between 1 and 2 or average 1.5. Alternately f can be measured by monitoring the microcracking process [11] or by adjusting it to fit experimental results [4]. Both these approaches show f to be between 1.2 and 1.5.

The advantage of finite fracture mechanics analysis of microcracking is that a single material property, G_c^*, can correlate data for different layups, such as the three layups in Fig. 1. To verify this important fracture mechanics result for more experiments, the microcrack formation criteria can be cast in a master plot form [4] by rewriting Eqs. (4) and (5) (and analogous expressions for laminates with surface 90° plies) as

$$\sigma_{red} = D_{red}\sqrt{G_c^*} + \Delta T \qquad (6)$$

where σ_{red} and D_{red} are reduced stress and crack density defined by

$$\sigma_{red} = -\frac{k_m}{k_{th}}\sigma_{app} \quad \text{and} \quad D_{red} = -\frac{1}{k_{th}\sqrt{G_{unit}^*}} \qquad (7)$$

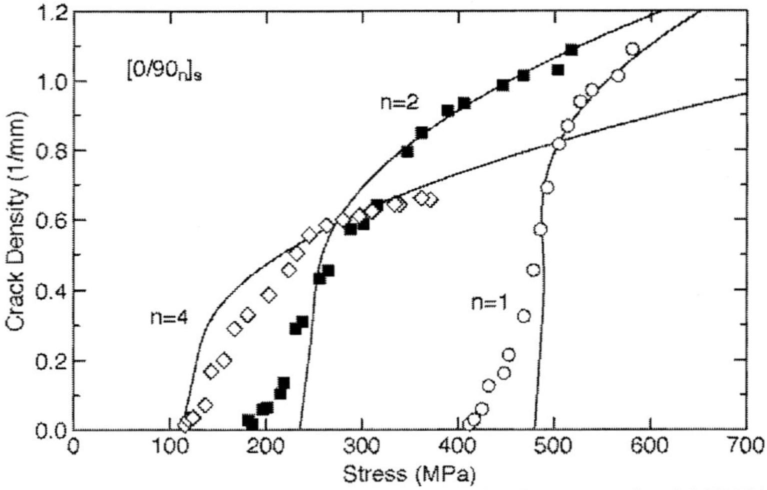

Fig. 1. Microcrack density as a function of applied stress for three Hercules AS4/3501-6 cross-ply laminates. The smooth lines are finites to finite fracture mechanics theory assuming G_c = 230±30 J/m^2 and $\Delta T = -100^\circ C$.

Here G^*_{unit} is the normalized energy release rate for unit σ_{90}. If finite fracture mechanics works well, a plot of σ_{red} as a function of D_{red} should be a straight line. The slope of the line gives the microcracking toughness and the intercept gives the thermal stress term. A sample master plot is given in Fig. 2 [4]. The results from 14 different layups, including both central (open symbols) and surface (filled symbols) cracked plies, all fall on the same line.

DISCUSSION

In typical microcracking experiments, microcracks form as a series of events. Experimental results consist of recording the number of microcracks as a function of loading conditions and specimen geometry. More detailed information could include the distribution of microcracking spacings [11]. If microcracks form in multiple ply groups (such as cracks on each surface of $[90_n/0_m]_s$ laminates), it is important to record correlations between damage states of the various plies. For transparent composites, full microcracks can be observed. For opaque composites, microcracks can be observed only on the edge. It is important to verify such microcracks correspond to a complete microcrack by observing opposite edges or by using x-ray methods. Microcracking experiments have been done by static tensile loading [1,2], mechanical fatigue [12], thermal cycling fatigue [13], and environmental exposure [14]. Static results are best analyzed by the methods above. Fatigue and thermocycling results can be analyzed by a modified Paris law where the rate of increase in crack density, D, per cycle, N, is related to the range in energy release rate, ΔG^*, by

$$\frac{dD}{dN} = K\left(\Delta G^*\right)^m \tag{8}$$

where K and m are fatigue toughness properties.

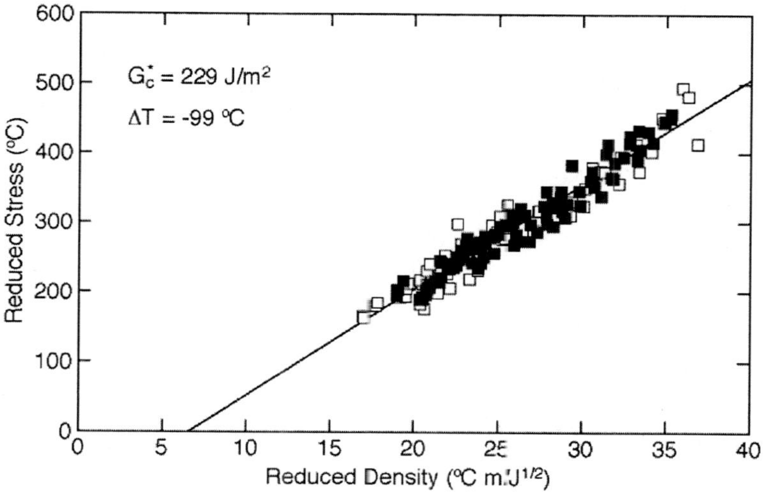

Fig. 2: Master plot results for 14 different layups of Hercules AS4/3501-6 laminates. The open symbols are for $[0_m/90_n]_s$ laminates; the filled symbols are for $[90_n/0_m]_s$ laminates. The straight line is a fit to all results; the slope and intercept give $G_c = 229\ J/m^2$ and $\Delta T = -99°C$.

The master plot in Fig. 2 shows that finite fracture mechanics of microcracking captures all major features of the microcracking process. There are minor details, however, and systematic effects that require further analysis. For example, the onset and initial rise in crack density is typically slower in experiments than as predicted by theory (see Fig. 1). This effect can be explained by introducing statistical variations in G_c^*. The early microcracks occur where the toughness is low; the later cracks are controlled by geometry (*i.e.*, they form midway between existing microcracks). Statistical variations can be included in the analysis by using simulations to fit experiments [2]. The results of such fits are that the mean toughness is similar to the toughness calculated by the simpler methods. The early crack density data gives new information about material variability. Simulations naturally account for the effect of microcracks to occur in larger microcracking intervals and thus do not need an f factor. Simulations results show that f is between 1.3 and 1.5 [11].

When G_c^* is found by fitting to individual layups, rather than a global master plot, there is a systematic trend towards lower toughness as the strain to initiate microcracking increases. This trend suggests that loading is causing diffuse damage in the microcracking plies that is causing a drop in G_c^*. Laminates that initiate microcracks at low strain (*e.g.*, laminates with thick 90° plies) have less damage and thus have a higher *apparent* toughness. Laminates that initiate at higher strain show a slightly lower toughness. This effect can be analyzed by continuum damage mechanics methods [10]. By using numerical methods and plotting G^* as a function of dimensionless crack spacing, ρ, it was observed that G^* is a property of the ply material but independent of the supporting plies. This observation made it possible to incorporate finite fracture mechanics methods into continuum damage mechanics models. By coupling a diffuse damage evolution law with microcracking toughness, a wider range of laminate structures can be predicted with greater accuracy [10].

REFERENCES

1. J. A. Nairn and S. Hu (1994) in *Damage Mechanics of Composite Materials*, eds., Ramesh Talreja, Elsevier Science, The Netherlands, 187-243.
2. J. A. Nairn (2000) in *Polymer Matrix Composites*, eds., R. Talreja and J.-Å. E. Manson, Elsevier Science, The Netherlands, 403-432.
3. L. N. McCartney (2000) *Comp. Sci. & Tech.* **60**, 2255-2279.
4. J. A. Nairn, S. Hu, and J. S. Bark (1993) *J. Mat. Sci.* **28**, 5099-5111.
5. Z. Hashin (1996) *J. Mech. & Phys. Solids* **44**, 1129-1145.
6. J. A. Nairn (1999) *5th Int'l Conf. on Def. and Fract. of Composites*, London, UK, March 18-19.
7. J. A. Nairn (2000) *Int. J. Fract.* **105**, 243-271.
8. Z. Hashin (1985) *Mech. of Mater.* **4**, 121-136.
9. K. W. Garrett and J. E. Bailey (1977) *J. Mat. Sci.* **12**, 157-168.
10. P. Ladevèze and G. Lubineau (2001) *Comp. Sci. & Tech.* **61**, 2149-2158.
11. D. Bal (2001) *M.S. Thesis*, MS&E Dept., University of Utah.
12. L. Boniface and S. L. Ogin (1989) *J. Comp. Mat.* **23**, 735-754.
13. C. Henaff-Gardin, M. C. Lafarie-Frenot, and J. L. Desmeuzes (1995) *Proc. ICCM-10*, Vancouver, BC, Canada, Aug 14-18.
14. M. H. Han and J. A. Nairn (2002) *Composites, Part A*, submitted.

The Application of Fracture Mechanics to Polymers, Adhesives and Composites
D.R. Moore (Ed)

FRACTURE MECHANICS OF THE MICROBOND AND PULL-OUT TESTS

JOHN A. NAIRN

INTRODUCTION

The properties of the fiber/matrix interface in composites have an influence on their overall performance. Common methods for studying the interface are micromechanical tests that load a single fiber/matrix interface to failure. Two popular tests are the microbond test [1] and the single-fiber, pull-out test [2]. In a microbond test a droplet of matrix is sheared off the fiber by pulling the fiber while restraining the droplet. In a pull-out test, the end of a fiber is embedded in a polymer and pulled until failure while restraining the matrix. A common interpretation of these tests is to reduce the load at failure to an interfacial shear strength (ISS) by dividing the applied load by the total interfacial area. Physically this term is the average interfacial shear stress at the time of failure. ISS has some use in qualitative work; it is less useful for fundamental characterization of the interface.

An alternative approach is use fracture mechanics methods. In each test, a crack initiates and propagates along the fiber matrix interface. By analyzing the interfacial cracking process with fracture mechanics, it is possible to determine an interfacial toughness instead of an ISS. The key analysis problem is to solve for the energy release rate of a propagating crack. The resulting equations can then be used to interpret experimental results. Two complicating features are residual stresses and friction. Residual stresses are always important in composites due to differential shrinkage between the fiber and the matrix. Friction is important in microbond and pull out tests because of the predominantly mode II loading conditions. This chapter presents a fracture mechanics model for both tests that includes all relevant effects. The model has been used to interpret experimental results.

FRACTURE MECHANICS THEORY

Detailed stress analysis of interface cracks is a difficult problem that has received much attention [3]. Fortunately, good results can be obtained for microbond and pull-out specimens by using an approximate, global energy analysis rather than a local, interfacial crack tip analysis. This section summarizes an analysis for energy release rate due to crack growth in microbond and pull-out specimens. More details can be found in Refs. [4-6].

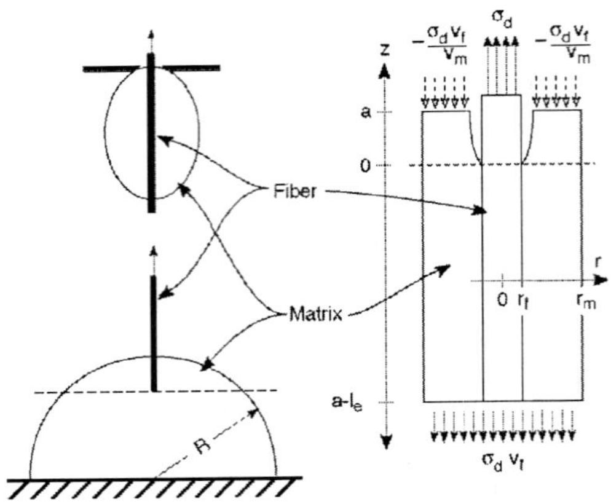

Fig. 1. The left side shows the microbond (top) and pull-out (bottom) specimen geometries. The right side shows the equivalent concentric cylinder model where the embedded length of fiber is in a cylinder of matrix. The fiber in both specimens is loaded with stress σ_d. The "dotted'" arrows are the remaining microbond specimen boundary conditions; the solid arrows on the bottom of the specimen are the remaining pull-out specimen boundary conditions.

Figure 1 shows a reduction of real specimens to an idealized geometry more amenable to analysis [4,5]. The simplification for the pull-out test is to replace the matrix region surrounding the embedded fiber by an equivalent cylinder of matrix. The length of the cylinder is equal to the embedded fiber length, l_e. The radius, r_m, is chosen to preserve the total fiber volume fracture, v_f, within the zone of the embedded fiber. The energy release rate analysis can thus focus on the concentric cylinder model on the right of Fig. 1 because neither the free fiber outside the matrix nor the matrix zone below the fiber end release energy during crack growth. The pull-out specimen is loaded by a fiber stress of σ_d. By force balance, the total stress on the bottom of the specimen is $\sigma_d v_f$. The simplification for the microbond specimen is to replace the elliptical matrix droplet by a matrix cylinder with length equal to the embedded fiber length and matrix radius chosen to preserve the total fiber volume fraction within the matrix. The microbond specimen is loaded by fiber stress of σ_d at the top of the specimen; that stress is balanced by a matrix stress of $-\sigma_d v_f/v_m$. The bottom of a microbond specimen is stress free.

Applying the general composite fracture mechanics methods from Ref. [6] to the idealized geometry in Fig. 1 with an interfacial debond of length a, the energy release rate for debond growth in both specimen types can be written as

$$\begin{aligned} G(a) = \frac{r_f}{2}\Bigg\{ & C_{33s}\bar{\sigma}^2 + 2D_{3s}\bar{\sigma}\Delta T + \left(\frac{D_3^2}{C_{33}} + \frac{v_m(\alpha_T-\alpha_m)^2}{v_f A_0}\right)\Delta T^2 \\ & - \left[\frac{m v_f \sigma_d}{2}\left(\frac{1}{E_A}-\frac{1}{E_m}\right) + D_{3s}\Delta T\right]\left[\frac{2\tau_f}{r_f}C_T(a) - \left(\bar{\sigma} + \frac{(1+m)D_3\Delta T}{C_{33}}\right)C_T'(a)\right]\Bigg\} \end{aligned} \qquad (1)$$

where m=0 or 1 is for pull-out or microbond tests, respectively. The term $\bar{\sigma}$ is a reduced debonding stress defined by

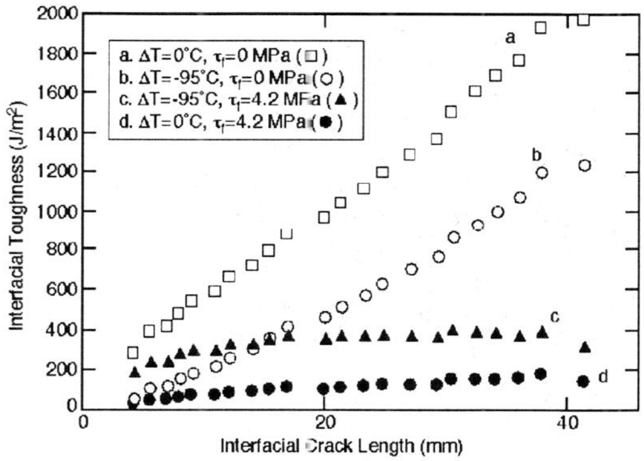

Fig. 2. Crack-resistance curves for a steel/epoxy model specimen analyzed four different ways. Curve *a* ignores residual stresses and friction; curve *b* includes residual stresses but ignores friction; curve *c* includes both residual stresses and friction; curve *d* ignores residual stresses, but includes friction.

$$C_T(a) = \int_0^{l_e-a} F(z)\,dz \tag{3}$$

where $F(z)$ is the solution for axial fiber stress in concentric cylinders of length l_e-a subjected to unit normal stress on the fiber and a balancing $-v_f/v_m$ stress on the matrix, both at $z=0$ in addition to zero stress on the other end at $z=l_e$-a. The functions $F(z)$ and $C_T(a)$ can be found by any analytical, numerical, or even experimental means and then substituted into Eq. (1) to find the energy release rate. One simple analytical approach, which was demonstrated to be accurate by comparison to finite element analysis [4,6], is to use shear-lag analysis for which it is easy to derive [4]:

$$C_T(a) = \frac{1}{\beta}\left[\coth\beta(l_e - a) - \operatorname{csch}\beta(l_e - a)\right] \tag{5}$$

where β is a shear-lag parameter as defined in Ref. [4] and elsewhere.

MICROBOND AND PULL OUT EXPERIMENTAL RESULTS

Verification of fracture mechanics methods for the microbond test was done using macroscopic model specimens of a steel wire embedded in a cylinder of epoxy (see Ref. [4] for specimen details). Steel/epoxy specimens are analogues of carbon/epoxy or glass/epoxy microscopic specimens with modulus ratio similar to carbon/epoxy but higher than glass/epoxy specimens. In macroscopic experiments, crack growth along the interface could be observed visually. After debonding was complete, it was further possible to observe and record frictional stress. The key experiments are force as a function of crack length for various specimen geometries. The verification experiments are summarized in Fig. 2 in which the raw data are interpreted by fracture mechanics four different ways. For fracture mechanics to be useful, one expects the measured toughness to be independent of geometrical factors and of crack length. Microbond experiments satisfy this requirement provided the data are analyzed correctly.

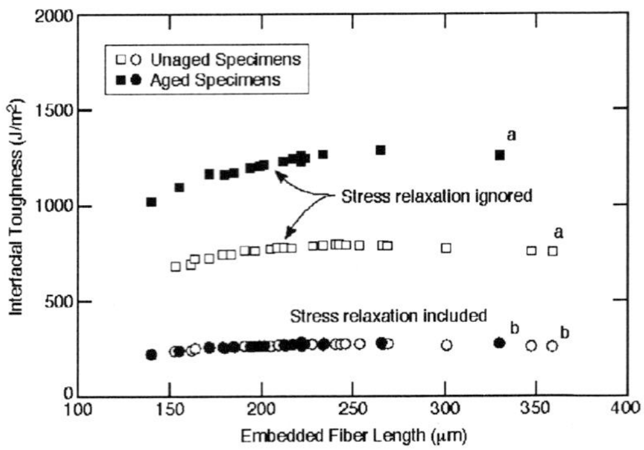

Fig. 3. Fracture toughness calculated from microbond specimens as a function of droplet length. Curves *a* include residual stresses but do not account for relaxation of those stresses during aging. Curves *b* account for relaxation of residual stresses.

Curve *a* in Fig. 2 is an analysis that ignores both residual stresses and interfacial friction. This curve is clearly a poor fracture mechanics result. Curve *b* includes residual thermal stresses by setting ΔT=-95°C, but still ignores friction; it is also a poor fracture mechanics result. Curves *c* and *d* both include friction effects by setting τ_f=4.2 MPa which was the measured interfacial shear stress after complete debonding. Curve *c* is an analysis that included both residual stresses and friction. This curve is a good fracture mechanics result. There is an initial rise in toughness at short debond lengths, but the results soon level out at an approximately constant toughness (360 J/m^2). Curve *d* includes interfacial friction, but ignores residual stresses. Although curve *d* looks relatively flat on the scale of Fig. 2, it actually never levels off and is a poor fracture mechanics result. The difference between curves *c* and *d* illustrates the magnitude of the contribution of residual stress to debonding. The magnitude is large; in fact, most of the energy released comes from residual stresses.

The experiments are more difficult with micro-specimens because it is difficult or impossible to observe crack growth. From the macroscopic specimen results, and in agreement with theory, it was found that interfacial crack growth is stable and the load continued to increase until complete debonding. A proposed fracture mechanics approach when there is no crack length data is to record the peak load and calculate interfacial toughness from Eq. (1) by taking the limit as *a* approaches droplet length [4-6]. This approach was verified by using it on the macroscopic experiments for which crack length could be observed. The toughness calculated by the peak-load method agreed reasonably well with a full analysis that included crack-length information. Figure 3 shows an experimental investigation of the effect of aging on interfacial toughness from micro-sized E-glass fiber/epoxy specimens [6,8]. These data were analyzed by the peak-load method. Fracture mechanics analysis should show a toughness that is independent of droplet length and differences between aged and unaged results should show the effect of physical aging on the interfacial toughness. The square symbols are an initial fracture mechanics analysis that used a simplistic calculation of residual stresses. These results gave the impression that the interfacial toughness increased with aging. Similarly, a non-fracture mechanics method suggested that aging affected the interface. The true results were revealed by new fracture mechanics calculations that included the actual level of residual stresses by accounting for stress relaxation during the aging process. The final interpretation (circular

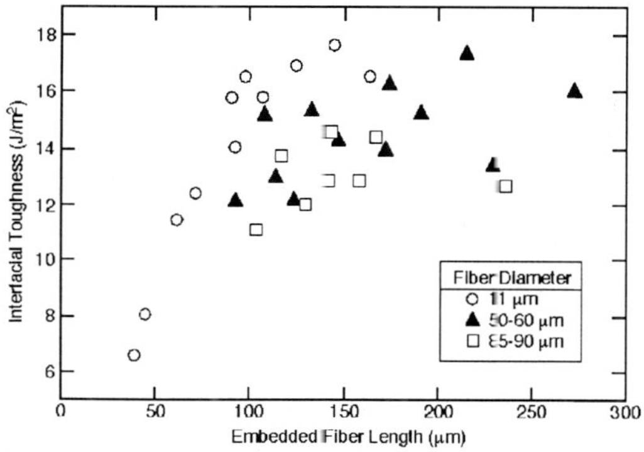

Fig. 4. Fracture toughness calculated from initiation of debonding in pull-out tests for glass fibers of three different diameters in a vinyl-ester matrix as a function of embedded fiber length.

symbols) showed that aging does not affect interfacial toughness; it only changes the level of residual stresses. These experiments illustrate how fracture mechanics interpretation can lead to the proper interpretation of experiments, but it is essential to have a good energy release rate analysis and to include all significant effects such as residual stresses and friction.

Fracture mechanics can analyze pull-out tests by similar methods. In the pull-out results for glass fiber/vinyl ester [9] analyzed in Fig. 4, it was possible to observe initiation of crack growth as a kink in the load-displacement curve. A toughness was calculated from each specimen using Eq. (1) by setting a=0 (or any suitably small a). Friction could be ignored because it had a negligible effect for initiation. The effect of friction only becomes significant as the debond length gets long. The calculated interfacial toughness for three fibers of different diameters is plotted in Fig. 4 as a function of embedded fiber length. After an initial rise, all results became roughly constant. In other words, as expected for good fracture mechanics results, the interfacial toughness was independent of both fiber length and fiber diameter.

RECOMMENDATIONS

Equation (1) provides an accurate, analytical result that can be used to calculate the energy release rate for interfacial crack growth in both the microbond and the pull-out test geometries. Hence, it is possible to derive fracture toughness information from these common interfacial test methods. The calculation of interfacial toughness requires sufficient experimental input for determination of all terms in Eq. (1). The key results needed are specimen geometry, current interfacial crack length, the load required to extend the crack, the actual level of residual stresses, and the magnitude of any friction on the debond surfaces. Residual stresses are particularly important for analysis of microbond specimens because they account for much of the energy released. Friction is especially important as the debond gets long. In some micro-sized specimens it can be difficult to observe crack growth. This situation can be handled by using indirect means to deduce crack length. For example, the crack length at the peak load in microbond tests is nearly equal to the droplet length. For pull-out tests, debond initiation can sometimes be observed as a kink in the force-displacement curve and the "kink" load corresponds to zero initial crack length.

REFERENCES

1. B. Miller, P. Muri, and L. Rebenfeld (1987) *Comp. Sci. & Tech.* **28**, 17-32.
2. L. S. Penn and E. R. Bowler (1981) *Surf. Interf. Anal.* **3**, 161-164.
3. M. L. Williams (1959) *Bull. of the Seismological Society of America* **49**, 199-204.
4. C. H. Liu and J. A. Nairn (1999) *J. of Adhes. & Adhesives* **19**, 59-70.
5. J. A. Nairn (2000) *Adv. Comp. Letts.* **9**, 373-383.
6. J. A. Nairn (2001) *Proc. 16th Ann. Tech. Conf. Amer. Soc. Composites*, Sept. 9-12, 2001, Blacksburg, VA.
7. J. A. Nairn (2000) *Int. J. Fract.* **105**, 243-271.
8. D.-A. Mendels, Y. Leterrier, J.-A. E. Månson, and J. A. Nairn (2003) *J. Comp. Mat.* in press.
9. S. Zhandarov, E. Pisanova, E. Mader, and J. A. Nairn (2000) *J. Adhesion Sci. & Tech.* **15**, 205-222.

The Application of Fracture Mechanics to Polymers, Adhesives and Composites
D.R. Moore (Ed)

FRACTURE ANISOTROPY

D.R. MOORE

INTRODUCTION

It is seldom that plastics exhibit homogeneous isotropic properties. For unfilled or unreinforced plastics, anisotropy can be linked to molecular orientation and processing stresses during fabrication. For filled and reinforced plastics, the nature of the reinforcement adds another layer of anisotropic characteristics. When the reinforcement is fibre-like then this dominates the anisotropy.

Fracture properties are no exception to these considerations of anisotropy. In fact, since fracture relates to toughness and strength and since anisotropy implies different properties in different directions, then there is bound to be a weak direction for strength that will need to be accommodated. In terms of fracture properties, application of linear elastic fracture mechanics (LEFM) has already been seen to be of value. For example, for unreinforced polymeric systems the measurement of K_c and G_c at slow and impact speeds can be used to measure and describe the anisotropy due to molecular orientation [1]. For fibre reinforced plastics the use of LEFM is not quite so straight-forward because the theory can only be rigorously applied for homogenous materials. Nevertheless, a pragmatic application to discontinuous fibre reinforced plastics has demonstrated that LEFM can be helpful in defining the fracture anisotropy but also in simply measuring the toughness [2].

The purpose of this article is to consider the application of LEFM to continuous fibre reinforced polymeric composites. For these materials, the nature of the anisotropy of the continuous fibre dominates the fracture characteristics.

FRACTURE MECHANISMS IN COMPOSITES.

There are three principal fracture mechanisms in continuous fibre reinforced composites. These are shown in Figure 1 and involve breaking a fibre (trans-laminar fracture), running a crack between fibres (intra-laminar fracture) and running a crack between the plies that make up the composite (inter-laminar fracture or delamination).

Each of these principal mechanisms has an associated fracture toughness, as indicated in Figure 1. Each mechanism will entail the crack, in part, being conducted through the matrix (whether thermoplastic or thermosetting polymer). Therefore, subsidiary mechanisms may also be involved and might include, tearing, debonding, crazing, cavitation, voiding and yielding.

A primary requirement to describe fracture anisotropy in these composites is to measure the three principal fracture toughness properties. In order to demonstrate how this can be achieved, carbon fibre (AS4) reinforced poly (ether-ether-ketone) (CF/PEEK) composite is used in the form of 40 ply unidirectional laminates $[0]_{40}$.

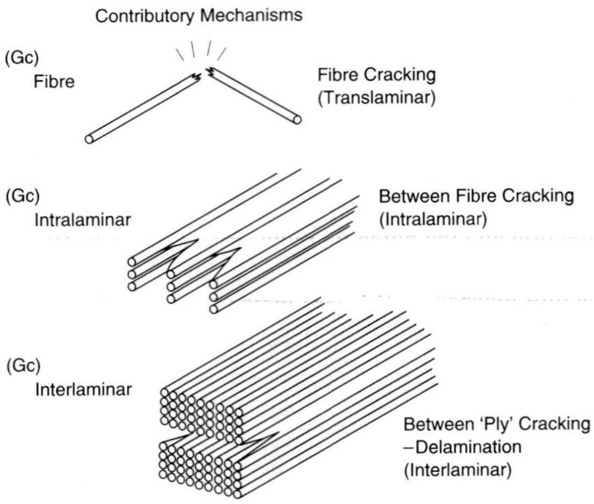

Figure 1 Fracture characteristics for continuous fibre reinforced composites.

FRACTURE TOUGHNESS FOR CF/PEEK.

Figure 2 shows how specimens can be obtained from a $[0]_{40}$ laminate in order to generate the three principal fracture mechanisms.

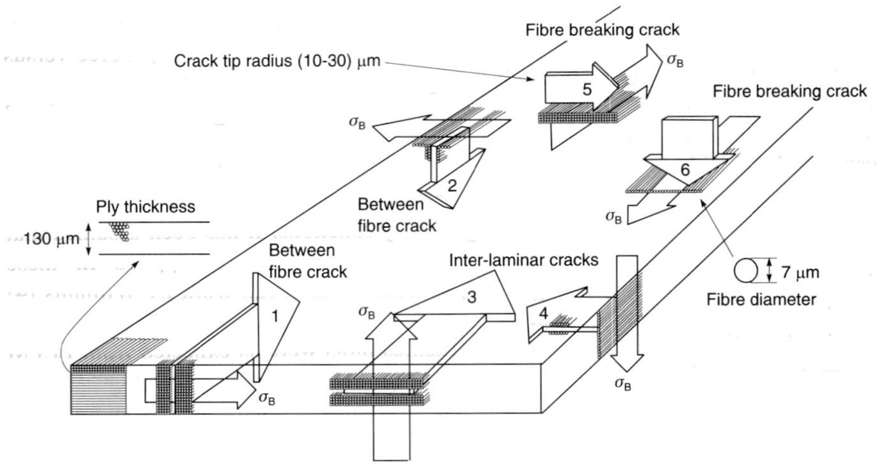

Figure 2 Specimen configuration for generating the three principal fracture mechanisms from a unidirectional composite.

The thickness of a 40 ply laminate of CF/PEEK is about 5.8 mm. LEFM experiments were conducted on single edged notched bend (SENB) specimens of nominal dimensions 60 mm x 10 mm x 5.8 mm, using a span of 40 mm. These specimens can be cut directly from the laminate for crack directions 1, 2, 5 and 6, as shown in Figure 2. However, crack directions 3 and 4 are more complicated. These specimens were prepared by gluing appropriate pieces together in order to achieve the required span for testing. Full details are given elsewhere [3]. Some 6-8 specimens were prepared for each crack direction and the notch tip radius was notionally 10 μm.

Linear elastic fracture mechanics (LEFM) methods are used for determining fracture toughness [1], namely, measurements of K_c and G_c at 1 mm/min and at 23 °C.

$$K_c = \sigma_F Y a^{\frac{1}{2}} \qquad (1)$$

$$G_c = \frac{U}{BW\phi} \qquad (2)$$

Where, σ_F is the stress at fracture that for three-point flexure is given by:

$$\sigma_F = \frac{3SP}{2BW^2} \qquad (3)$$

Y and ϕ are geometry functions [1], a is the crack length, B is the specimen thickness, S the specimen span and W is the specimen width. U is the total energy absorbed during fracture. B and W were nominally 5.8 mm and 10 mm, but because of the nature in cutting specimens from the $[0]_{40}$ laminate the values for B and W had to be interchangeable (W was 12 mm for crack direction 1, 5.6 mm for crack direction 2, 10 mm for crack directions 3, 4 and 5 and 5.8 mm for crack direction 6). Force versus displacement curves were recorded during the fracture test but were seldom the conventional triangular shape. Often cracks were initiated and grew in a convoluted manner. Therefore fracture stress was based on peak load and but energy to fracture was associated with the total energy absorbed.

Perhaps the most serious dilemma in the calculation of fracture toughness for composites is the determination of the geometry functions. It has been assumed that the values for isotropic and homogenous materials can be applied in these circumstances. This is without rigorous justification, although work by Williams [4] has shown that to a certain extent that this is not always an unfounded assumption. Despite these limitations, the fracture mechanics plots were as expected from LEFM theory. For example, Figure 3 shows plots for crack directions 1 and 2 where force is plotted against the function $(\frac{2BW^2}{3SYa^{1/2}})$ (derived from combining equations 1 and 3) and energy plotted against $(BW\phi)$. The slopes of these curves give K_c and G_c, respectively. Table 1 summarises all values of fracture toughness obtained.

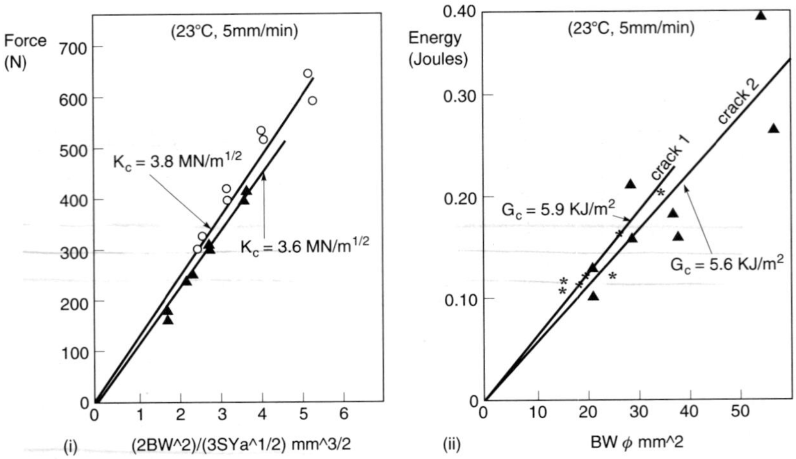

Figure 3 Fracture mechanics plots for cracks 1 and 2. (i) as K_c and (ii) as G_c

Apparent crack mode	**Crack direction**	**Kc (MPam $^{1/2}$)**	**Gc (kJ/m^2)**
Between fibre cracks	1	3.8	5.9
	2	3.6	5.6
Delamination cracks	3	3.7	3.4
	4	4.0	3.8
Fibre-breaking cracks	5	42	406
	6	38	128

Table 1 Summary of fracture toughness values for six crack directions.

It is apparent that the fracture toughness values for crack directions 1 to 4 are the same, implying that the cracking mode is between fibres in all cases. In other words, crack directions 3 and 4 are not delamination cracks. Fracture morphology of the fracture surfaces confirms this view [3]. Bearing in mind that cracks are machined into the SENB specimens (rather than being initiated by insert films) then it is unlikely that the machining operation for locating the crack can strike the inter-laminar region (see the dimensions in Figure 2).

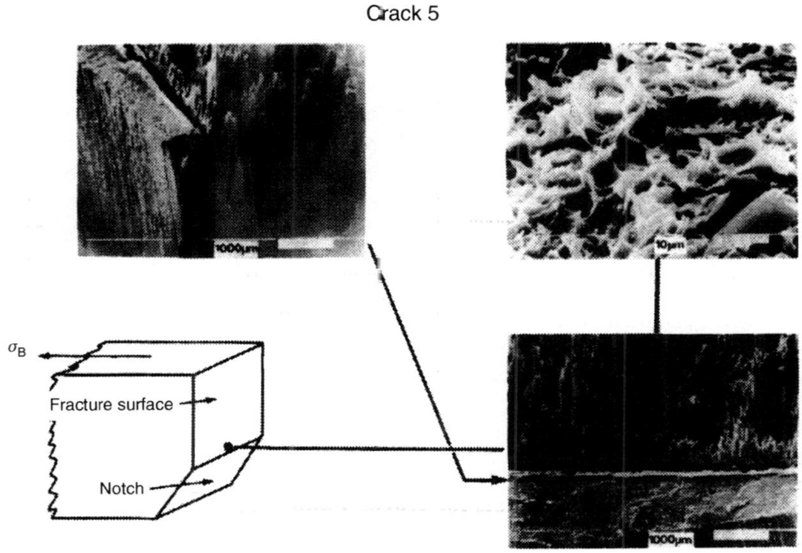

Figure 4 SEM micrographs of the fracture surface for crack 5 at several magnifications.

The fibre breaking mode gives the highest values of fracture toughness as seen by the data for crack directions 5 and 6. The K_c data are consistent but the G_c data are quite different. This is due to additional crack mechanisms occurring for crack direction 5 as shown in Figure 4. The SEM micrographs confirm that fibre fracture is occurring. However, a magnified side-on view of the crack shows the presence of shear cracks. The combination of the fibre-breaking and shear cracks is creating too large a value for the total energy recorded which in turn leads to the value for G_c for fibre-breaking being too large for crack 5 compared with that for crack 6. On the other hand, K_c is similar in both cases because it is just based on peak load.

Delamination fracture toughness is believed to be the smallest toughness for these types of composites. Therefore, it is necessary to find other means of its measurement. This is achieved by the use of double-cantilever beam (dcb) geometry, where an insert film laid between the central plies is used as the crack starter.

A protocol for the determination of mode I fracture toughness (G_{Ic}) has been established by ESIS TC4 [5]. Several methods for conducting the test and analysing the data have been researched and documented. However, at the time of exploring the fracture anisotropy for CF/PEEK an early version of the analysis was all that was available, based on an area method analysis [3]. Although an area method analysis is completely acceptable, it would nowadays include corrections for large displacements of the beam, corrections for the stiffening of the beam associated with the end-blocks and also a correction for the crack length due to root rotation of the beam [5]. However, these early results will be included now in order to maintain consistency with the samples already discussed. The net effect in not using the various corrections incorporated in the protocol leads to the values quoted for G_{Ic} being up to 40 % too large. Nevertheless, these results enable us to complete the discussion of the fracture anisotropy.

The dcb specimen comprised a $[0]_{40}$ laminate with a starter crack incorporated between the centre plies during moulding. Propagation of the crack then passes along the inter-ply region and provides a measurement of delamination toughness. Results are presented in Figure 5 in the form of G_c versus crack length (an R-curve) and an average value is obtained for the delamination fracture toughness of 2.4 kJ/m^2. An equivalent value for K_c cannot be determined by an analytical method, although numerical methods could be used. Detailed microscopy confirmed that the fracture occurred by a delamination crack [3].

As previously discussed, the G_c value would be in the region of 1.7 kJ/m^2 had the superior analytical methods [5] been used, but an exact sample was not available for this.

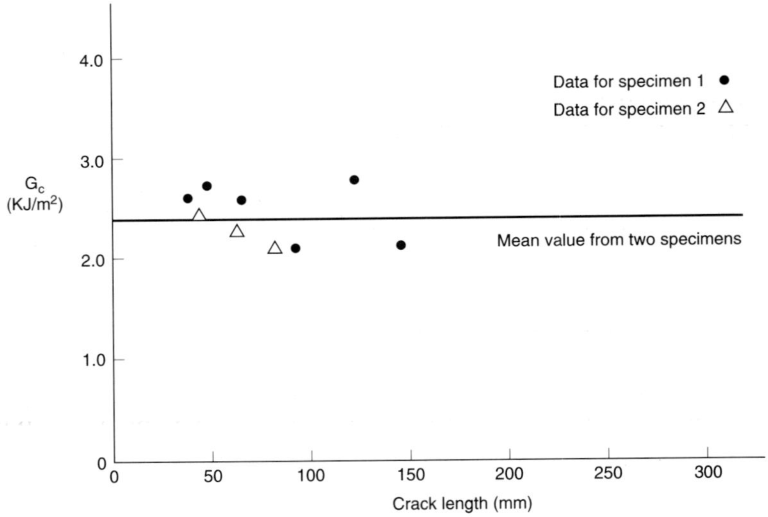

Figure 5 An R-curve of Gc versus crack length from dcb specimens of CF/PEEK

SUMMARY OF FRACTURE ANISOTROPY

The principle fracture toughness (G_c) values, for CF/PEEK, can now be summarised. These toughness values represent a high anisotropy for fracture. Of course, in most practical laminates the fibres are not unidirectional but are multi-axial e.g $[0,\pm45,90]_{4s}$ etc. Therefore, it can be expected that a combination of the three principal fracture mechanisms will occur. However, laminates are a combination of plies and therefore delamination cracking will remain a possibility. This is perhaps an important consideration because delamination fracture toughness is the smallest energy absorbing mechanism.

Fracture mechanism	**Gc (kJ/m²)**
Delamination or inter-laminar fracture	2.4 (more likely to be 1.7)
Between crack or intra-laminar fracture	3.5-6.0
Fibre breaking or trans-laminar fracture	c 130

Table 2 Fracture anisotropy for CF/PEEK.

A key benefit in using fracture mechanics to describe the anisotropy is that it places the fracture toughness measurements on a common scale. This is achieved through the need to adopt different geometry forms in order to make the measurements.

REFERENCES

1 Moore D.R., Pavan A, Williams J.G. ed *Fracture Mechanics Testing Methods for Polymers Adhesives and Composites* Williams, J.G., *K_c and G_c at Slow Speeds for Polymers* ESIS Publication 28, Elsevier 2001, Ch 1 p11

2 Moore D.R., Pavan A, Williams J.G. ed *Fracture Mechanics Testing Methods for Polymers Adhesives and Composites,* Moore, D. R., *The Measurement of K_c and G_c at Slow Speeds for Discontinuous Fibre Composites* ESIS Publication 28, Elsevier 2001, Ch 1 p 59

3 Leach D. C., Moore D.R., *Composites Science and Technology* **23** (1985), 131-161

4 Friedrich, K., ed *Application of Fracture Mechanics to Composite Materials* Williams J.G., *Fracture Mechanics of Anisotropic Materials* p3-38 Elsevier1989.

5 Moore D.R., Pavan A, Williams J.G. ed *Fracture Mechanics Testing Methods for Polymers Adhesives and Composites,* Brunner, A.J., Blackman, B.R. K., Davies, P., *Mode I Delamination* ESIS Publication 28, Elsevier 2001, Ch 4 p277

The Application of Fracture Mechanics to Polymers, Adhesives and Composites
D.R. Moore (Ed)

DETERMINATION OF INTERFACE ADHESION IN SINGLE-FIBRE PULL-OUT AND MICROBOND EXPERIMENTS

B. LAUKE, T. SCHÜLLER AND W. BECKERT

INTRODUCTION

The quality of adhesion between reinforcing fibres and matrix in composites is important for the macroscopic mechanical properties of these materials. For example, the strength of short-fibre reinforced polymers increases with the adhesion strength at the interface while fracture toughness improvement demands an optimised (not maximum) adhesion quality. Consequently the determination of micromechanical adhesion parameters is important for the understanding and improvement of macromechanical properties. The characterisation of adhesion between different materials generally follows two concepts: determination of adhesion strength or determination of fracture mechanics parameters. The application of the strength concept at the interface between two materials often involves fundamental problems. Because of the non-uniform stress distribution in most of the applied test methods the normalisation of the applied critical force with debonded interface area provides only a rough approximate measure of the composite quality. To circumvent stress singularities it is necessary to develop new tests, see for example Ref. [1]. The single fibre pull-out (SFPO) and microdroplet or microbond (MB) tests, however, show stress singularities and crack propagation along the interface because of the geometry and loading situation. In such cases only the fracture mechanics approach of bimaterial composites leads to reasonable answers to the question: How can adhesion (rather than toughness or load transfer) between a fibre and matrix be characterised? The aim of our contribution is to show how fracture mechanics can be applied to derive adhesion parameters for these tests.

THE EXPERIMENTAL AND MODELLING PROBLEM

The single fibre pull-out test was introduced decades ago by Kelly [2] and the microbond test was proposed by Miller [3]. Both tests, shown in Fig. 1, have been the subject of numerous fundamental experimental and theoretical investigations.

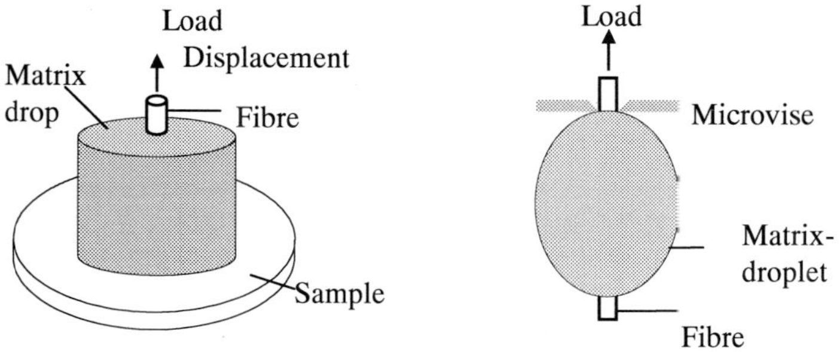

Fig. 1 Schematic representation: (left) single-fibre pull-out test (SFPO); (right) microbond test (MB).

Despite the simple geometry, the calculation of the stress and strain distributions is not simple [4]. It becomes even more complicated if a crack has been initiated during loading of the fibre and propagates along the interface. Then friction due to residual stresses or caused by the microvice in the MB-test and roughness must be considered [5]. In ideally elastic models the maximum interface stresses are infinite and this mathematical exact result is not obtained by most of the analytical approaches, as for example shear lag analysis [6]. Such approximations remain finite because they are approximations. In reality, by contrast, the stresses remain finite because of inelasticity.

Atkinson et al. [7] were among the first authors who have applied successfully the theory of bimaterial fracture mechanics to this problem. They calculated the energy release rate variation with the propagating crack, without consideration of friction and residual stresses. One of the most valuable experimental test set ups was constructed by Hampe [8]. The main advantages of the equipment are the high stiffness of the machine and the short free fibre length used. Both facts are necessary to achieve a stable crack propagation at the interface which is a requirement for data reduction to obtain adhesion parameters. The debonding initiation is shown in the force-displacement curve by a kink in the slope. This reveals that debonding starts long before the maximum force is reached. Quite similar experimental results can be obtained for the microbond test [9]. These experiments provide the following quantities: critical load for debonding initiation at the entry point of the fibre to the matrix, force-displacement and force- crack length curve (at least for transparent materials).

In the following we try to show how this experimental information can be used as input data in the fracture mechanics models for the experiments to draw conclusions about the adhesion quality. The analysis must include other influencing factors, for example friction, in order to give hints how these other factors can be separated from adhesion related parameters.

BASIC RELATIONS OF INTERFACE CRACK PROPAGATION

Consider two isotropic elastic solids joined along the x-axis as indicated in Fig. 2.

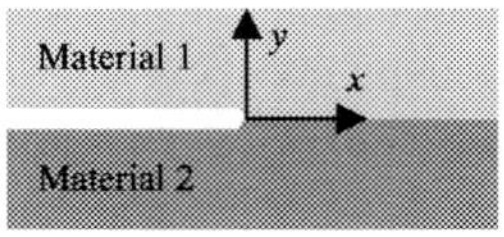

Fig 2. Sketch of an interface crack with co-ordinate system.

The interface is a low-toughness fracture path. The energy release rate G for crack extension can be derived from elastic strain energy U using $G = \partial(W_{ext} - U)/\partial A$ where W_{ext} is the external work and ∂A the new-formed crack area.

A more detailed description of interface failure involves the stress fields near the interface crack tip. The dominant stress singularity for $x \to 0$ and $y = 0$ is of the form

$$\left(\sigma_{yy} + i\sigma_{xy}\right) = \frac{\left(K\rho_0^{i\varepsilon}\right)\left(\frac{x}{\rho_0}\right)^{i\varepsilon}}{\sqrt{2\pi x}} \text{ with } K_1^{\rho 0} = \mathrm{Re}\left[K\rho_0^{i\varepsilon}\right] \;\; K_2^{\rho 0} = \mathrm{Im}\left[K\rho_0^{i\varepsilon}\right] \;\; \Psi^{\rho 0} = \arg\left[K\rho_0^{i\varepsilon}\right] \quad (1)$$

where K is the complex stress intensity factor which uniquely characterises the singular stress field and Ψ^{ρ_0} is the mixed-mode angle. The bimaterial constant ε depends on the material properties. Since x is not a dimensionless quantity it has to be normalised by a reference length ρ_0. The choice of ρ_0 is arbitrary at this point. Changing from one reference length to another causes a rotation in the K-space. The term $x^{i\varepsilon}$ prevents also a simple relationship between the remote loading mode and the local mixed-mode state at the crack tip from being deduced. That's why Arabic numbers are used for mode 1 and 2 instead of I and II used for homogeneous materials. The energy release rate is related to the near crack tip fields by

$$G_{tot} = \frac{1}{E^*}\left(K_1^2 + K_2^2\right) = G_1\left(K_1^{\rho_0}\right) + G_2\left(K_2^{\rho_0}\right). \tag{2}$$

with E^* defined according to Hutchinson and Suo [10].

Crack extension criterion

If the energy release rate G for a given geometry (including a crack) exceeds a critical value G_c the crack will grow. The parameter G_c is a material property of the interface. However, also the local mode mixity at the crack tip controls interfacial failure. Therefore, a more realistic crack propagation criterion must include more details about the load state at the crack tip. The local stress fields which control the local fracture mechanisms are characterised completely by the complex stress intensity factor K. Consequently a curve in the K-space represents the fracture criterion. A reasonable and common approximation for the actual fracture envelope is an ellipse [11]. Because of the particular nature of interfacial cracks its orientation in the K-space towards the K_1, K_2-axes does not need to be parallel in general, a transformation to a principal axes representation is possible by using the rotational effect of reference length ρ_0 and choosing an appropriate value ρ_c:

$$(K_1^{\rho_c}/K_{1c})^2 + (K_2^{\rho_c}/K_{2c})^2 = 1. \tag{3}$$

The parameters K_{1c}, K_{2c} and ρ_c are properties of the interface.

FINITE ELEMENT ANALYSIS AND CRACK GROWTH

The simulation of crack propagation requires two things. First, a finite element model establishes the connection between the remote load and the local stress state at the crack tip for a fixed crack length. Second, a method must be found to implement a growing crack which brings a changing geometry into the model. Both steps are independent of each other in the case of linear-elastic fracture mechanics. The crack growth is path independent. Therefore, the course of cracking can be assembled from a series of independent finite element models, each one for a particular crack length. Each model gives the local tip stress connected to a certain remote load or, conversely, the critical remote load belonging to a critical tip stress defined by a certain crack extension criterion. The parametric plot of these critical remote loads and displacements as functions of crack length results in the simulated load-displacement curve. Fig. 3 shows three examples from the series of the finite element models used. They take contact and friction in the debonded region behind the crack tip into account. The extraction of the tip stresses depends on the finite element code. ANSYS® provides special singular elements with direct calculation of stress intensities. However, this method turned out to be less reliable. Therefore a more complicated technique was developed. A very fine mesh at the crack tip (compare Fig. 3) provides the accurate linear-elastic near tip stress field.

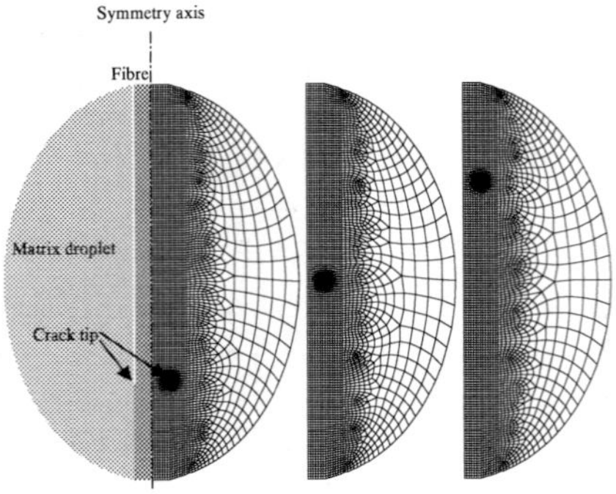

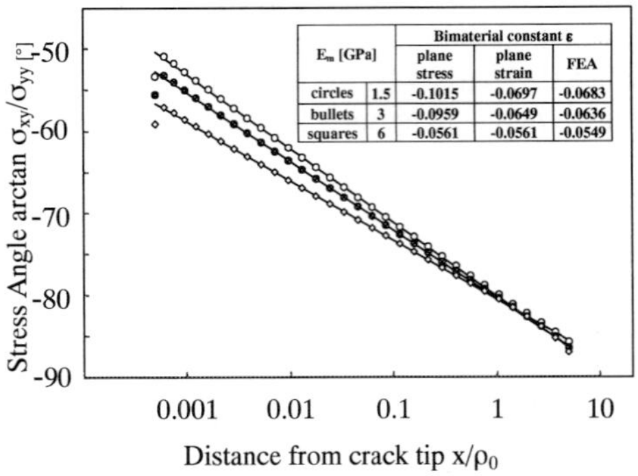

Fig. 3. (left) Finite element models for different crack lengths.
Fig. 4 (right) Least square fit of the near tip solution (solid lines) to the finite element results (symbols) for different Young's-moduli of the matrix and glass fibre, ρ_0=1μm.

Then the desired values for the bimaterial constant ε, the stress intensity $|K|$ and the mixed-mode angle Ψ^{ρ_0} result from a least square fit of

$$\left|\sigma_{yy} + i\sigma_{xy}\right| = \frac{|K|}{\sqrt{2\pi x}} \quad \text{and} \quad \arctan\frac{\sigma_{xy}}{\sigma_{yy}} = \Psi^{\rho_0} + \varepsilon \ln\frac{x}{\rho_0} \tag{4}$$

to the numerical near tip solution for stresses.
Fig. 4 shows examples of curve fitting with the second equation of Eq. (4). The finite element results agree very well with the expected behaviour except the results of the node next to the crack tip. The obtained bimaterial constant ε only slightly differs from the expected plane strain value. Thus, fitting the stress intensity factors to numerical near tip solutions is a very stable and accurate technique.

RESULTS AND DISCUSSION

The total energy release rate is dominated by the fibre's contribution, which is approximately $G_{fibre} = P/(4\pi^2 E_f r_f^3)$ where r_f and E_f are the fibre diameter and Young's modulus respectively. Noticeable deviations occur only for very short and very long cracks in the case of the SFPO-Test (Fig 5, left) whereas a noticeable deviation remains also for medium crack lengths in the case of the Microbond-Test (Fig 5, right). The wide plateau of the energy release rate for medium crack lengths leads to a stable and self similar crack propagation which can be used to extract the critical energy release rate from experimental data. This is the desired interface property.

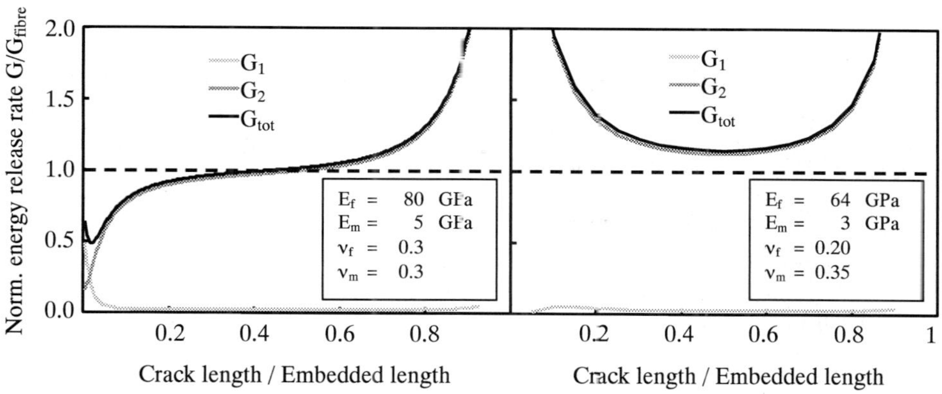

Fig. 5. Energy release rate normalised by the fibre contribution vs. crack length, (left) SFPO-test, (right) MB-test.

However, as Fig. 5 reveals, it is obtained for a dominant mode 2 loading of the interface, which is not expected to be the most serious one. This is probably the biggest disadvantage of pull-out tests.

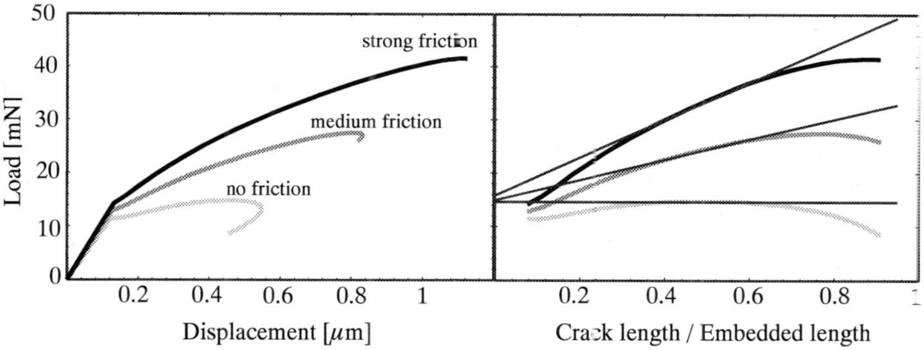

Fig. 6. Simulated load vs. displacement curves and load vs. crack length curves for the Microbond-Test.

Experimental load-displacement-curves show an almost linearly increasing force during stable crack propagation which can be explained by the frictional contact of debonded surfaces. Due to the linearity an extrapolation of the measured curve for medium length cracks down to zero crack length separates the frictional part and gives the force contribution that drives the crack. Fig 6 illustrates this procedure for simulated curves with different levels of friction. The extrapolation reveals a critical force much smaller than the maximum force. The data reduction scheme then follows three steps: Find the critical force P_c from experimental curves by extrapolating the stable crack propagation down to zero crack length. Calculate the fibre's contribution of the critical energy release rate G_c from P_c by $G_c = P_c/(4\pi^2 E_f r_f^3)$. In the case of the SFPO-test the contribution of the matrix is negligible and the fibre contribution is equal to the total energy release rate. In the case of the MB-test, however, the contribution of the matrix must be considered. The easiest way is to apply a correction factor of about 1to 1.25 which depends on the geometry and follows from the finite-element analysis, as given in Fig. 5b.

CONCLUSIONS

The SFPO- and MB-tests are appropriate to determine the critical energy release rate as an adhesion parameter. However, the mixed mode angle Ψ is near 90° and the local load is mainly shear. Consequently the tests do not provide the critical adhesion parameter for material design, because under normal local loads the interface would fail at lower critical energy release rates. As long as linear elastic material behaviour is a reasonable assumption for the materials tested it can be stated that at present the benefits but also the limits of the SFPO- and MB-tests are known. The existence of the plateau region of energy release rate, i. e. a self-similar crack growth, guarantees an extended validity of the modelling even for cases where small a plastified matrix regions may exist.
For the evaluation of the interfacial fracture toughness, no matter which mechanisms are contributing to it, then the integration of the whole force-displacement curve would be appropriate to get this quantity.

REFERENCES

1. Schüller, T., Beckert, W., Lauke, B., Friedrich, K., Single-fibre transverse debonding: tensile test of a necked specimen, (2000) *Composites Sci. and Technology* **60,** 2077-2082.
2. Kelly, A. (1979) *Proc. R. Soc.* **A319**, 95
3. Miller, B., Muri, P. and Rubenfeld, L. (1987), Composites Sci. and Technology **28**, 17.
4. Marotzke, Ch., The elastic stress field arising in the single-fibre pull-out test (1994) *Composite Sci. and Technology* **50**, 393-405.
5. Kerans, R.J., Theoretical analysis of the fibre pull-out and pushout tests, (1991) *J. Am. Ceram. Soc.* **74**, 1585-1596.
6. Zhou, L.M., Kim, J.K., Mai, Y.W. Mai, Interfacial debonding and fibre pull-out stresses, (1992) *J. of Materials Science* **27,** 3155-3166.
7. Atkinson, C., Avila, J., Betz, E. and Smelner, The rod pull-out problem, theory and experiment, (1982) *J. Mech. Phys. of Solids* **30**, 97-120.
8. Hampe, A., Faser-Matrix Haftung: Weiterentwicklung der Einzelfaser-Auszugs-Methode, (1993) *Materialprüfung* **35**, 269-275.
9. Schüller, T., Beckert, W., Lauke, B. and Perche, Analytical and Numerical Calculation of the Energy Release Rate for the Microbond-Test, (1999) *J. Adhesion* **70**, 33-56.
10. Hutchinson, J.W., Suo, Z., Mixed mode cracking in layered materials, (1992) *Advances in Applied Mechanics* **29**,63-191.
11. Beckert, W., Lauke, B., Critical discussion of the single-fibre pull-out test: does it measure adhesion?, (1997) *Composites Science and Technology* **57**, 1689-1706.

The Application of Fracture Mechanics to Polymers, Adhesives and Composites
D.R. Moore (Ed)

MICROSTRUCTURE DEPENDENT FRACTURE AND FATIGUE CRACK PROPAGATION BEHAVIOR IN INJECTION-MOLDED SHORT FIBER-REINFORCED THERMOPLASTICS

J. KARGER-KOCSIS

INTRODUCTION

Injection-molded short glass (GF) and carbon fiber (CF) reinforced composites are potential candidates for being used in technical domains, so far occupied by metallic and ceramic materials. Short fiber reinforcements provide high stiffness, strength, heat and dimensional stability to the related thermoplastic matrices rendering them suitable for various applications which usually require high resistance to fracture (at both dynamic and static conditions) and to fatigue crack propagation (FCP). Note that the related response can be assessed by the linear elastic fracture mechanics (LEFM) for which a recommendation was released by the ESIS TC-4 group [1]. The LEFM approach is usually valid for injection-molded composites except for some systems with supertough matrices. Nowadays, fracture mechanics tests are preferred instead of standardized ones as the former yield material parameters which can be used for design and construction purposes. This work surveys how the fracture and fatigue performance of composites with discontinuous fiber reinforcements depend on their microstructural details (fiber volume fraction, layer structure, fiber orientation, fiber aspect ratio and its distribution) generated by the injection molding process.

MICROSTRUCTURE DEVELOPMENT

The molding-induced microstructure both of the unfilled matrices and their short fiber reinforced grades can be approached by the flow model of Tadmor [2] and Rose [3]. This model considers the shear and elongational flow field developed during the cavity filling process. The reinforcing fibers may adopt a complex alignment due to the dominating flow field. As a result, several layers with different planar fiber orientation can be distinguished across the specimen thickness. Nevertheless, the fiber layering can well be approximated by a 3-ply laminate model composed of one central (C) and two surface (S) layers. The fiber are aligned in the mold filling direction (MFD) in the S-, whereas they adopt a transverse orientation to the MFD in the C-layer. In each layer the mean fiber orientation (f_p) can be characterized by a modified Hermans-type planar orientation parameter. It has been also shown that the fiber layering and orientation depend on the fiber volume fraction (V_f). Increasing V_f causes fiber attrition due to which the fiber aspect ratio (l/d) reduces. The microstructure of the short fiber reinforced composites with respect to the loading direction can be described by the reinforcing effectiveness parameter (R) [4-7]:

$$R = \left(\frac{2S}{B}\cdot f_{p,eff,S} + \frac{C}{B}\cdot f_{p,eff,C}\right)\cdot V_f\left(\frac{1}{d}\right)\cdot\left(\frac{1_n}{1_w}\right) \quad (1)$$

where B is the thickness of the molding, $f_{p,eff}$ is a complex function of f_p and the loading direction, and l_w/l_n represents the fiber length distribution (the subscripts w and n relate to the weight- and number-averages, respectively) and all other terms were defined before. Note that with increasing mold thickness the C/B ratio increases, too. Increasing V_f is usually accompanied with increasing f_p and decreasing l/d. However, injection-molding may cause fiber enrichment or depletion across the thickness and thus V_f is no longer constant. Further, often more than 3 layers can be distinguished. Therefore it was necessary to give a universal description for R [5-7]:

$$R = \sum_i T_{rel,i} f_{p,eff,i} \phi_{f,i} \left(\frac{l}{d}\right)_{equ,i} \frac{(l/d)_{n,i}}{(l/d)_{m,i}} \quad (2)$$

where $T_{rel,i}$ is the relative thickness of the i-th layer normalized to the sample thickness B, $f_{p,eff,i}$ is the effective orientation in the i-th layer calculated using the function of planar orientation, f_p vs. $f_{p,eff}$, introduced by Friedrich [4], $V_{f,i}$ is the fiber volume fraction in the i-th layer; $(l/d)_{equ,i}$ is the equivalent aspect ratio in the i-th layer; and $(l/d)_{w,i}$ and $(l/d)_{n,i}$ are the mean weight- and number-average aspect ratios in the i-th layer, respectively.

Accordingly, R considers the effects of fiber structuring (layering and orientation with respect to the crack path), molding induced local change in the reinforcement content, and the aspect ratio (affected by fiber bunching and bending) and its distribution (stress concentration effects of fiber ends). Note that the aspect ratio and its distribution consider implicitly the effect of fiber/matrix adhesion (interphase). The empirical nature of R is given by the fact that its constituent terms are interrelated, but they are considered in Equations 1 and 2 as independent parameters. It is worth noting that terms of the reinforcing effectiveness could be defined in other ways, as well (e.g. defining a three-dimensional fiber orientation parameter instead of a two-dimensional one).

MICROSTRUCTURE DEPENDENT FRACTURE TOUGHNESS

The microstructural efficiency (M) concept, credited to Friedrich [4], seems to be a powerful tool to estimate the change in fracture toughness due to reinforcement. According to this approach the fracture toughness of the composite related to the matrix at a given testing condition linearly depends on R (cf.Figure 1):

$$\frac{K_{c,c}}{K_{c,m}} = M = a + n \cdot R \quad (3)$$

where $K_{c,c}$ and $K_{c,m}$ are the fracture toughness of the composite and matrix, respectively; "a" is the matrix stress condition factor, and "n" is the energy absorption ratio.

The relative improvement in the toughness of the composites is plotted against R for composites based on polyphenylenesulfide (PPS), polyethernitrile (PEN) and polyetheretherketone (PEEK) in Figure 1.

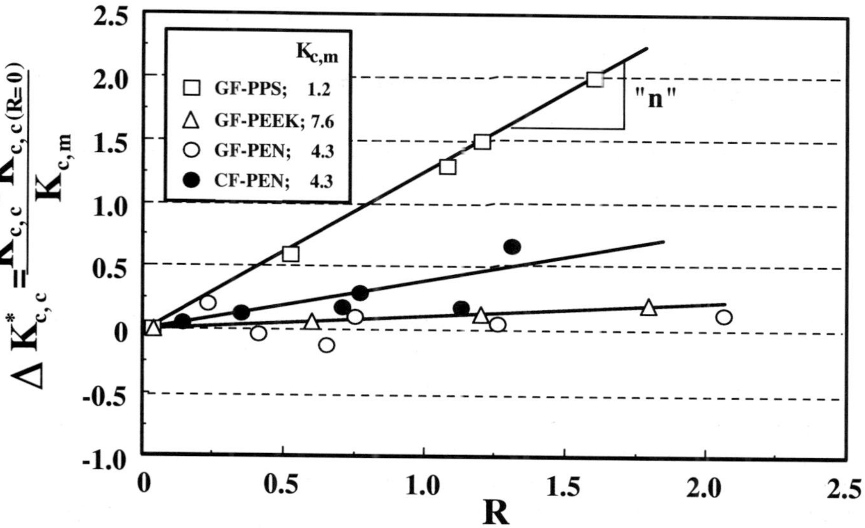

Fig.1. Relative toughness improvement by discontinuous fiber reinforcement in some high temperature-resistant polymers. (Testing conditions: compact tension specimens with both L-T and T-L notches, T=20 °C, v=1 mm/min crosshead speed). Notes: $K_{c,m}$ is given in $MPa.m^{1/2}$; according to ASTM E 616 "T-L" indicates that the loading is transversal, whereas the notching is longitudinal to MFD – for "L-T" specimens the opposite holds. [8]

Figure 1 shows that the efficiency of reinforcement is the higher the lower the matrix toughness is. That is the reason why all brittle polymers are commercialized only in filled, reinforced versions. The sign of "n" is usually positive which indicates that the energy absorption owing to fiber-related failure events (i.e. fiber fracture, fiber pull-out and fiber/matrix debonding) is higher than that of the matrix (crazing, cavitation, shear deformation). For GF-reinforced PEEK for example $n \approx 0$ and for GF-"reinforced" polycarbonate (PC) $n<0$ were found. As a consequence the failure of the PEEK-composite is matrix-dominated (incorporation of fibers does not change the failure mode due to the high matrix ductility), whereas GF addition "weakens" instead of reinforcing PC. The validity of Equation 3 was shown for many discontinuous short and long fiber reinforced injection-molded composites which were tested both in static and dynamic (impact) conditions ([6] and references therein).

MICROSTRUCTURE DEPENDENT FATIGUE CRACK PROPAGATION (FCP)

Under cyclic loading (tensile-tensile) conditions final breakdown of the short fiber reinforced composites is controlled by FCP instead of a crack initiation process. This is due to the fact that inherent "flaws" in form of structural inhomogeneities are always present [5,9]. The stable acceleration FCP range generally can be described by the Paris-Erdogan power law [10]:

$$\frac{da}{dN} = A \cdot (\Delta K)^m \tag{4}$$

where A and m are constants, da/dN is the FCP rate, and ΔK is the amplitude of the stress intensity factor.

Figure 2 shows the validity of Equation 4 on the example of injection-molded polypropylene (PP) reinforced with various amount of short (SGF) and long glass fibers (LGF). One can clearly see that with increasing V_f and l/d the FCP resistance is strongly improved.

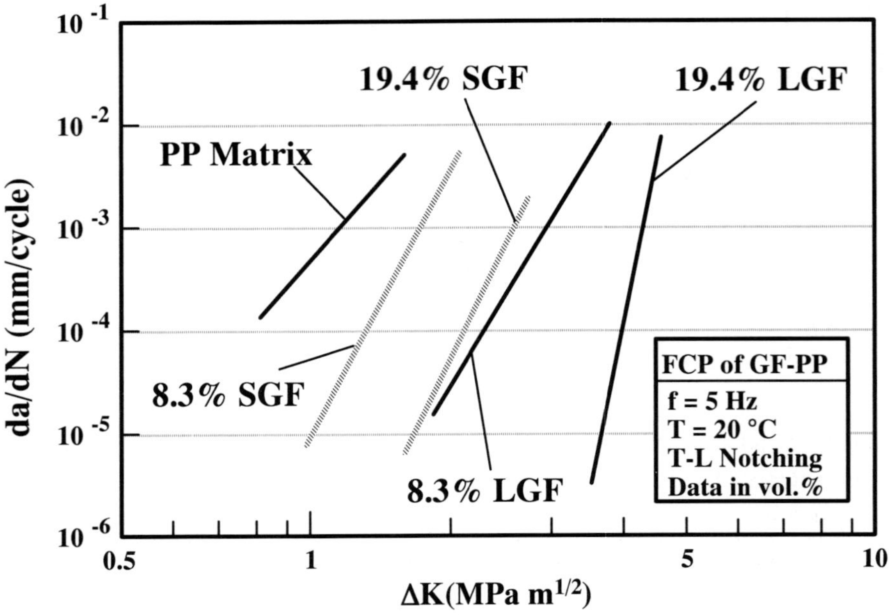

Fig.2. Simplified FCP response of SGF- and LGF-reinforced PP composites.
Notes: the minimum/maximum load ratio was 0.2; compact tension specimens were notched in T-L direction; experimental results published in Ref. [11].

The lessons from the FCP studies performed on various injection-molded composites can be summarized by the scheme in Figure 3. Note that below a threshold ΔK (ΔK_{th}) no crack growth occurs. Albeit this value is of great importance for design purpose, it can not be always determined. It should be mentioned that the stable acceleration (Paris-Erdogan range) is often preceded by a stable deceleration range. Its onset is traced to the microstructure and to the formation of an "equilibrium" damage zone ([11-12] and references therein). Figure 3 demonstrates that fast (catastrophic) fracture during FCP tests takes place at the static fracture toughness values.

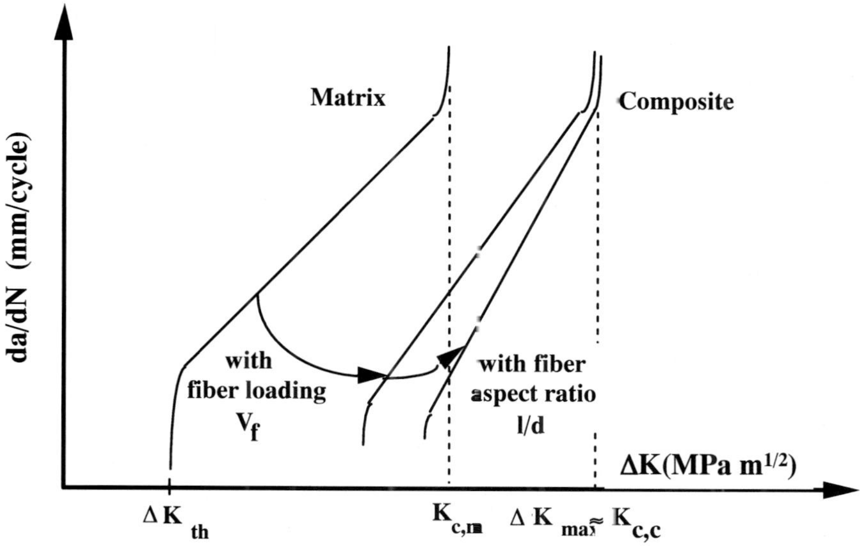

Fig.3. Effects of selected microstructural parameters (V_f and l/d) on the FCP response of injection-molded thermoplastic composites

Figure 4 shows the effects of microstructure on the preexponential (A) and exponential (m) parameters of Equation 4 on the example of PP reinforced by SGF and LGF in various amounts.

Since the upper bound of the validity of the Paris-Erdogan range is at about the critical fracture toughness ($K_{c,c}$ - cf. Figure 3) and the related failure events are also very similar at subcritical (FCP) and critical loadings (static fracture), the fatigue and fracture results are interrelated. Thus increasing R results in improved resistance to FCP of these materials. Further, the M concept (cf. Equation 3) is valid also for FCP [13-14]. Thus Equation 4 can be modified assuming that its constants are functions of M:

$$\frac{da}{dN} = \left(\frac{A}{M}\right) \cdot (\Delta K)^{m(M)} \quad (5)$$

$$\log\left(\frac{da}{dN}\right) = \log A - \log(a + n \cdot R) + m(M) \cdot \log(\Delta K) \quad (5)$$

Equation 6 simplifies further as m(M) or m(M).log(ΔK) are considered as constants. This occurs when "m" only slightly changes with "M", which is the case when "n" tends to 0. This prerequisite is the easier fulfilled the lower is the energy absorption ratio (n), or in other words, the higher is the ductility of the matrix. As a consequence the log(da/dN) vs. log M data pairs lay on a straight line, that was reported for PEEK- [13-14] and PP-composites [11-12]. It is worth mentioning, that the simplified version of Equation 6 works well, provided that the valid Paris-Erdogan range on the ΔK-scale is broad enough.

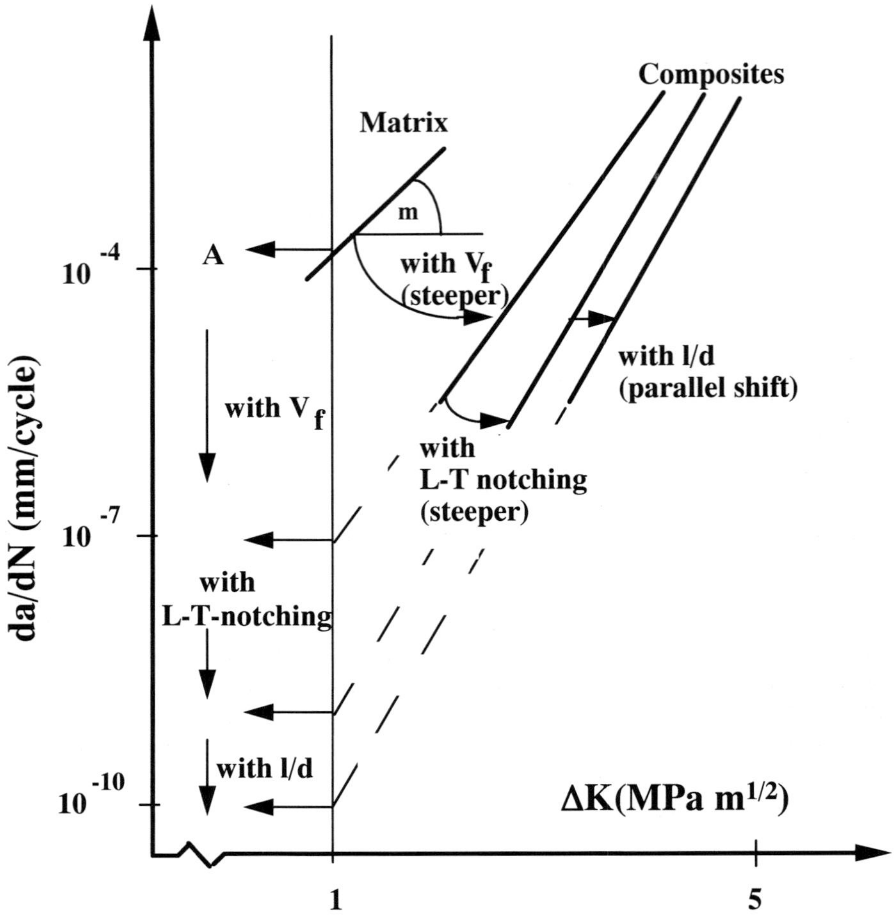

Fig.4. Effects of microstructural factors on the FCP behavior (viz. values A and m) of injection-molded PP composites with SGF and LGF reinforcements. Note: L-T notching at the usual thickness of injection-molded parts (ca. 3mm) results in a higher proportion of reinforcing fibers aligned in the loading direction as 2S > C holds.

CONCLUSIONS

Improvement in fracture toughness and fatigue crack propagation (FCP) behavior achieved by incorporation of discontinuous fibers highly depends on the matrix characteristics. The related effect is the stronger the lower is the matrix toughness. Both the fracture toughness and the FCP response can be correlated to the microstructure of the composites by means of the microstructural efficiency factor (M). This "M" term counts for the effects of the reinforcing fibers incorporated, by considering their relative effectiveness with respect to the loading direction (R), the change in the matrix stress state (a) and the variation in the energy absorption balance between fiber- and matrix-related failure events (n) induced by the fibers present.

Fractographic analysis revealed that the dominant failure events are practically the same in static fracture and FCP tests (if no hysteretic heating occurs in the latter case). This is a further support for the reliability of the microstructural efficiency concept and for the common treatise of the fracture and FCP responses using the LEFM.

REFERENCES

1 Moore, D.R. (2001). In: *Fracture Mechanics Testing Methods in Polymers Adhesives and Composites*, ESIS Publ.28, pp. 59-71, Moore, D.R., Pavan, A. and Williams, J.G. (Eds.), Elsevier Sci., Oxford
2 Tadmor, Z.(1974) *J.Appl. Polym. Sci.*, **18**, 1753.
3 Rose, W.(1961) *Nature,* **191**, 242.
4 Friedrich, K. (1985) *Compos. Sci. Technol.*, **22**, 43.
5 Karger-Kocsis, J. (1989). In: *Application of Fracture Mechanics to Composite Materials*, pp. 189-247, Friedrich,K. (Ed.), Elsevier, Amsterdam
6 Karger-Kocsis, J. (2000). In: *Polymer Blends, Vol.2: Performance*, pp. 395-428, Paul D.R. and Bucknall, C.B. (Eds.), Wiley, N.Y.
7 Karger-Kocsis, J. and Friedrich, K.(1988) *Compos. Sci. Technol.*, **32**, 293.
8 Karger-Kocsis, J. (1991) *Polym. Bull.*, **27**, 129.
9 Karger-Kocsis, J.(1991). In: *International Encyclopedia of Composites*, Vol. 5, pp.337-356, Lee, S.M. (Ed.), VCH Publ., N.Y.
10 Paris, P.C. and Erdogan, F.(1963) *J. Basic Eng.*, **85**, 528
11 Karger-Kocsis, J., Friedrich, K. and Bailey R.S. (1991) *Adv. Compos. Mater.*, **1**, 103.
12 Karger-Kocsis, J. (1995). In: *Polypropylene: Structure, Blends and Composites*, Vol.3 Composites, pp.142-201, Karger-Kocsis, J. (Ed.), Chapman and Hall, London
13 Karger-Kocsis, J., Walter, R. and Friedrich, K.(1988) *J. Polym. Eng.*, **8**, 221.
14 Friedrich, K., Walter, R., Voss, H. and Karger-Kocsis, J. (1986) *Composites,* **17**, 205.

The Application of Fracture Mechanics to Polymers, Adhesives and Composites
D.R. Moore (Ed)

A STIFFENER DEBONDING STUDY USING FRACTURE MECHANICS DATA

Peter DAVIES

INTRODUCTION

Although fracture mechanics tests are frequently used to characterise adhesives, both on metal and composite substrates, there are few examples of the application of the measured data to predict the response of bonded structures. Kinloch and colleagues have developed a methodology to predict the service life of bonded joints using mode I fracture data [1].

The aim of the present study was to examine whether the values of G_{Ic} obtained from simple test specimens could be used to predict the debonding of a stiffener in a structural composite element. Previous results of test-prediction correlations for lap shear specimens (see chapter 3.2) appeared promising and the failure mode in tests on stiffened panels often involves the development and propagation of an interface crack, which is well-suited to a fracture mechanics approach.
The element chosen for this study is representative of assemblies widely used in marine structures, in which a structure such as a boat hull or a bulkhead is manufactured, and subsequently stiffened by overlaminated top-hat sections. Such elements have been extensively studied by Shenoi and coworkers [2] and some attempts have been made to use a fracture mechanics approach to look at delamination in the curved part of the overlaminate [3]. In the present work it is another damage mechanism, debonding at the interface between panel and stiffener, which is addressed.

MATERIALS AND FABRICATION

Two materials have been studied, a woven glass fabric reinforced polyester and a stitched glass reinforced epoxy. Table 1 presents the material characteristics.

	Glass/Polyester	Glass/Epoxy
Resin	Isophthalic polyester (Cray Valley S70390TA)	Epoxy/amine hardener (Sicomin SR1500/2505)
Reinforcement	Woven fabric 0/90°, 500 g/m²	Multi-axial stitched fabric, QX, (0/90/+45/-45°), 1034 g/m²
Fibre content Mass %	58	61
Tensile modulus, GPa	21	15
Mode I fracture toughness, J/m² (COV in brackets)	180 (6%)	330 (19%)

Table 1. Materials studied

Double cantilever beam (DCB) and top hat specimens were manufactured by hand layup at IFREMER. For the latter the sandwich panels with a 40mm thick balsa core are produced first. Then a foam former is placed on the panel and the top hat is produced by overlaminating. Either 8 layers of woven fibres or 4 layers of QX are used to make the overlaminate. The same number of layers makes up each arm of the DCB specimens (about 4mm thickness).

FRACTURE TESTS ON DCB SPECIMENS

The mode I fracture toughness of the two materials was determined by tests on specimens of the same thickness as the top hat elements, Figure 1. The test procedure was that developed within the ESIS TC4 group [4]. Specimen dimensions are larger than those usually employed for aerospace materials, as the weave cell dimensions are several millimetres. Specimen width is 50 mm, length is 300 mm and starter film is 8 micron thick polypropylene 75 mm long. The initiation values from the fracture tests, defined as non-linearity on the load-displacement plots, are given in Table 1.

Figure 1. Mode I test

TOP HAT PULL-OFF TESTS

The test set-up is shown in Figure 2. Two digital cameras are used to record crack initiation and propagation, and two acoustic emission transducers are placed on the stiffener webs.

Figure 2. Top hat pull-off test.

Specimens were prepared with and without implanted 8 micron thick polypropylene defects. The latter were nominally 20, 40 or 60 mm long, actual lengths were measured on fracture surfaces after test. Four specimens were loaded to failure for each condition at a machine cross-head displacement rate of 5 mm/minute. Figure 3 shows examples of load-displacement plots for specimens with and without defects.

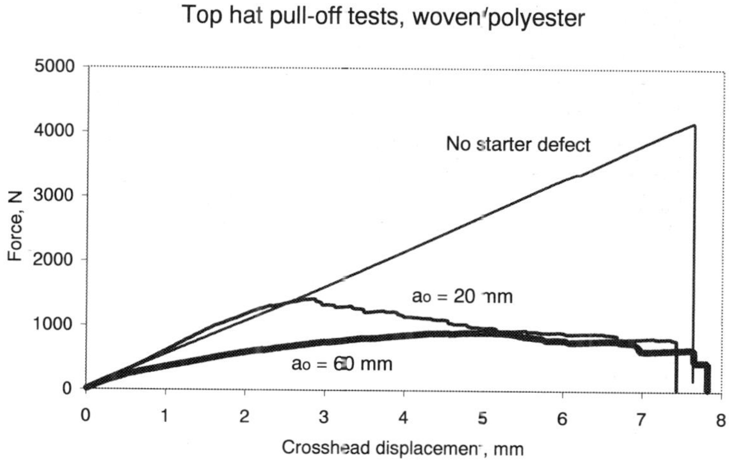

Figure 3. Examples of load-displacement plots, glass/polyester pull-off tests

PULL-OFF LOAD PREDICTION

The pull-off load for these tests was estimated using a simple beam analysis. This involves a number of assumptions. Loading is assumed to be pure mode I, which appears justified given the very stiff sandwich base. Linear elastic behaviour is also assumed. This may be justified for the short (20 mm) defects but is clearly more debatable for the longest (60mm) starter cracks.

The expression for pull-off load is based on the analysis of Hashemi et al. [5] ;

$$F = 2\left[\left(\frac{4B^2h^3EG_{Ic}}{21a_o^2}\right)\right]^{0.5} \qquad (1)$$

with B the specimen width, h the thickness of the overlaminate, E the tensile modulus, and a the crack length.

Different initiation criteria were applied, here a non-linearity criterion (NL) is presented.

This is clearly a very simple model, and numerical modelling has also been examined to look at more complex cases such as stiffener ends, but the test geometry considered here is very close to the mode I DCB test.

TEST-PREDICTION CORRELATION

The correlation between predictions and test results is shown in Figure 4 below.
It may be noted that the failure loads for the QX/epoxy are significantly higher than those measured for the woven/polyester specimens. This difference is also found in the mode I tests, and suggests that those tests can be used to examine parameters such as surface preparation. The agreement between tests and predictions is quite good for the polyester composite. For the epoxy the measured values are higher than predictions for the longer cracks. There are several areas of uncertainty, concerning both the model and the input data, and these have been discussed elsewhere [6]. One point to bear in mind is that these materials are made by hand lay-up and that the fabrication procedure involves a delay between panel fabrication and stiffener overlamination. This has two important consequences. First, the mode I fracture toughness was initially taken to be 240 J/m², based on results from tests on unidirectional glass/epoxy specimens presented elsewhere [7], but the tests on QX fibre reinforced specimens laminated with the same delay gave values of 330 J/m².

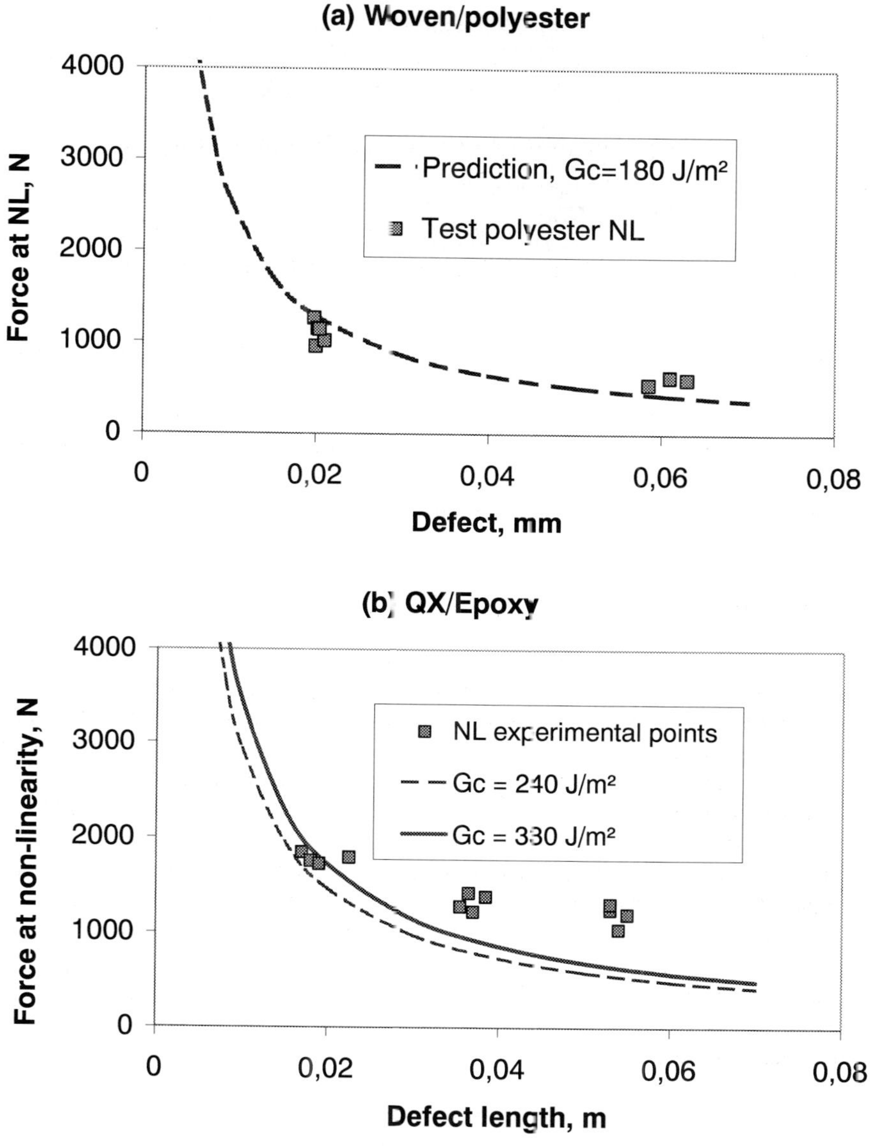

Figure 4. Test-prediction correlation, top hat pull-off tests.

Predictions using both values are shown in Figure 4. Second, the predictions were based on nominal properties and geometry but the actual thickness may vary in practice, Figure 5, particularly for the epoxy with the higher surface weight reinforcement which is harder to form into a tight radius. The simple model is very sensitive to thickness variations and an improved correlation can be obtained by adjusting this parameter, but this reduces the usefulness of the model as a predictive tool. Nevertheless the failure load estimations are quite reasonable, particularly for the short starter cracks, even with the nominal thickness values.

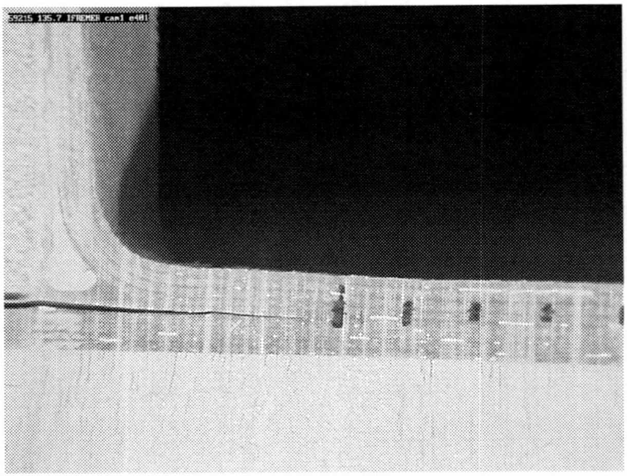

Figure 5. Detail of radius of stiffener/plate joint

CONCLUSIONS

The first results from this study show that:

- Fracture mechanics tests can be used to give a good qualitative indication of stiffener/panel interface performance.
- Simple analyses can give a reasonable estimation of failure loads for simple structures.
- Application to industrial structures requires great care to ensure that the input values are appropriate. This will require determination of mixed mode fracture data for materials in many cases.

REFERENCES

1. Curley AJ, Hadavinia H, Kinloch AJ, Taylor AC, Predicting the service life of adhesively bonded joints, Int. J. Fracture, 103, 2000, 41-69.
2. Shenoi RA, Hawkins GL, An investigation into the performance characteristics of top-hat stiffener to shell plating joints, Composite Structures, 30, 1995, 109-121.
3. Phillips HJ, Shenoi RA, Moss CE, Damage mechanics of top-hat stiffeners used in FRP ship construction, Marine Structures, 12, 1999, 1-19.
4. Moore DR, Pavan A, Williams JG, eds., Fracture Mechanics Testing methods for Polymers, Adhesives and Composites, ESIS Publication 28, 2001, Elsevier.
5. Hashemi S, Kinloch AJ, Williams G, Mixed mode fracture in fiber-polymer composite laminates, ASTM STP 1110, ed O'Brien TK, 1991, 143-168

6. Davies P, Sargent JP, Fracture mechanics tests to characterise bonded glass/epoxy composites, Proc. 3rd ESIS TC4 conference, September 2002.
7. Ducept F, Gamby D, Davies P, A mixed mode failure criterion from tests on symmetric and asymmetric specimens Comp. Sci & Technology, 59, 1999, p609-619.

The Application of Fracture Mechanics to Polymers, Adhesives and Composites
D.R. Moore (Ed)

THE INFLUENCE OF EPOXY-METAL INTERPHASE PROPERTIES ON FRACTURE TOUGHNESS

A.A. ROCHE and J. BOUCHET

INTRODUCTION

Understanding the interphase is important for both fundamental and practical aspects of adhesion. Indeed, for composite systems made of two components (the substrate and the organic layer) the interphase and its properties determine the final properties (practical adhesion, corrosion resistance, and durability). Liquid epoxy-diamine mixtures are extensively used as adhesives or paints in many industrial applications. When they are applied onto metallic substrates and cured, epoxy-amine liquid monomers react with the metallic oxide and/or hydroxide to form chemical bonds [1, 2] increasing practical adhesion between the epoxy polymer and the substrate surface [3, 4]. Different studies report the influence of the nature of the metallic substrate on the prepolymer cross-linking [5]. Moreover, when epoxy resins are applied onto metallic substrates and cured, intrinsic and thermal residual stresses develop within the entire organic layer [6-8]. Whatever their source, these residual stresses reduce the practical adhesion and may induce cracks in coating materials [9-11] resulting in a drop of the overall performance of adhesives or paints. To gain a better understanding of the fundamental epoxy/metal adhesion requires a full knowledge of chemical and physical reactions that take place at the epoxy/metal interface [12,13] and the related diffusion phenomena [14]. The organic layer in the vicinity of the substrate surface has to be considered as an interphase containing gradients of residual stresses and Young's modulus [15] resulting from structural rearrangement, intermolecular and inter-atomic interactions and diffusion phenomena [12] in which some new chemical species may be formed. When epoxy/metal systems failed, it was possible not only to correlate the residual stresses at the interphase/metal interface to practical adhesion as defined in this contribution but also to correlate the fundamental adhesion and durability to the presence or absence of some chemical species [16]. The aim of this paper is to establish the influence of the interphase formation on the fracture toughness. Then, fracture mechanics (i.e., a version of a G_{Ic} measurement) is used to distinguish the effects of different interphase properties (gradients of Young's modulus, residual stress profiles,...) on adhesive layers produced by special preparation and curing as well as after ageing.

EXPERIMENTAL

The metallic substrate used was a 0.5 mm thick commercial rolled aluminium alloy (5754 from Pechiney) prepared by die-cutting to provide identical sized strips (50x10 mm or 150x10 mm). Before any polymer application, aluminum substrate surfaces were ultrasonically degreased in acetone for 10 min. and wiped dry with a soft absorbent paper. Some aluminium panels were also chemically etched (immersed in a solution of 250g/l of sulfuric acid, 50g/l chromic acid and 87 g/l aluminium sulphate octadecahydrate at 60°C for 20 min., rinsed in running tap water for 1 min., immersed in deionized water for 5 min. and wiped dry with a soft absorbent paper.). After surface treatment, all substrates were kept in an air-conditioned room (22±2°C and 55±5% R.H) for 2 h. The bifunctional epoxy prepolymer used was a liquid diglycidyl ether of bisphenol A (DGEBA, M=348 g/mole, DER 332 from Dow Chemical).

The liquid diamine curing agent used was isophorone-diamine (IPDA or 3-aminomethyl-3,5,5-trimethylcyclohexylamine from Fluka). Monomers were used without further purification. Assuming a functionality of 4 for diamine and 2 for the epoxy monomer, a stoichiometric ratio (a/e = aminohydrogen/epoxy) equal to 1 was used throughout this work. Homogeneous mixtures of DGEBA and diamine were achieved by stirring under vacuum (1 Pa) at room temperature for 1 h (Rotavapor RE211 from Büchi, Switzerland) to avoid air bubble formation. To allow chemical reactions between the substrate surface and liquid monomers to take place, leading to the full interphase formation, liquid monomers were kept in contact with the metallic surface for 3 hours at room temperature (under N_2 flow to prevent monomer oxidation and/or carbonation) before commencing the polymer curing cycle thereby achieving the maximum conversion of monomer and the highest glass transition temperature ($T_g^{\infty} = 163°C$): (20→60°C (2°C/min); 2 hours at 60°C; 60→140°C (2°C/min); 1 hour at 140°C; 140→190°C (2°C/min); 6 hours at 190°C; cooling (8 hours) down to 20°C in the oven). In the following, we have denoted (i) 3 hours at room temperature plus the curing cycle (see figure 1). Conversely, when we did not want the interphase formation, immediately after the epoxy-diamine application (less than 2 min.) the coated specimens were placed in a preheated oven at 190°C, held for 6 hours and cooled down to 20°C in the oven for 8 hours. This curing cycle is denoted (ii) (see figure 1). To obtain the desired coating thickness, an automatic film applicator (from Sheen) was used. The coating layer and the substrate have the same widths. After the curing cycle and cooling down, the coating thickness was determined using a Digital Linear Gauge (Model EG-100) having a ± 5 μm sensitivity.

RESULTS AND DISCUSSIONS

When liquid epoxy-diamine prepolymers were applied onto metallic substrates, interphases between the coating part having the bulk properties and the metallic surface were created. Chemical, physical and mechanical properties of the formed interphase depend on the natures of both the substrate surface and the diamine hardener. In previous works, we have pointed out that the interphase formation mechanisms result from a dissolution and a diffusion phenomena [17,18]. When the pure DGEBA monomer was applied onto the metallic surfaces or when pure diamine monomers were applied onto gold coated substrates, no chemical reaction was observed. On the contrary, when pure diamine monomers were applied onto either titanium or aluminum metallic surfaces, chemical reactions occurred. Following the amine chemical sorption onto oxided or hydroxided metallic surfaces, a partial dissolution of the surface oxide (and/or hydroxide) metallic substrate was observed according to the basic behavior of diamine monomers [2].

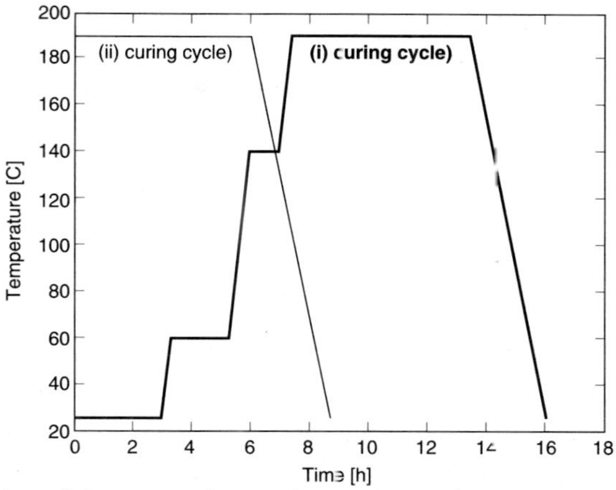

Fig. 1. Representation of the two curing cycles used for DGEBA-IPDA system. ((i) with interphase formation, (ii) without interphase formation)

Metallic ions diffuse within the liquid monomer mixture (epoxy-diamine) and react by coordination with amine groups of the diamine monomer to form organo-metallic complexes (or chelates). When the complex concentration was higher than its solubility limit, complexes (or chelates) crystallized as sharp needles. During the curing cycle, organo-metallic complexes react with the epoxy monomer leading to a phase separation corresponding to the formation of a new epoxy network having a lower Tg. Moreover, crystals not dissolved after the curing cycle, acted as short fibers in the organic matrix leading to an increase of the Young's modulus. The same mechanisms were observed for Sn, Zn, Fe, Cr, Cu metallic substrates covered by their oxide or hydroxide layer and E-glass substrate. Since dissolution and diffusion phenomena were expected, the interphase formation should be related to the liquid-solid contact duration between liquid prepolymers and metallic substrates. This means that it is possible to control the interphase formation by limiting the contact duration between the liquid prepolymers and metallic substrates. That can be easily achieved. Thus, immediately after the application of the DGEBA-IPDA mixtures onto treated metallic substrates (less than 2 min.), panels were introduced in a oven at 190°C and kept in for 6 hours before cooling down (Curing cycle (ii)). In table 1, we have reported the final physical and mechanical properties of both bulk polymers and films according to the two different curing cycles (i and ii). We observed that bulk mechanical properties (E) are identical irrespective of the curing cycle. However the bulk physical properties (Tg) are slightly different due certainly to a slight diamine evaporation during curing cycle (ii). For the 30 μm film on aluminum with curing cycle (i), we observed, as explained previously, that the mechanical and physical properties were quite different from those of the bulk inferring interphase formation. However, for the 30 μm film on aluminum with curing cycle (ii), mechanical and physical properties were the same as those of the bulk suggesting that the interphase was not formed or that the interphase is so thin that we are unable to observe it.

Table 1. Mechanical (E) and physical (Tg) properties of DGEBA-IPDA systems.

Material	Curing cycle	E [GPa]	Tg [°C]
Bulk	(i)	3.2	163
	(ii)	3.1	156
30 μm thin film on Al	(i)	14	110
30 μm thin film on Al	(ii)	3.3	154

As the practical adhesion measurement is related to an adhesional failure (or failure within the interphase region), it is obvious that the interphase properties have to be considered. In recent works, we have reported that the consideration of the mechanical interphase properties is of prime importance to understand the mechanical behavior of bonded structures [19-21]. To point out the role of interphase formation on fracture toughness, we used the model developed in this volume in order to calculate both residual stresses and the critical strain energy release rate. For three-layer systems (i.e. when curing cycle (i) was used), residual stresses were calculated using the Young's modulus gradient, within the interphase, observed experimentally [19]. Obviously, for bi-layer systems (i.e. when curing cycle (ii) was used), residual stresses were calculated using the Young's modulus of the bulk coating. Figures 2 and 3 represent the profile of the residual stresses in the tri and bi-layer system, respectively. The maximum stresses within the tri-layer system were at the interphase/metal interface while the ones observed in the bi-layer system were at the bulk coating/metal interface irrespective of the surface treatments. In mechanical terms this means that the failure of such systems should take place where residual stresses are at their maximum, i.e. that adhesional failures should be observed. Moreover, we can observe, for tri-layer systems (i.e. those which contain an interphase), an increase of the maximum stress intensity compared to the bi-layer systems (i.e. without interphase). Thus the presence of the interphase may favor an increase of the residual stresses. Those results are consistent with the assumption that the practical adhesion will decrease as soon as the interphase is formed. By using the model previously developed in this volume, we have calculated the quantity of energy required to initiate the failure.

To point out the role of the interphase formation on the $G_{I_c}^{flexure}$ value, the calculation was made for systems with or without interphase before and after hydrothermal aging (immersion of bonded specimen in distilled water at 40°C for 12h). The results obtained for both degreased and chemically etched aluminum are listed in Tables 2 and 3 before and after hydrothermal aging respectively. From Table 2, we can observed that $G_{I_c}^{flexure}$ decreases when the interphase is formed (curing cycle (i)) whatever the surface treatment is. To consider residual stresses for the $G_{I_c}^{flexure}$ calculation leads to an increase of the $G_{I_c}^{flexure}$ values whatever the surface treatment is. Residual stresses within painted or bonded systems act as a potential deformation energy. Obviously, as soon as this potential energy is higher than the intrinsic adhesion energy, a spontaneous coating delamination (or buckling) is expected and experimentally observed [8].

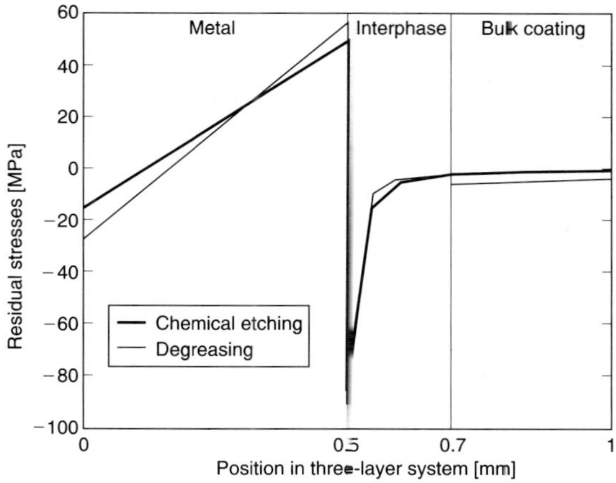

Fig. 2. Variation of residual stresses as a function of position in the three-layer system for both chemically etched and degreased aluminium.

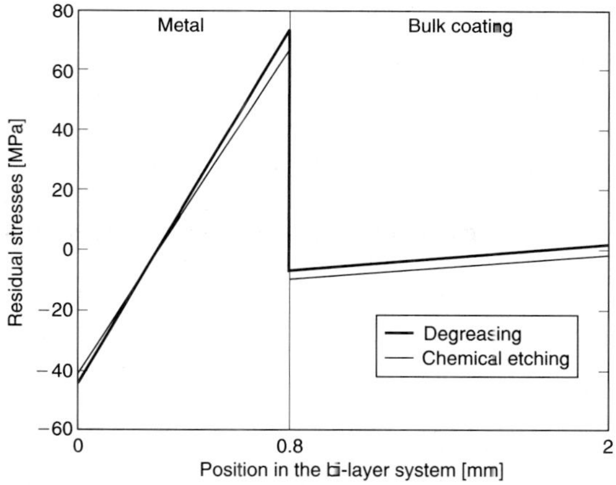

Fig. 3. Variation of residual stresses as a function of position in the bi-layer system for both chemically etched and degreased aluminium.

Table 2. $G_{I_c}^{flexure}$ results obtained for both degreased and chemically etched aluminium before aging.

Surface treatment	**Calculation assumptions**	**Without interphase** $G_{I_c}^{flexure}$ **[J/m²]**	**With interphase** $G_{I_c}^{flexure}$ **[J/m²]**
Degreased	Without stresses	84±17%	40±13%
	With stresses	115±14%	73±8%
Chemically etched	Without stresses	116±14%	74±12%
	With stresses	138±15%	82±13%

According to residual stress profiles shown in Fig 1 and 2, this potential energy appears to be higher for degreased samples than chemically etched ones. Thus the $G_{I_c}^{flexure}$ values for chemically etched samples are more significant than the degreased ones. As preliminary work, some samples were hydrothermally aged by immersion in distilled water at 40°C for 12 h. Results obtained are listed in Table 3.

Table 3. $G_{I_c}^{flexure}$ results obtained for both degreased and chemically etched aluminium after aging.

Surface treatment	**Calculation assumptions**	**Without interphase** $G_{I_c}^{flexure}$ **[J/m²]**	**With interphase** $G_{I_c}^{flexure}$ **[J/m²]**
Degreased	Without stresses	52±18%	66±14%
	With stresses	81±14%	92±13%
Chemically etched	Without stresses	87±16%	105±14%
	With stresses	98±13%	115±12%

After hydrothermal aging, a decrease of critical strain energy release rate was observed for samples without an interphase (curing cycle (ii)). On the contrary, when the interphase was formed, the critical strain energy release rate increased slightly with the duration of the liquid-solid contact time. According to this result, we can assume that the interphase and more particularly the organo-metallic (chelate) complexes act as a barrier material or as a water corrosion inhibitor within the interphase region. The corrosion inhibiting behavior of amine organo-metallic complexes has been reported previously [22].

CONCLUSION

When epoxy-diamine prepolymers were applied onto metallic substrates, interphases between the coating part having the bulk properties and the metallic surface were created. The mechanisms of the interphase formation are dominated by dissolution and diffusion phenomena. Since dissolution and diffusion phenomena were observed, the interphase formation should be related to the liquid-solid contact duration between liquid prepolymers and metallic substrates. Using an appropriate curing cycle for prepolymers it is possible to control the interphase formation. We observed that the interphase formation decreases practical adhesion by increasing residual stresses and increases the hydrothermal durability of coated samples. Thus, it has been clearly shown that the practical adhesion represented here by the critical strain energy release rate depends on the residual stresses. Samples that contain interphase are subjected to a higher residual stress level compared to the ones without interphase. However, the presence of the interphase and more particularly the organo-metallic (chelate) complexes act as a barrier material or water corrosion inhibitor within the interphase region. Nevertheless, these organo-metallic complexes certainly seem to be responsible for the increase of the residual stresses. For otherwise identical systems, overall properties of coatings, including the interphase ones, depend on the contact duration between metallic surfaces and liquid prepolymers and can explain the different findings observed in the literature.

Obviously to characterize practical adhesion using ultimate parameters such as ultimate load before failure and/or assuming a perfect interface between the organic coating and the substrate is not sufficient and/or useful in order to correlate practical adhesion to fundamental adhesion. The interphase formation and resulting Young's modulus gradient and residual stress profile have to be considered in applying relevant fracture mechanics concepts to this goal.

REFERENCES

1. Peillex, E. (1996). *Ph.D. thesis*, Université de Lyon 1, France.
2. Fauquet, C. (1992). *Ph.D. thesis*, Université Paris 6, France.
3. Walker, P. (1980) *J. Coating. Technol.* **52**, 49.
4. Guminski, R.D. and Meredith, F.M.P. (1961) *J. Oil. Colour Chem. Assoc.* **44**, 93.
5. Hine, P.J., Muddarris, S.E.L. and Packham, D.E. (1987) *J Adhesion Sci. Technol.* **1**, 69.
6. Entenberg, A., Lindberg, V., Fendrock, L., Hong S.K., Chen, T.S. and Horwath, R.S. (1989). In: *Metallized Plastics 1: Fundamental and Applied Aspects*, pp. 103-113, K.L. Mittal and J.R. Susko (Eds.), Plenum Press, New York.
7. Von Preissing, F.J. (1989) *J. Appl. Phys.* **66**, 4262.
8. Scafidi, P. and Ignat, M. (1998) *J. Adhesion Sci. Technol.* **12**, 1219.
9. Orsini, H. and Schmit, F. (1993) *J. Adhesion* **43**, 55.
10. Thouless, M.D. and Jensen, H.M. (1994) *J. Adhion Sci. Technol.* **8**, 579.
11. Mulville, D.R. and Vaishnav, R.N. (1975) *J. Adhesion* **7**, 215.
12. Mittal, K.L. (1978). In: *Adhesion Measurement of Thin Films, Thick Films and Bulk Coatings*, pp. 5-17, K.L. Mittal (Ed.), ASTM, Philadelphia.
13. Sharpe, L.H. (1972) *J. Adhesion* **4**, 51.
14. Gaillard, F., Hocquaux, H., Romand, M. and Verchère, D. (1992). Proceedings of Euradh'92, pp. 122-127, Karlsruhe, Germany.
15. Safavi-Ardebili, V. Sinclair, A.N. and Spelt, J.K. (1997). *J. Adhesion* **62**, 93.
16. Finlayson, M.F. and Shah, B.A. (1990). *J. Adhesion Sci. Technol.* **4**, 431.
17. Bouchet, J. and Roche, A.A. (2002) *J. Adhesion* **78**, 799.
18. Bentadjine, S., Roche, A.A. and Bouchet, J. (2001). In: *Adhesion Aspects of Polymeric Coatings*, Vol. 1, pp. 239-260, Mittal, K.L. (Ed.). VSP, Utrecht, the Netherlands.
19. Bouchet, J., Roche, A.A. and Jacquelin, E. (2002) *J. Adhesion Sci. Technol.* **15**, 321.
20. Bouchet, J., Roche, A.A., Jacquelin, E. and Scherer, G.W. (2001). In: *Adhesion Aspects of Thin Films*, Vol. 1, pp. 217-237. Mittal, K.L. (Ed.).VSP, Utrecht, The Netherlands.
21. Bouchet, J., Roche, A.A. and Jacquelin, E. (2001). *J. Adhesion Sci. Technol.* **15**, 345.
22. Duprat, M. and Dabozi, F. (1981). *Corrosion*, **37**, 89.

The Application of Fracture Mechanics to Polymers, Adhesives and Composites
D.R. Moore (Ed)

FRACTURE TESTS TO OPTIMISE MARINE COMPOSITE MANUFACTURING

P. DAVIES and H. LOAEC

INTRODUCTION

Most composite manufacturing processes involve the assembly of composite components, either in one or multiple steps. For example a complete structure with integral stiffeners may be fabricated in one shot in an autoclave or a structure such as a boat hull may be produced in a mould and then stiffeners over-laminated in a second step. In both cases there is a need to optimise the mechanical strength of interfaces between components, and fracture mechanics can be useful to evaluate surface preparations, adhesives and cure cycles without having to produce large structures each time. This article will describe one example in which fracture mechanics tests have been used to quantify the influence of the manufacturing procedure on assembly performance. The case considered is that of over-lamination. This is a procedure widely used in the marine industry (for both pleasure boats and military ships) and involves wet lay-up of glass fibres and resin onto a partly- or fully-cured substrate. There are many parameters which can influence the behaviour of the interface but two which have been seen to be critical are:

- the delay period between completion of the substrate and over-lamination, and
- the preparation of the substrate surface.

It should be emphasised that the polyester resins most frequently used in the marine industry are not post-cured and will continue to cure naturally for days or weeks after fabrication. Thus a long delay before over-lamination is likely to affect the interaction with the new laminate. This situation may be worsened if additives, such as those present in some low styrene emission resins, can migrate to the surface. There are different surface preparations available. The simplest is to apply a peel ply, which protects the surface and is removed just before over-lamination to reveal a new surface. Additional surface preparations include abrasion, solvent cleaning or even grinding to ensure that a clean, uncontaminated surface is offered to the overlaminate.

METHODS OF EVALUATION OF THE INTERFACE

There are relatively few practical options for characterising interfaces such as these. The most widely used method is the standard short beam shear (ILSS) test. This requires little material and can produce useful qualitative results if care is taken, but it may also produce failure modes other than interface failure and results cannot be used in design calculations. For this reason values obtained are usually termed “apparent shear strength”. In order to obtain more valid shear strength data, more complex and expensive specimens such as the Iosipescu beam are required. There are also through-thickness tensile test procedures, but again they are complicated and not very satisfactory. The idea of using fracture mechanics tests is therefore quite attractive.

MATERIALS

The materials tested in the present study are all woven glass reinforced polyester composites. The reinforcement consists of 16 plies of 500 g/m^2 woven E glass (balanced

0/90°), about 8 mm total thickness, in an isophthalic unsaturated polyester (Cray Valley 703) resin. Seven panels were produced over a two month period. The first 8 plies of six of the panels were impregnated the same day. Three of these panels were protected with peel plies (referred to as P), the other three were left exposed to the air (referred to as N). After one day, one week and 6 weeks the second series of 8 plies were laminated onto the first 8 plies, after removing the peel plies when present. An 8 micron thick polypropylene starter film was included at mid-thickness. No additional surface preparation was performed before overlamination. A seventh panel was laminated continuously from the same resin and fibres, as a reference material (C) and in order to determine how the aging of the composite with time affected ILSS and fracture behaviour.

TESTS PERFORMED

A first series of tests was performed on specimens from the seven panels. The conditions for these tests were the following (Table 1).

Time after substrate fabrication	Surface preparation	Period between overlamination and testing (days)
0	Continuous	3
1 day	No peel ply Peel ply	56 56
1 week	No peel ply Peel ply	50 50
6 weeks	No peel ply Peel ply	14 14

Table 1 First series of tests. Direct comparison of fabrication conditions.

These tests were performed two months after the completion of the first two overlaminated panels but only three days after the manufacture of the continuously laminated panel.

A third parameter which will affect the results obtained is the time between the fabrication of the interface (overlamination) and testing. As all tests were performed on the same day, this will not affect the comparison between the two tests. However, it is well known that the cure of polyester resins is not instantaneous, so this parameter is of interest for practical applications. In particular it is important to know how long it takes for stable properties to be attained. Further tests were performed therefore after different periods of time, Table 2. All specimens were stored in a controlled temperature (20°C) and humidity (50% RH) laboratory before testing.

Time after substrate fabrication	Surface preparation	Period between overlamination and testing (days)
0	Continuous	3, 7, 15, 21, 28, 56
6 weeks	No peel ply Peel ply	14, 28, 56 14, 28, 56

Table 2 Second series of tests. Influence of time after fabrication.

Tests were performed on mode I DCB, and standard short beam shear (ILSS) specimens. A small number of mode II 4ENF and three point flexure specimens were also tested. DCB specimens were 35 or 50 mm wide with 60 mm long 8 micron thick PP starter films at mid-thickness. ILSS tests were performed with 40 mm between supports (l/h = 5).

Additional physico-chemical measurements were also made in order to follow the cure of the composite matrix with time. Both Differential Scanning Calorimetry (DSC) and Dynamic Mechanical Analysis (DMA) measurements were made, using TA Instruments modulated DSC and DMA in three point flexure, but due to the limited space available here these results will be presented elsewhere.

EXAMPLES OF RESULTS

Figure 1 shows the results from the short beam shear tests, 4 or 5 specimens for each condition. The differences between the highest and lowest of the mean values is quite small, the range is only from 25 to 33 MPa.

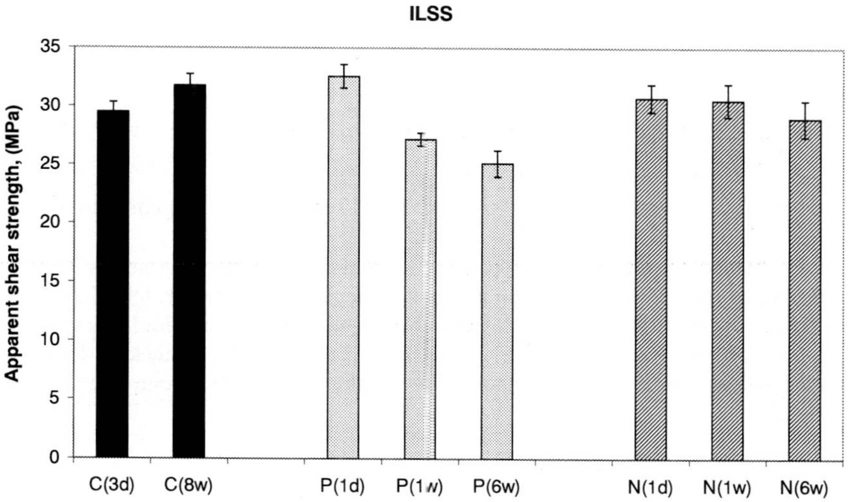

Figure 1. Short beam shear test results, mean and standard deviations
C: Continuous fabrication (test time after fabrication in brackets), P: Peel ply, N: No peel ply (delay between fabrication of first and second halves in brackets), tests on P and N specimens performed 8 weeks after fabrication started.

There do appear to be differences which correspond to the different fabrication conditions, notably lower values for the overlamination onto a surface protected by a peel ply when the delay is one week or more, compared to an unprotected surface, and higher values for overlaminations after shorter delays, but they are quite small. This test does not appear very sensitive to manufacturing conditions.

Mode I DCB tests were performed on 1 to 3 specimens for each condition. This is a small number of specimens, but this was a preliminary study to establish whether the fracture mechanics tests could provide useful information rather than a full characterisation campaign.

Figure 2 shows examples of R-curves for the different manufacturing conditions. Initiation values are similar but there are very significant differences for propagation values.

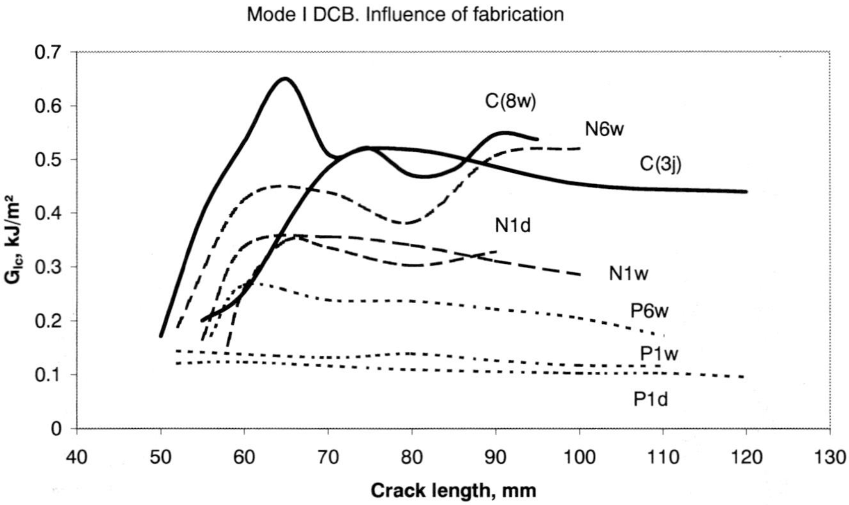

Figure 2. Examples of mode I R-curves for different fabrication conditions

The crack resistance of the central interface in a panel produced by continuous lamination is higher than all those produced in two lamination steps, even when the delay between the steps is only one day. It is also much easier to propagate a crack at the mid-plane interface when a peel ply is used than when the surface has not been protected. This rather surprising result will be discussed further below. Delays of one day or one week produced similar results.

Some additional flexural tests were performed. These included three-point loading (distance between supports 120 mm, crosshead displacement rate 5 mm/min.) and four point loading with a starter film (4ENF), (distance between lower supports 230 mm, between loading points 150 mm, loading rate 5 mm/min.). Neither of these tests showed significant variations with fabrication conditions. Figure 3 shows one example of results.

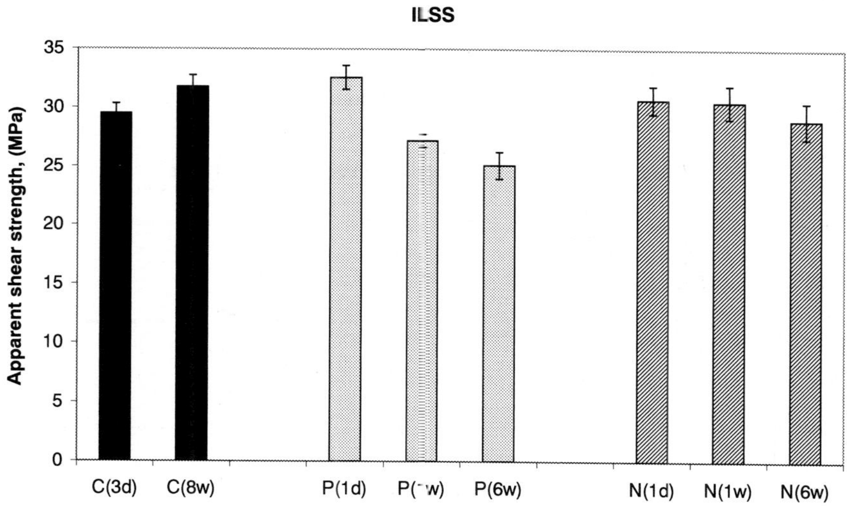

Figure 3. Flexural strengths, two tests for each condition

These limited results suggest that neither of those tests is particularly sensitive to the change in manufacturing conditions. If these conditions are to be optimized with respect to crack propagation, the mode I DCB test appears the most useful. Further tests were therefore run using DCB and ILSS tests, to look at how properties of the continuously produced material evolved with time after fabrication. Figure 4 shows the latter, and there is no significant evolution of ILSS values after the first week after fabrication.

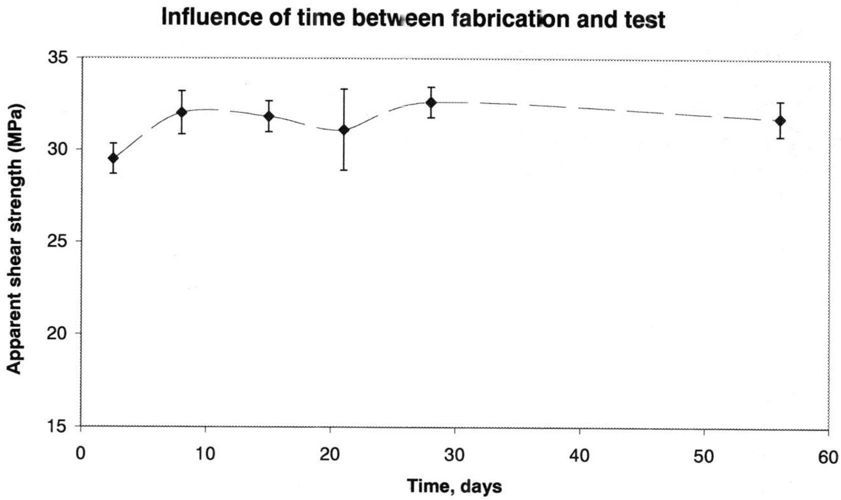

Figure 4. ILSS tests, influence of time after fabrication, continuously laminated specimens (C), mean values and standard deviation error bars.

Two DCB specimens were tested at each time and propagation values of G_{Ic} were determined (taking the mean of all values except the first 10mm of crack advance). The mean and standard deviations are shown below in Figure 5 and here there does not appear to be a significant evolution in fracture resistance either. This suggests that the comparisons in Figure 2 are valid in spite of the different delays between fabrication and testing, as the material is quite stable once the first week after fabrication has passed.

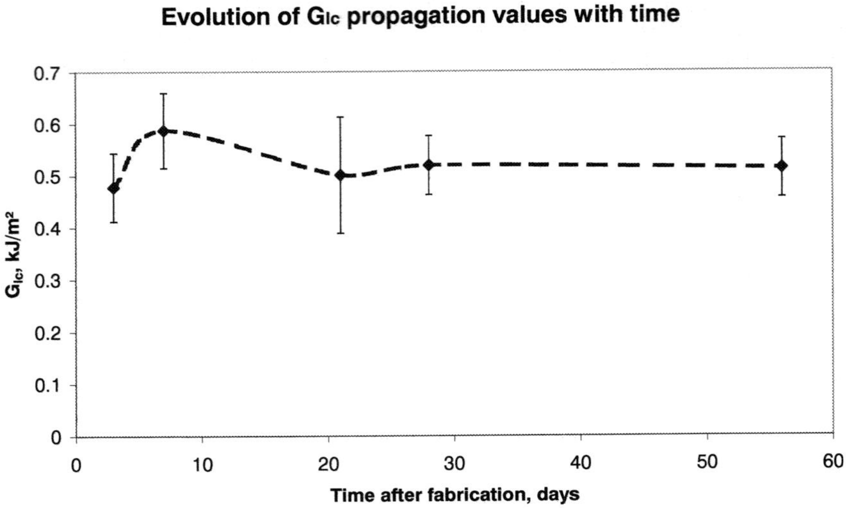

Figure 5. Mode I DCB test results, propagation values, influence of time after fabrication, continuously laminated specimens (C), mean values and standard deviation error bars.

FURTHER CONSIDERATIONS

The initial aim of these tests was to establish whether there were advantages to using fracture mechanics tests rather than the conventional ILSS or three-point flexure to reveal effects related to changes in manufacturing conditions. The results shown above certainly suggest that the mode I DCB tests are very sensitive to such changes, at least for the type of glass reinforced polyester system cured at room temperature, commonly used in marine structures. If the results are to be analysed further, there are several factors which need to be considered.

One consideration is that, although all the specimens contain the same number of reinforcement layers, the mid-plane lamination region thickness is not the same for the three conditions (continuous, overlaminated unprotected and with a peel ply). Thickness measurements and examination under the microscope reveal that the continuously laminated specimens are thinner than the overlaminated specimens due to resin rich regions being produced at the mid-plane when the lamination is restarted after a delay. This difference is a consequence of the fabrication process, and hence a real effect, but it may explain some of the differences observed. A second parameter is the change in the physico-chemical behaviour of the surfaces with time, with or without the presence of a peel ply, and this requires more detailed surface analysis to establish why the fracture resistance of both series of overlaminated specimens increases with delay before overlamination.

CONCLUSIONS

The conventional interlaminar shear strength test, frequently used for quality control puposes, is shown to be relatively insensitive to significant differences in manufacturing conditions. The mode I fracture mechanics test appears more sensitive to these conditions and may offer a better potential for the optimisation of fabrication parameters. Significant differences were noted between the mode I fracture resistance of specimens fabricated under different conditions. Further tests are being performed to investigate these differences.

The Application of Fracture Mechanics to Polymers, Adhesives and Composites
D.R. Moore (Ed)

SUPPRESSION OF INITIATION OF DELAMINATION CRACKING IN UNIDIRECTIONAL COMPOSITES BY SELF-SAME RESIN INTERLEAVING

I.K. PARTRIDGE and D.D.R. CARTIÉ

INTRODUCTION

Delamination cracking in continuous fibre reinforced laminated composites is a universal concern in materials selection and structural design. In aerospace composite structures, the 'no crack' design philosophy has prevailed for many years. The present case study limits itself to the treatment of suppression of initiation of delamination cracking in aerospace grade thermosetting composites and does not address related issues in adhesively bonded joints or indeed in thermoplastic matrix composites.

Early attempts at a solution of the classical brittleness of thermosetting composites concentrated on the toughening of the matrix resin itself, by blending with different polymeric modifiers. Requirements of thermal stability above 120°C dictate the use of 'high temperature' thermoplastics as the toughening agents, in preference to reactive rubber modifiers commonly used in toughened adhesive formulations. Largely due to the very high crosslink densities in the high temperature thermosets, the improvements in toughness achievable by this technique are limited. The best of commercially available aerospace grade resins still fail to achieve fracture toughness (G_{IC}) levels much above 500 J/m^2 [1-3].

Even if considerably tougher resin systems were to be produced, all the indications are that above about 400 J/m^2, the translation of any further resin toughness improvement into a toughness increase in the laminate made from it is very poor [4,5]. The first indications of this limitation of toughness transfer came in the work by Hunston et al [4] and at the time there was considerable speculation about the nature of the apparently complicated correlation between the Mode I toughness values measured in UD composites and their corresponding resin matrix G_{Ic}. The difficulties in resolving this issue were connected to a large extent with the unavailability of standardised fracture mechanics testing procedures for the determination of Mode I toughness, both in the unreinforced polymer and in the composite form.

The argument which became generally accepted is that the toughening mechanisms in the resin, effectively crack tip blunting by plastic yielding in the resin, tend to be inhibited by the proximity of the stiff fibre plies. Moving the stiff fibre plies further apart, allowing the plastic zone to develop in the resin rich region around a crack tip, thus looked a promising route to achieving the desired toughening and gave rise to the so-called 'interleaf technology'.

RESIN INTERLEAVING

The first documented use of resin interleaving concerned very tough thermoplastic layers used with a thermosetting prepreg [6]. Very impressive compression–after-impact performance of such composites sparked off early interest, but the drawbacks of the use of relatively compliant resin films for this purpose soon became clear. As might be expected, the low resistance to microbuckling of the reinforcement fibres into the compliant interleaf led to disappointing performance under 'hot-wet' compression conditions.

The use of well toughened adhesive films as the composite interleaves was also explored by several researchers [7-10]. However, the dissimilar interfaces created at the borders between the 'adhesive resin' and the 'prepreg resin' led to frequent premature interfacial failures in such specimens, limiting the effectiveness of this particular approach.
A compromise solution has been the use of polymeric particles, sprinkled onto the outer layers of prepreg plies, acting as fixed 'spacers' and defining the minimum separation of composite plies and hence of the (regularly spaced) resin rich layers. Several such prepreg systems have been very successful in recent years, both in the aerospace and in the automotive industries [11, 12]. Fig.1 is a micrograph of one such system, clearly showing the polymeric spacer particles. High increases in delamination initiation resistance and compression-after-impact resistance can be achieved in composites made in this way, but at the expense of reduced specific stiffness.

Fig.1 Scanning Electron Micrograph of a fracture surface of (unreinforced) resin containing solid polymeric spacer particles

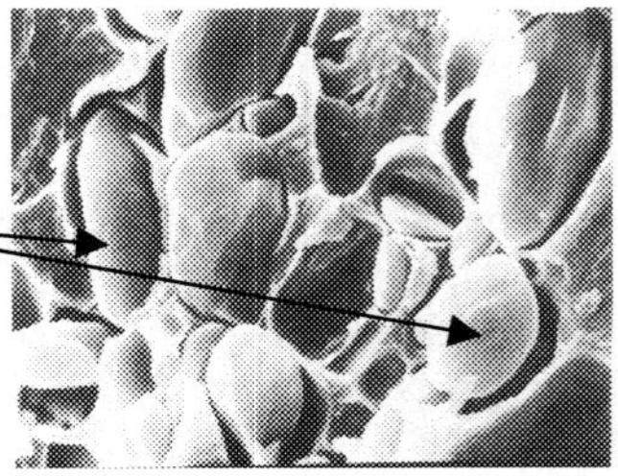

------- 10 μm

The final option then is to use a film of the same resin as the resin originally used to impregnate the reinforcing fibres in order to create the desired interleaf, in a specific position within a laminate or structure. The term 'self-same' resin interleaving is used to describe this version of the approach and is the topic of the remainder of this article.

SELF-SAME RESIN INTERLEAVING

Sample manufacture
Fig.2 is a schematic of a unidirectionally reinforced (UD) DCB specimen lay-up, of standard dimensions as specified by ISO 15024 but containing several layers of matrix resin films in its mid-plane. Whilst the number of resin films can be used as an approximate control of the resin layer thickness, the final thickness can only be determined after the laminate has been compacted during cure. The boundaries of the resin rich layer are uneven, and the measured layer thickness value can only be regarded as an average value over the specimen (see Fig.3)

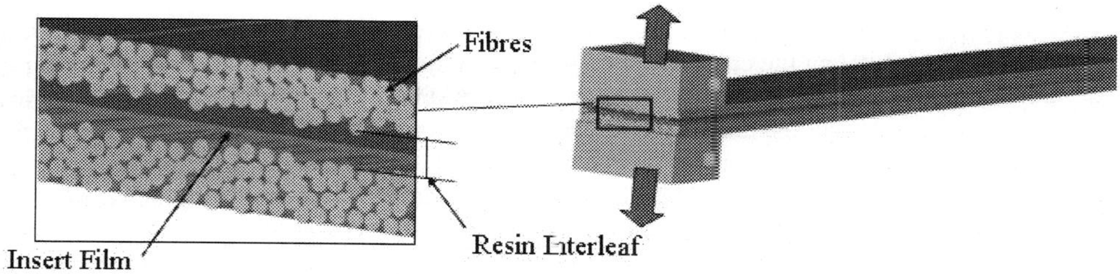

Fig.2 Unidirectional double-cantilever-beam specimen containing a resin layer in its mid-plane

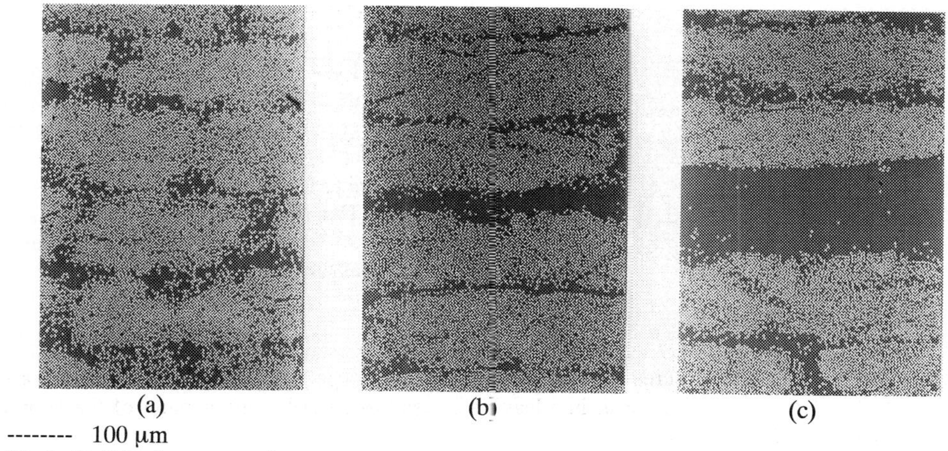

Fig.3 Polished cross-sections of the central layers in (a) standard lay-up DCB, DCB samples containing (b) 4 and (c) 8 resin film layers in the mid-plane of the (UD) sample

Delamination fracture testing of interleaved composites

The presence of a central resin layer in a UD composite delamination specimen introduces specific complications which need to be considered. Firstly, as the cross section of the beam is inhomogeneous, a correction factor for the beam compliance may need to be applied to the standard beam equation calculations on which the data analysis is based. While the equation for G_{Ic} does not require modification when applied to interleaved materials, the equation for G_{IIc} does [8]. In the case of composites containing a relatively compliant thermoplastic interleaf layer (interleaf stiffness typically 2.5GPa c.f. matrix resin stiffness of between 3.5 and 4GPa), the homogeneous beam theory significantly underestimates the Mode II fracture values. However, for the case of self-same resin interleaving using an aerospace grade epoxy resin, the correction required was found to be less than 2% of G_{IIc}, when compared with the conventional experimental compliance calibration method, even in the case of interleaf thickness of nearly 200μm [13]. Given the existing status of Mode II delamination testing, with an absence of an agreed standard on test procedures or analysis, it seems unnecessary to apply any additional correction factors to the results obtained from our interleaved specimens.

What is probably more of a reason for concern when interpreting data from fracture of interleaved composites is the fact that the crack is not constrained to run in a globally self-similar manner, as is the case for delamination cracks propagating between closely spaced plies. In an interleaved composite, we can expect the crack path to change direction frequently, being bound to stay within the interleaf but generally favouring a direction which is out of the plane of the interleaf. On a microscopic level the crack may not propagate under the same loading mode as that which had been applied externally. Scanning electron micrographs from such specimens show evidence of complex crack paths, which in themselves might be contributing to the delamination resistance of the material (see Fig.4) However, these considerations are unlikely to be important in the earliest stages of crack growth, i.e. initiation of a delamination crack, and will therefore be neglected for the purposes of the present discussion.

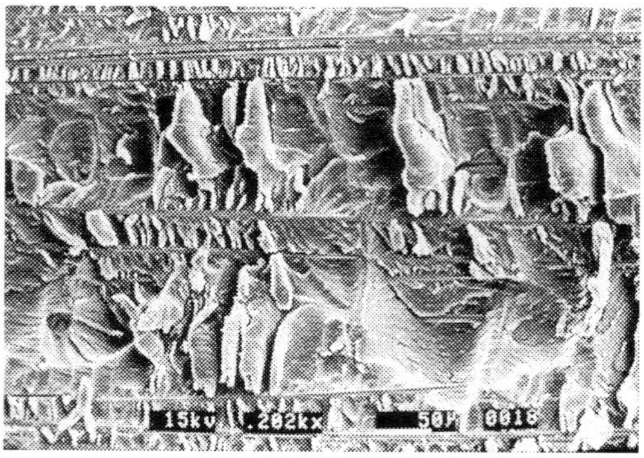

Fig.4. Scanning Electron Micrograph of a Mode II delamination fracture, grown through UD CF/977 specimen containing a thick central resin interleaf.
The extensive plastic deformation in the resin evidences the high energy absorption by the specimen.

Delamination test results from interleaved DCB specimens
The graphs shown in Fig.5 (a) and (b) summarise the relationship between the crack initiation resistance of the specimen and the thickness of the interleaf.

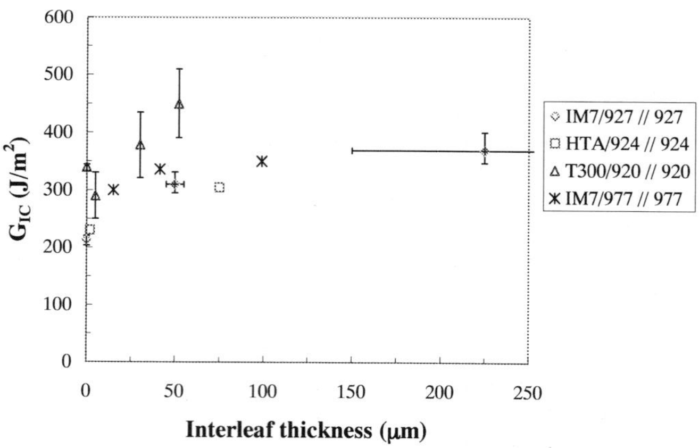

Fig.5 (a) Resistance to delamination crack initiation under Mode I loading, against the interleaf thickness, for four different self-same resin interleaved carbon fibre reinforced composites. Original data from refs. 13 and 14.

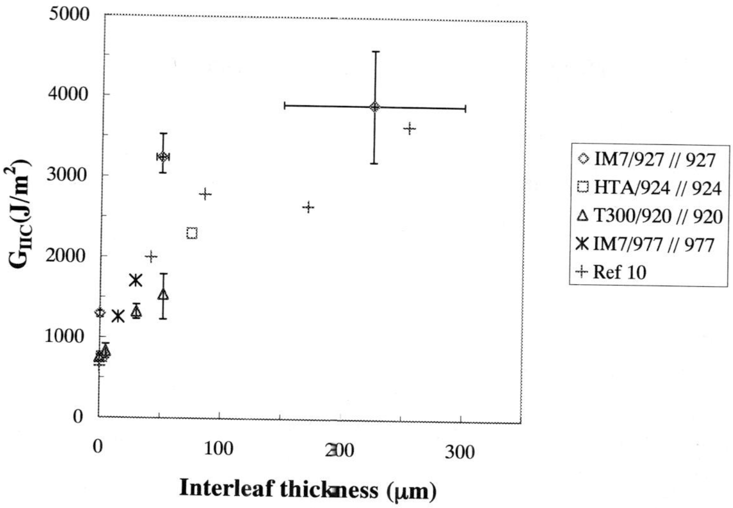

Fig.5 (b) Graph equivalent to Fig. 5 (a), but under Mode II loading. Additional data are presented from the literature (ref.10), where a toughened aerospace adhesive film, FM300, was used as the interleaf.

Fracture resistance values were also determined on samples of the unreinforced resins having the same glass transition temperature and morphology as the resin within the corresponding composite. Allowing for measurement uncertainties, these values range between 300 and 350 J/m^2 for the aerospace grade 927, 924 and 977 resin systems. The 920 system, based on a lower crosslink density epoxy resin (a '120°C resin'), exhibits fracture toughness in excess of $500J/m^2$.

PLASTICITY CONSTRAINT CONSIDERATIONS

The results shown in Fig. 5 appear to indicate that the Mode I delamination toughness of the interleaved specimens increases as the thickness of the central resin rich layer increases, tending to an asymptote at the value of G_{Ic} of the unreinforced resin itself. Hence the fracture toughness of the appropriate resin may be used to estimate the minimum thickness of interleaf which would allow an unconstrained plastic zone to develop in the composite as it fails under Mode I loading. The yield stress values used in the estimate have to be obtained from uniaxial compression tests on the unreinforced resin as tensile tests inevitably result in premature failure.

The specific example of the IM7/927 is taken for further analysis, as the largest range of interleaf thickness has been tested for this material. A simple elastic-plastic analysis, using the von Mises yield criterion and plane strain assumptions, gives the height of an unconstrained plastic zone developing under mode I loading in unreinforced 927 resin as 90 μm. This is expected to be an underestimate of the plastic zone size; application of a finite element analysis method developed by

Ozdil & Carlsson [15] gives an estimate of the plastic zone height as 135μm for this material [13]. The Mode I delamination resistance of interleaved IM7/927 composite would therefore not be expected to increase once the thickness of the interleaf exceeds this value.

An equivalent elastic-plastic analysis to estimate the height of an unconstrained plastic zone in an isotropic resin layer under Mode II loading [16] gives 2 mm for the unconstrained plastic zone height in the 927 resin based composite. However, it seems from Fig. 5(b) that the Mode II toughness values may be indicating an asymptote, at about 4.5 kJ/m^2, for interleaf thicknesses well below this theoretical limit. This is most probably caused by the tendency of the crack front to change direction so as to propagate in Mode I, in the isotropic central layer. In any case, a 10% drop in the flexural modulus has already been incurred by placing the 200μm resin layer in the centre of the composite beam; it is unlikely that any further significant stiffness reduction penalty would be acceptable in practice.

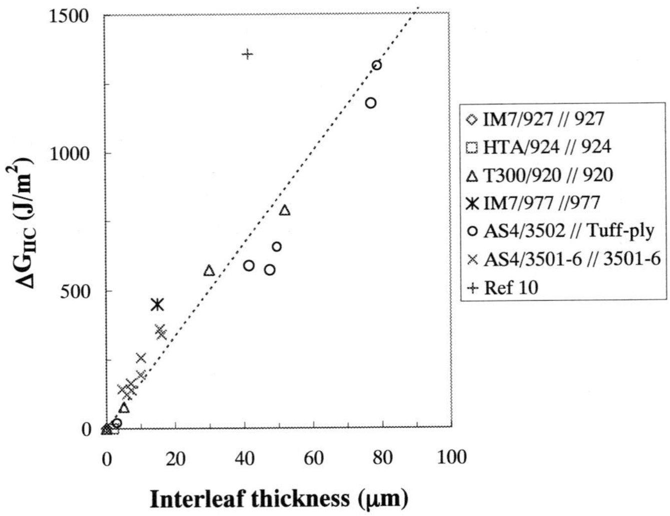

Fig.6 Absolute increase in the Mode II toughness for a range of interleaved composites, against interleaf thickness.

Further analysis is thus limited to composites with relatively thin resin interleaves, below 100 μm. Fig.6 shows how the improvement in the crack initiation resistance under Mode II loading depends upon the thickness of the interleaf. The existence of an apparent master-curve, containing data from a wide range of interleaved composite systems [data from 8,10,17] , would seem to suggest a purely geometric dependence, with no influence of the type of material or even the temperature of test [17]. It is likely that the geometry of the characteristic resin cusp formation, that accompanies shear failures in UD composites, is responsible for this observation.

CONCLUSIONS

The implications of the observed strong relationship between the crack initiation resistance and the spatial arrangement of resin and fibre plies in a UD specimen can be considered on different levels. Results of measurements on laboratory coupons should be accompanied with information on the cross-sectional structure of the sample, particularly in the case of Mode II loading measurements. There is a danger of the data generated being an un-conservative representation of the delamination resistance of a structure manufactured under different conditions of resin flow and bleed than the test specimen. Similarly, it is unlikely that correct interpretation of the failure mechanisms in complicated specimens, such as the CAI specimen, will be reached without the mesoscopic structure information being available.
Despite the fact that the understanding of the influence of the mesostructure upon the delamination resistance of composites is still incomplete, the industrial practice of selective interleaving of composites with toughened resin layers is becoming established technology. Some of the best examples can be found in the lay-up detail of composite T-stiffeners in thin-walled aerospace structures. The barriers to more extensive use of this technology in large complex structures are the need for reliable predictions of the locations of regions of crack initiation and crack growth and an inevitable increase in the manufacturing complexity.

The advantage in using a fracture mechanics approach for this type of a study is the fact that the data obtained can be regarded as a material property, independent of specimen dimensions. The applicability of the alternative methods for characterisation of failure in these materials is likely to remain limited by the practical difficulty of determining accurate values of tensile failure stress in thermosetting resins.

REFERENCES

1. Bucknall C.B. and Partridge I.K. (1983)*Polymer*, **24**, 639-644
2. Yee, A.F. and Pearson R.A., NASA Contractor Report 3718 (1983) and 3852 (1984)
3. Pascault J-P and Williams, R.J.J. (2000) "Formulation and Characterisation of Thermoset-Thermoplastic Blends" Ch 13 in 'Polymer Blends', Vol.1, eds. D R Paul and C B Bucknall, John Wiley&Sons,
4. Hunston, D.L. (1984) *ASTM Composites Technology Review*, **6**, 176
5. Bradley W. L. (1989) "Relationship of matrix toughness to interlaminar fracture toughness", Ch 5 in 'Application of Fracture Mechanics to Composite Materials' ed. K Friedrich, Elsevier
6. Krieger R. B. (1985) *Proc. 6th Int. Eur. Chapter Conf. of SAMPE, Scheveningen, May 28-30* pp. 189-199 eds. G Bartelds and R J Schliekemann, Materials Science Monographs 29, Elsevier
7. Ishai O., Rosenthal H., Sela N. and Drukker E. (1988),, *Composites*, **19**, 49-54
8. Sela N., Ishai O. and Banks-Sills L. (1989)*Composites*, **20**, 257-264
9. Ozdil F. and Carlsson L.A. (1992) *J. Comp. Mater.*, **26**, 432-459
10. Aksoy A. and Carlsson L.A. (1992) *Comp. Sci. Tech.*, **43**, 55-69
11. Odagiri N., Kishi H. and Nakae T. (1991) *6th Tech. Conf. of Am. Soc. for Comps.*, p.43
12. Altstädt V., Gerth D., Stängle M. and Recker H.G (1993) *Polymer*, **34**, 907-909
13. Singh S. and Partridge I.K. (1995) *Comps Sci and Tech*, **55**, 319-327
14. Jaussaud J.A.M. (1992) "Toughness transfer between unreinforced matrix and fibre composites" PhD thesis, Cranfield University

15. Ozdil F. and Carlsson L.A. (1992) *Eng. Fract. Mech*, **41**, 645-658
16. Aksoy A. and Carlsson L.A. (1991) *Eng. Fract. Mech.*, **39**, 525-534
17. Russell, A. J. (1987) *Polymer Composites*, **8**, 342-351

The Application of Fracture Mechanics to Polymers, Adhesives and Composites
D.R. Moore (Ed)

SUPPRESSION OF PROPAGATION OF DELAMINATION CRACKING IN UNIDIRECTIONAL COMPOSITES BY Z-Fiber® PINNING

D.D.R. CARTIÉ and I.K.PARTRIDGE

INTRODUCTION

About Z-pinning

The presence of Z-Fibers® (Z-pins) as the through-the-thickness reinforcement in continuous fibre/polymer matrix composites results in dramatic increases in the resistance to crack propagation under Mode I, Mode II and mixed mode loadings. The Z-pins can be made from metals, glass or ceramic but those currently finding the greatest range of commercial applications are made by fully curing thin pultruded carbon fibre tows impregnated with bismaleimide resin. They are supplied as a preform of a certain areal density of Z-pins in a foam sandwich (see Fig. 1) and are inserted into an uncured prepreg stack with the aid of an ultrasonic gun. Detailed account of the raw materials, manufacturing processes and sample preparation relevant to Z-pinned laminates is presented elsewhere, together with an up-to-date review of the currently available mechanical property data [1].

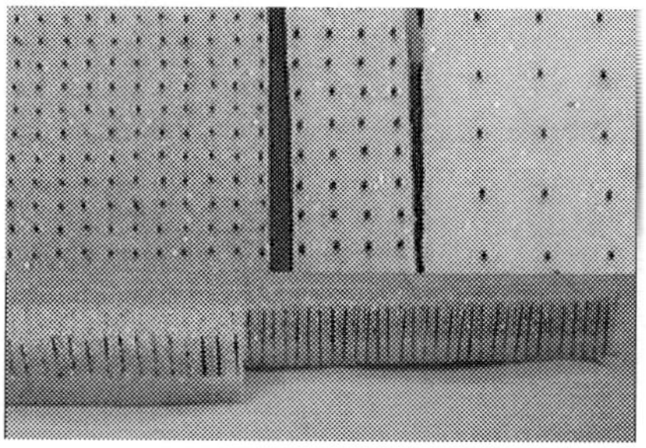

Fig.1 Top view of three different Z-Fiber® preforms, areal densities of reinforcement 2, 1 and 0.5% going from L to R; side view of a 2% and a 4% reinforcement density preform (L to R) showing the double foam layer

APPLYING FRACTURE MECHANICS TO Z-PINNED LAMINATES

The base data on the delamination resistance of Z-pinned beams have been obtained from standard unidirectional composite beams [2] containing a single 25 mm long block of Z-pins in a regular square array, placed some 5mm beyond the edge of the thin crack starter film. This geometry arrangement has ensured that the test determines the resistance of the specimen to a moving delamination.

The purpose of this case study is to illustrate how a combination of fracture mechanics data, the beam theory and FE analysis is used to design a new type of specimen, intended to quantify experimentally the crack stopping potential of a block of Z-pins for a standard Mode I delamination fracture test beam. The 25 mm block geometry is maintained, and the individual blocks have been made with different pinning densities, namely 0.5% and 2%. The block length of 25 mm is above the ACK limit (steady state crack propagation) [3], for both reinforcement densities used. The basic idea is to grow a crack through a first block of Z-pins, artificially enhancing the elastic energy stored in the sample, which then results in an unstable crack growing from the end of the first Z-pin block into another Z-pin block, as shown in Fig.2. A specimen with the first block with areal Z-pin density X followed by Z-pin block of areal density Y is designated as SX/Y.

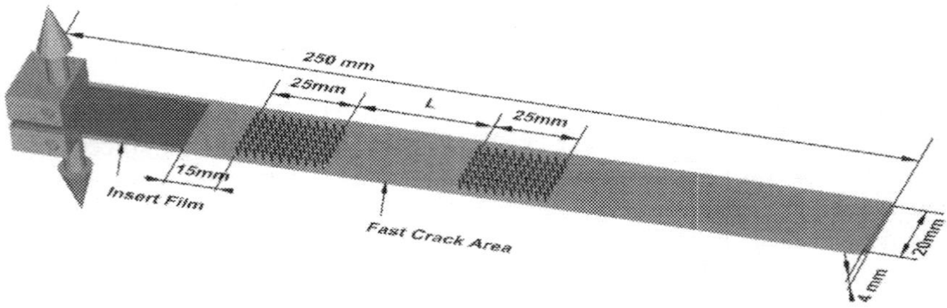

Fig.2: Double Z-pin block specimen configuration

SPECIMEN DESIGN

The primary aim is the determination of a suitable distance L between the Z-pin blocks, long enough to avoid stress interference from the first block and short enough to ensure that a rapidly propagating crack is driven into the second block of Z-pins. The model used in the design progresses through (a) a micro-model which expresses the effect of the Z-pins in terms of the traction law and (b) a local-model which considers the deformation in the standard beam containing a central delamination subject to the specified traction law and uses previously determined delamination resistance data as the basis for the crack growth / no growth decisions.

Micro-Model

The micro-model expresses the characteristics of the crack bridging forces and energies associated with the presence of the Z-pins. These have been determined experimentally by single pin pullout experiments. Fig.3 illustrates the main features of the bilinear traction law: It commences with an elastic stretch and a de-bonding phase up-to a maximum force F_{MAX}, signifying the onset of pin pullout. The second phase of the law is characteristic of frictional pullout. Experimentally, for the pullout of a 0.51mm diameter Z-pin from a 2mm thick laminate, F_{MAX} = 42N. The energy absorbed by a single Z-pin during the pullout phase can be estimated by integration of the area under the curve, as shown. As an example, the energy absorbed by a single 0.28mm Z-pin in pulling out of a 4mm thick UD specimen is 23J. In the subsequent local-model, the exact number of Z-pins per row is taken into account, as well as the laminate thickness.

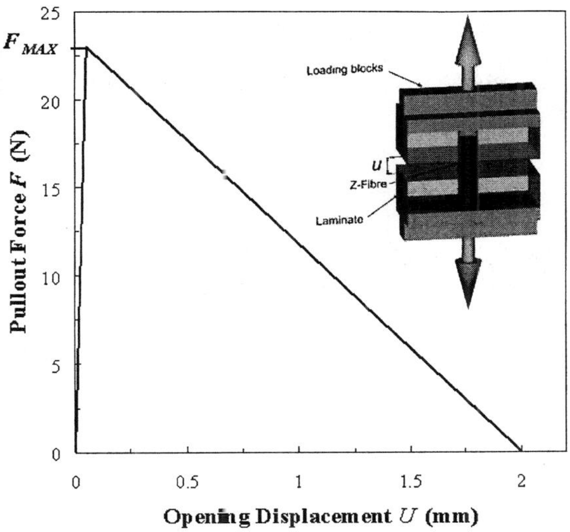

Fig.3 Bilinear load-displacement curve of Z-pin stretch/de-bond and pullout. The laminate material is unidirectional IM7/977-2. The integrated area under the lines represents the energy absorbed in the pullout. The insert indicates the experimental arrangement of the test.

Local-Model

The crack bridging actions of the Z-pins are represented by forces applied at nodes representative of their location. 2D Finite Element Analysis was chosen to compute the deformation of the beam, using quadratic shell elements (Timoshenko element) representative of the geometry. Use of this type of element ensures exact displacement results and the result is not mesh size sensitive. The alternative possibility is the use of beam theory analysis, but FEA methods are relatively easy to automate and this *local* FE model can be inserted directly into a more *global* structural FE model.

The approach taken is to model a (infinitely) long DCB containing one block of Z-pins only, to estimate how far the crack would propagate after breaking through the Z-pin block, as a function of the areal density of Z-pinning within the block. The results indicated that L=40mm was suitable for practical experimental configurations. Further details of the model and its validation can be found in a separate publication [4].

TEST RESULTS AND DISCUSSION

The above determined value of L=40 mm was used to prepare a number of DCB specimens containing two Z-pin blocks. These were tested under Mode I loading conditions, at constant cross-head speed of 2 mm/min. Examples of a load-displacement curve and of an R-curve are shown in Fig.4, and the resultant fracture surface is depicted in Fig.5.

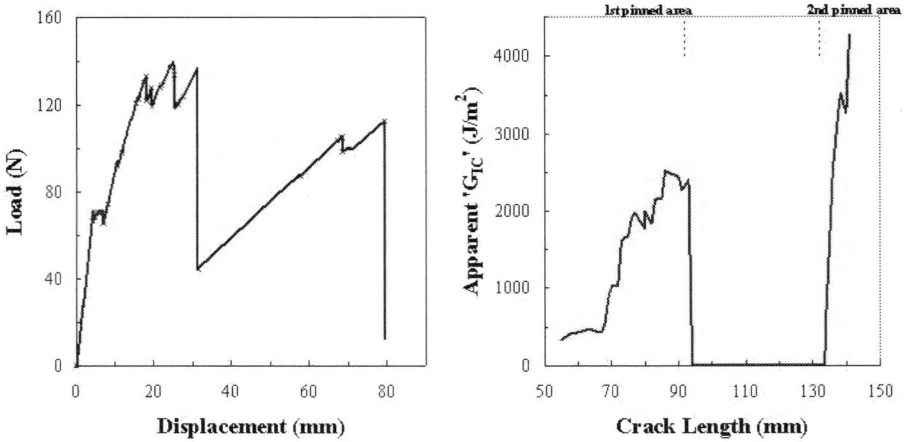

Fig. 4 Load-displacement curve and an R-curves obtained from testing of a S0.5/2 specimen. The term Apparent 'G_{Ic}' is used, as the validity conditions specified for the test [2] cannot be respected fully.

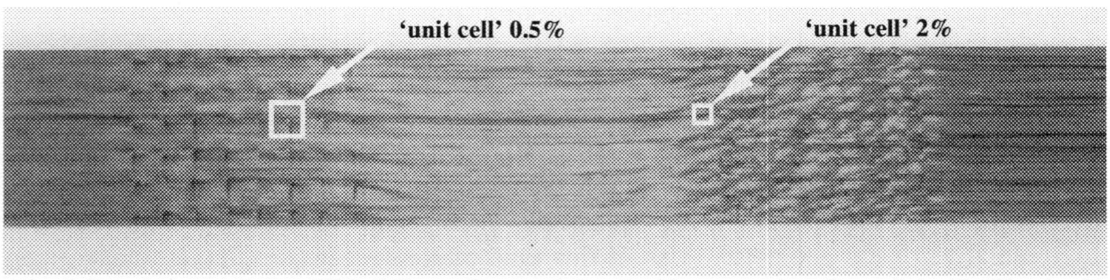

Fig.5 Fracture surface of S0.5/2 specimen, delamination growing from L to R. The 'unit cell' areas appropriate to the different pinning densities are indicated.
(The specimen width is 20 mm)

Fig.6 shows comparison of the R-curves for the experimental and model simulated results for two samples with different Z-pin block combinations. For clarity of presentation, the delamination resistance is considered equal to zero during the unstable crack growth. The good agreement between the FE analysis and the experimental results demonstrates that the bridging laws determined experimentally from single pin pullout tests are indeed representative of the actions of the Z-pins during beam delamination testing under Mode I loading conditions.

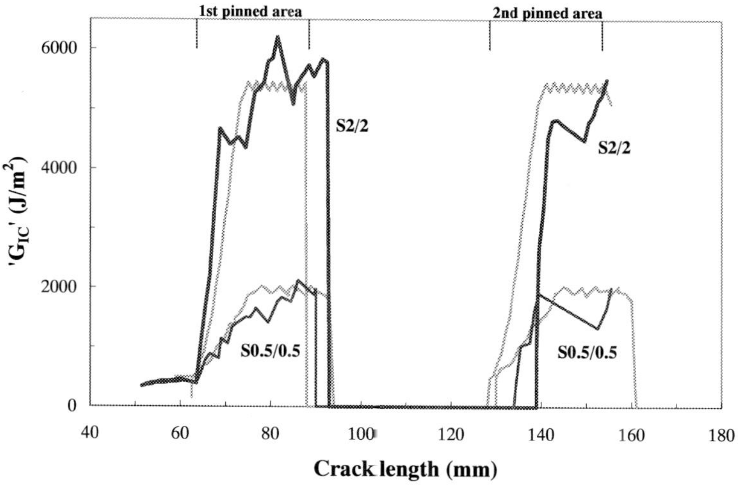

Fig.6. Comparison of FE simulation (grey lines) to experimental results (black lines)

The crack propagation resistance of such specimens could be estimated using a simple energy argument. In conditions of steady state crack propagation (i.e. crack propagation length within Z-pinned zone greater than the ACK limit), it can be assumed that a constant number of Z-pins bridge the crack at any given time. The average energy per unit area absorbed by the pin deformation and pull-out is therefore the energy absorbed by one pin divided by the 'unit cell' area. Typically, for 0.5% Z-pinning density, 0.28mm diameter Z-pins, the pullout energy density is 1.8kJ/m^2. Adding the energy required to propagate the crack per unit area of unpinned specimen, the global energy release rate of the specimen is 2.2kJ/m^2. The good agreement between this value and the experimental toughness level indicates that this method could be used for determining, to first approximation, the toughness of a Z-pin reinforced structure failing by Mode I dominated delamination.

In the graphs shown in Fig.6 it is particularly noteworthy that once the plateau value of crack resistance is reached, these levels are independent of the crack length. This suggests that a fracture mechanics based approach to designing against delamination fracture of Z-pinned composites structures is applicable.

REFERENCES

1. Partridge I.K., Cartié D.D.R., Bonnington T. (2003) "Manufacture and performance of Z-pinned composites", Ch 3 in *Advanced Polymeric Materials: Structure-property relationships*, Advani S.and Shonaike G. (Eds.) CRC Press
2. ISO 15024 "Determination of Mode I (Critical Strain Energy Release Rate or Fracture Toughness, G_{IC}) of unidirectional Fibre reinforced Polymer Laminates using the Double Cantilever Beam (DCB) Specimen", International Standards Organisation

3. Massabò R, Mumm D.R. and Cox B.N. (1998) *International Journal of Fracture*, **92**, 1-38
4. Cartié D.D.R. and Partridge I.K.(2003) "Prediction of the delamination behaviour of the Z-Fiber® reinforced laminates" *Procs. 44th AIAA/ 44th AIAA/ASME/ASCE/AHS/ ASC Structures, Structural Dynamics, and Materials Conference*, Norfolk, Virginia, USA, 7- 11 April 2003.

AUTHOR INDEX

Balika, W. 83
Beckert, W. 227
Béguelin, Ph. 55
Blackman, B.R.K. 143
Bouchet, J. 163, 249

Cartié. D.D.R. 273, 265
Castellani, L. 71
Colombarini, S. 71

Davies, P. 157, 241, 257

Forster, St. 171
Frassine, R. 71

Greenshields, C.J. 103
Grein, C. 55

Huang, D.D. 121

Jaworek, T. 171

Karger-Kocsis, J. 25, 233
Kausch, H.-H. 55
Kinloch, A.J. 143, 149, 187

Lang, R.W. 39, 47, 83
Lauke, B. 227
Leevers, P.S. 63, 103
Li, S. 121
Loaec, H. 257

Major, Z. 39
Marshall, G.P. 109
Moore, D.R. 7, 93, 97, 131, 137, 149, 201, 219

Nairn, J.A. 193, 207, 213

Partridge, I.K. 265, 273
Pinter, G. 47, 83

Ramsteiner, F. 31, 171, 179
Reed, P.E. 1
Roche, A.A. 163, 249

Sargent, J.P. 157
Sautereau, H. 17
Schüller, T. 227

Taylor, A.C. 187

Weber, M. 171
Williams, J.G. 109, 121

SUBJECT INDEX

ABS (acrylonitrile butadiene styrene) 109, 172, 173, 175
Adhesive bonding 191
Adhesive fracture energy 145, 188
Adhesive fracture toughness 131, 134, 135, 149–151, 154
Adhesive joint 157, 163, 187, 193, 197
Adhesive layer 144, 145, 153, 158, 160, 161
Adhesive–substrate interface 144
Adhesively bonded joints 143, 145, 146
Adiabatic heating 65, 66, 103
Anisotropy 219, 223, 224
AS4/3501 210, 270
AS4/3502 270

Beam theory 166, 195
Blunt 205
Blunt crack 203
Blunting 50, 176
Boat hull 257
Bond-line thickness 149–151, 154
Bonded joint 187, 192
Bridging laws 276

C Ring test 115, 116
CAI (compression after impact) 271
Carbon-fibre reinforced composite 143
Catalysts 19
Cavitation 20, 219, 235
CCT (center-cracked tension) 85, 87
CF/PEEK (carbon-fibre reinforced PEEK) 219, 221, 223, 224
Chain scission 80, 81
Chair base 124
Characteristic ratio 27
Charpy 65, 66, 71, 110, 171, 201
Chromic-acid etching 145
Classical beam theory 165
Coating 163–166, 171, 173, 175, 176, 182, 196, 197, 249, 250, 252
COD (crack opening displacement) 21, 48, 64
Cohesive failure 145, 146
Cohesive layer 144
Cohesive stress 64, 68
Cohesive zone 65, 67
Cold drawing 25
Compliance 18, 31, 63, 65, 73, 74, 163, 180, 195, 267
Component toughness 93, 96
Compression-after-impact resistance 265, 266
Conditioning 8
Continuous fibre reinforced laminated composites 265
Continuous fibre reinforced polymeric composites 219
Core–shell particles 17
CPE (chlorinated polyethylene) 117
Crack bridging 274
Crack initiation 17, 50, 53, 83, 98, 152, 157, 161, 165–167, 179, 187, 202, 203, 205–207, 217, 228, 235, 265 268, 270, 271
Crack propagation 17, 18, 23, 31, 33, 35, 66, 73, 74, 80, 83, 88, 103, 105, 157, 163, 165, 167, 173, 176, 180, 187, 193, 202, 203, 207, 227–229, 231, 241, 261, 273, 277
Crack resistance 117
Crazing 14, 27, 29, 64, 65, 67, 68, 77–81, 84, 85, 88–90, 1 8, 219, 235
CRB (cracked round bar) 43
Creep 34, 50, 52, 77, 97, 98, 101
Creep rupture 8, 11–13, 15, 97, 98, 101
CT (compact-tension) 18, 31, 48, 55, 85, 86, 152, 179–184, 187, 202, 205, 206
CTBN (carboxyl terminated butadiene acrylonitrile copolymer) 17, 19, 21, 22

DCB (double cantilever beam) 143–145, 157, 179, 187, 195–197, 223, 224, 241, 266, 275
Debonding 20, 228, 229, 231, 235, 241
Delamination 175, 196, 207, 219, 222–225, 241, 252, 265–268, 271, 273, 274, 276
DENT (double edged notched tension) 43
Design stress 97, 99, 100
DGEBA (diglycidyl ether of bisphenol A) 17, 19
Disentanglement 7, 15, 31, 33–36, 51, 65, 80, 81, 103
DKC (dynamic key curve) 45
DT (double torsion) 71, 73, 74
Ductile tearing 83
Ductile–brittle transition 7, 8, 10, 13, 55, 59, 60, 113, 138
Ductility factor 94–96
Dugdale cohesive zone 64, 67
Dugdale line length 14
Dugdale zone 88
Durability tests 146

E-glass 257
E-glass substrate 251
EBA (energy-based analysis) 36
Elastic corrections 134, 135
Embrittlement 171, 176
Embrittlement transition 40, 41
4ENF (four point end notched flexure) 260
Entanglement 25–27, 29, 32, 35, 50, 52, 53, 80, 81
Environmental stress cracking 99
Epoxy 17, 19, 21–23, 137, 144, 145, 149, 157, 163, 166, 215, 241, 244, 245, 249–251, 254, 267
ESC (environmental stress cracking) 111
EWF (essential work of fracture) 25, 55

Fatigue 8, 11, 15, 17–19, 21–23, 31, 33, 35, 36, 83, 85, 86, 88, 91, 97, 98, 101, 114, 125, 144–146, 149, 180, 187–189, 191, 193, 207, 210, 233, 237
Fatigue crack growth 187
FBA (force-based analysis) 36, 40, 45
FCP (fatigue crack propagation) 18, 20–23, 28, 83, 85–90, 233, 235–239
FEM (finite-element methods) 36, 97, 157, 166, 187, 189–192, 195, 215, 229–231, 269, 275
Fibre pull-out 235
Fibre reinforcement 93
Fixed arm peel 131–133
Fracture propagation 29
Free volume 8

GFR (glass-filled resin) 124
Glass fibre 204
Glass-fibre reinforcement 201, 205
Global energy analysis 131, 197, 208

Hardener 19
HDPE, *see* PE-HD
Hermans-type planar orientation parameter 233
HIPS (high-impact polystyrene) 35, 36, 184
HTA/924 269, 270
Hysteretic crack tip heating 84
Hysteretic heating 83, 239
Hysteretic specimen heating 85

IM7/927 269, 270
Impact 63–66, 68, 110, 122, 137, 140, 171, 219
Indentation correction 8
Initiation 122
Initiation, cracks 17, 50, 53, 83, 98, 152, 157, 161, 165–167, 179, 187, 202, 203, 205–207, 217, 228, 235, 265, 268, 270, 271
Initiation, fracture 71
Initiators 19
Interfacial toughness 213, 216, 217, 232
Interlaminar fracture 219
Interlaminar shear strength test 263
Interleaf technology 265–271
Interphases 250–252, 254, 255
Intralaminar fracture 219
Intrinsic viscosity 35
Iosipescu beam 257
iPP/EPR (ethylene–propylene rubber toughened isotactic polypropylene blends) 55–60
ISS (interfacial shear strength) 213
Izod number 121, 124, 203–205
Izod test 201, 202, 206

$J_{0.2}$ value 122
J integral 122, 125, 126, 158, 190

Lap shear 159–161, 163
Lap-shear joint 157, 158
Load amplitude 18
Load ratio 189

Marine industry 257
MB (microbond) test 213–215, 217, 227, 228, 230–232
MDPE (medium-density PE) 95, 96, 117, 118
Mesoscopic structure 271
Microcracking 208–211
Microcracks 207, 209
Minimum/maximum load ratio 85
Mixed-mode fracture 157, 158
Mixed-mode loading 273
Mode-I crack growth 194, 195
Mode-I DCB test 244, 258, 259, 261, 262
Mode-I deformation 183
Mode-I delamination 269, 270, 277
Mode-I failure 270
Mode-I fracture 163, 223, 244, 263
Mode-I loading 273, 276
Mode-I toughness 265
Mode-II crack growth 194, 195
Mode-II deformation 183
Mode-II 4ENF specimen 258
Mode-II fracture 163, 267
Mode-II loading 213, 270, 271, 273
Mode-III crack growth 194, 195
Modulus 7, 17–19, 63, 66, 73, 104, 122, 125, 132, 150, 158, 164, 166–169, 189, 196, 201, 203, 209, 215, 230, 244, 249, 251, 252, 255, 270
Molecular agents 182
Molecular mass 48, 50–52, 71, 77, 81
Molecular mass distribution 47
Molecular weight 7, 8, 13, 15, 17, 21, 25, 31–33, 35, 100, 107, 181
MPVC (modified PVC) 115

Necking 25
Nylon 122, 126
Nylon 6 (N6, PA6) 107, 121, 122
Nylon 6,6 (N66) 121

PA [poly(amide); *see also* Nylon] 91, 175
Paris law 20, 143, 188, 191, 192, 210
PB [poly(1-butene)] 85, 89–91, 107
PC [poly(carbonate)] 40, 43–46, 85–91, 107, 137, 140, 171, 174, 176
PE-HD (high-density polyethylene) 31–35, 47, 85–87, 89–91, 115, 118
PE [poly(ethylene)] 41, 63, 66, 68, 95, 97, 98, 100, 101, 106, 107, 109, 111–116, 118, 184
PEEK (polyetheretherketone) 144, 219, 234, 235, 237
Peel 131, 157, 259, 260
Peel strength 131–134
PEN (polyethylene naphthalate) 26, 27
PES (polyethersulfone) 7, 8, 11, 14–16, 201–203
PETP (polyethylene terephthalate) 26, 27, 124, 131–135
PETP/PVC (polyethylene terephthalate/polyvinyl chloride) 131
Phenol formaldehyde resin 201
Phenolic resin 121
Phenolics 122, 201–203
Phosphoric-acid anodising 145
Pipes 35, 47, 63, 66, 93–96, 103–107, 109–116, 118, 119
Plane strain 8, 14, 26, 41, 43, 46, 56, 63, 84, 103, 138, 151, 154, 203, 230
Plane stress 41, 43, 46, 63, 84, 88, 89, 106, 107, 138, 140, 150, 151
Plastic deformation 132, 136, 173, 191
Plastic zone 8, 10, 14, 26, 50, 51, 53, 55–58, 60, 84, 85, 87–91, 94, 95, 138, 140, 151, 154, 204, 205, 265, 269, 270
Plastic zone site 150
PMMA [poly(methyl methacrylate)] 85–91, 103, 179, 183
Polydimethylsiloxane (PDMS) 17, 19
Polyester 241, 245, 257, 258
Polymeric coating 137
POM [poly(oxymethylene)] 41, 43–45
Power law 110, 235
PP [poly(propylene)] 41, 43–45, 91, 236, 237, 243, 258
PPco (polypropylene copolymer) 107
PPS (polyphenylene sulfide) 234
Propagation, cracks 17, 18, 23, 31, 33, 35, 66, 73, 74, 80, 83, 88, 103, 105, 157, 163, 165, 167, 173, 176, 180, 187, 193, 202, 203, 207, 227–229, 231, 241, 261, 273, 277
Propagation, fatigue cracks 87
Propagation, fracture 29, 71
PS (polystyrene) 32, 33, 35, 71, 76–80, 179, 181–183
Pull-out test 214, 217, 277
PVC [poly(vinylchloride)] 40, 46, 112, 116, 117, 131–133, 135
PVC-U (unplasticized PVC) 85–87, 90, 93, 95, 96, 106, 107, 109–116
PVDF (polyvinylidene fluoride) 106

R curve 117, 118, 203, 259, 275, 276
R-ratio 88 89, 91
RCP (rapid crack propagation) 63–65, 67, 68, 104, 105, 109–111, 114
Reinforced plastics 219
Residual stress 164, 167–169, 193–197, 209, 213, 216, 217, 228, 249, 252, 254
Rubber-modified epoxy 152
Rubber-modified material 36
Rubber modifiers 265
Rubber-toughened acrylic 154
Rubber-toughened nylon 6,6 (RTN66) 121, 122

SAN (styrene acrylonitrile) 172, 173, 183
Scission 81
Scratch resistance 171, 176
SEN (single-edge notched) 85, 87
SENB (single-edge notched beam) 18, 35, 153, 202, 221, 222
SFPO (single-fibre pull-out) test 213, 227, 230–232
Shear bands 89
Shear lip 88, 89
Short-fibre reinforced composites 235
Short-fibre reinforcement 233
Single-lap joint 189, 190, 192
Single-pin pull-out 274
Specific heat 65
Standard short beam shear (ILSS) test 257, 258, 261, 262
Stress concentration 119

T300/920 269, 270
T-peel 131
TDCB (tapered double cantilever beam) 143, 145, 149, 152, 163, 164, 167, 169, 187
Tearing 25, 219
Thermal conduction 65
Thermal diffusivity 65
Thermal expansion 194, 195, 207
Thermal stress 164
Thermoelasticity analysis 193–195
Thermoplastics 25, 31, 63, 66, 109, 111, 201, 204, 219, 233, 265
Thermosets 17, 25, 219, 265
Tie molecules 28, 32, 33, 50–53, 180
Toughened acrylic 153
Toughened epoxies 17
Toughened nylons 121
Translaminar fracture 219
Triaxial stress field 118

UHMWPE (ultrahigh molecular weight PE) 121, 125, 126
Uniaxial compression 203
Uniaxial compression tests 269

Water storage tanks 97, 101
Welding 180–183
Welding experiments 179
Wöhler curves 17, 23, 83
WOL (wedge-open loading) 85
Work hardening 27

Yield 84, 103, 112, 113, 116, 117, 163, 173, 175, 176, 233
Yield strength 96, 138, 175, 203
Yield stress 7, 8, 18, 19, 41, 56, 84, 113, 116–118, 125, 138, 150, 269
Yielding 14, 25–27, 29, 36, 40, 41, 71, 84, 88, 89, 91, 111, 112, 116–118, 158, 205, 219, 265

Z-pins 273–277

AUTHOR AFFILIATIONS

W. Balika
Polymer Competence Center Leoben
A-8700 Leoben
Austria

W. Beckert
Fraunhofer-Institut
Dresden
Germany

Ph. Béguelin
Ecole Polytechnique Fédérale de Lausanne
Switzerland

B.R.K. Blackman
Research Lecturer
Department of Mechanical Engineering
Imperial College London
Exhibition Road
London SW7 2BX

J. Bouchet
INSA Lyon
Lyon
France

D.D.R. Cartié
Cranfield University
UK

L. Castellani
Senior Scientist, Polymers Characterization
Enichem
via Taliercio 14
46100 Mantova
Italy

S. Colombarini
Università di Ferrara-Ingegneria
Ferrara
Italy

P. Davies
Research Engineer, Materials & Structures Group
IFREMER
Centre de Brest
BP 70
29280 Plouzané
France

C.J. Greenshields
Nabla Ltd
Surrey
UK

C. Grein
Borealis GmbH
Linz
Austria

S.T. Forster
BASF
Germany

R. Frassine
Politecnico di Milano
Milano
Italy

D.D. Huang
E.I. Du Pont de Nemours Inc
Wilmington
USA

T. Jaworek
BASF
Germany

J. Karger-Kocsis
University of Kaiserslautern
Germany

H.-H. Kausch
Ecole Polytechnique Fédérale de Lausanne
Switzerland

A.J. Kinloch
Imperial College London
Exhibition Road
London SW7 2BX

R.W. Lang
University of Leoben
Leoben
Austria

B. Lauke
Institut für Polymerforschung Dresden
Dresden
Germany

P.S. Leevers
Imperial College London
Exhibition Road
London SW7 2BX

S. Li
Medical Devices Testing and Innovations
Sarasota
Florida
USA

H. Loaec
IFREMER
Centre de Brest
BP 70
29280 Plouzané
France

Z. Major
University of Leoben
Leoben
Austria

G.P. Marshall
Pipeline Developments Ltd
Salford
UK

D.R. Moore
Imperial College London
Exhibition Road
London SW7 2BX

J. Nairn
University of Utah
Utah
USA

I.K. Partridge
Cranfield University
UK

G. Pinter
University of Leoben
Leoben
Austria

F. Ramsteiner
BASF
Germany

P.E. Reed
University of Twente
Enschede
The Netherlands

A.A. Roche
INSA
Lyon
France

J.P. Sargent
BAE Systems
Filton
UK

H. Sautereau
INSA
Lyon
France

T. Schüller
Institut für Polymerforschung Dresden
Dresden
Germany

A.C. Taylor
Imperial College London
Exhibition Road
London SW7 2BX

M. Weber
BASF
Germany

J.G. Williams
Imperial College London
Exhibition Road
London SW7 2BX

18817